C++语言程序设计

（进阶篇）

［美］梁勇（Y. Daniel Liang） 著

张丽 译

Introduction to C++ Programming and Data Structures
Fifth Edition

机械工业出版社
CHINA MACHINE PRESS

Authorized translation from the English language edition, entitled *Introduction to C++ Programming and Data Structures, Fifth Edition*, ISBN: 9780137391448, by Y. Daniel Liang, Copyright © 2022, 2018, 2014 by Pearson Education Inc. or its affiliates.

All rights reserved. No part of this book may be reproduced or transmitted in any form or by any means, electronic or mechanical, including photocopying, recording or by any information storage retrieval system, without permission from Pearson Education, Inc.

Chinese simplified language edition published by China Machine Press, Copyright © 2024.

Authorized for sale and distribution in the Chinese Mainland only (excluding Hong Kong SAR, Macao SAR and Taiwan).

本书中文简体字版由 Pearson Education（培生教育出版集团）授权机械工业出版社在中国大陆地区（不包括香港、澳门特别行政区及台湾地区）独家出版发行。未经出版者书面许可，不得以任何方式抄袭、复制或节录本书中的任何部分。

本书封底贴有 Pearson Education（培生教育出版集团）激光防伪标签，无标签者不得销售。

北京市版权局著作权合同登记　图字：01-2022-2397 号。

图书在版编目（CIP）数据

C++ 语言程序设计 . 进阶篇：原书第 5 版 /（美）梁勇（Y. Daniel Liang）著；张丽译 . -- 北京：机械工业出版社，2024. 9. --（计算机科学丛书）. -- ISBN 978-7-111-76346-8

I. TP312.8

中国国家版本馆 CIP 数据核字第 2024JC7211 号

机械工业出版社（北京市百万庄大街 22 号　邮政编码 100037）
策划编辑：曲　熠　　　　　　　责任编辑：曲　熠
责任校对：薄萌钰　王　延　　　责任印制：常天培
北京科信印刷有限公司印刷
2024 年 12 月第 1 版第 1 次印刷
185mm×260mm · 25 印张 · 637 千字
标准书号：ISBN 978-7-111-76346-8
定价：99.00 元

电话服务　　　　　　　　　　网络服务
客服电话：010-88361066　　　机　工　官　网：www.cmpbook.com
　　　　　010-88379833　　　机　工　官　博：weibo.com/cmp1952
　　　　　010-68326294　　　金　书　网：www.golden-book.com
封底无防伪标均为盗版　　　机工教育服务网：www.cmpedu.com

译者序

Introduction to C++ Programming and Data Structures, Fifth Edition

C++ 是一种强大且应用广泛的程序设计语言。尽管计算机技术一直在快速发展，但调查表明 C++ 仍然是目前使用最多的语言之一。在当今计算机向各行各业迅速渗入的情况下，C++ 正在重新变得更加重要。

C++ 继承了 C 语言运行效率高的优势，而对于面向对象的支持，增强了其代码的可复用性和可维护性，因此适用于构建大型软件项目。C++ 有丰富而强大的库和框架，支持操作系统、嵌入式系统、游戏、图形和计算机视觉、高性能科学计算、桌面应用，以及网络通信和服务器端程序的开发。通过学习 C++ 而掌握的程序设计语言基本概念以及面向过程和面向对象的编程范式，也可以让你轻松地切换到其他编程语言，例如 Java、C#、Python 等。

本书作者采用"基础优先"的方法，在设计自定义类之前介绍基本的编程概念和技术。选择语句、循环、函数和数组的基本概念和技术是编程的基础。建立这个坚实的基础可为学习面向对象编程和高级 C++ 编程做好准备。本书的程序设计以问题驱动的方式呈现，侧重于解决问题而非语法，并使用了许多不同领域的示例，包括数学、科学、商业、金融、游戏、动画和多媒体，通过实例说明了基本编程概念。

《C++ 语言程序设计》中文版分为基础篇和进阶篇。基础篇（前 16 章）介绍基本的 C++ 程序设计，而进阶篇则关注栈、二叉树等数据结构以及排序、查找等经典算法。本书为进阶篇。

于芷涵、李宇博、刘家炜 3 位学生翻译了本书图表中的文字，并对全书进行了校对，特此表示感谢。

译者
2024 年 3 月

前 言
Introduction to C++ Programming and Data Structures, Fifth Edition

教学特色

- 每章的开头列出学习目标，明确学生应该从这一章中学到什么。这份简洁的列表有助于学生在完成学习后，判断自己是否达到了学习目标。
- 要点提示强调了每节中所涵盖的重要概念。
- CodeAnimation 模拟程序的执行，它引导学生逐行浏览代码、要求学生提供输入并立即展示这些输入对程序产生的影响。
- LiveExample 让学生能够在类似于 IDE 的环境中练习编码。给学生提供填写缺失代码的机会要求他们编译和运行程序，提交内容后能立即获得反馈。LiveExample 引导学生逐步接近正确答案，帮助他们坚持下去，并保持不断尝试的动力。
- 交互式流程图、算法动画和 UML 图可以提升解决问题和逻辑思维能力，有助于理解操作流程，并在学生开始编码之前帮助他们可视化程序中正在发生的事情。

本版新增内容

本版在细节上进行了全面修订，旨在改善清晰度、呈现方式、内容、示例和练习。主要的改进包括：

- 更新 1.2 节，包括云存储和触摸屏的内容。
- 更新 4.8.4 节，讨论基于元组的输入与基于行的输入。
- 在 C++17 中不再支持异常说明符。因此，在第 5 版中删除了第 4 版的 16.8 节。对所有使用异常说明符的代码都进行了修订。
- 18.11 节是全新的。它介绍了三种字符串匹配算法：暴力法、Boyer-Moore 算法和 KMP 算法。
- 21.11 节也是全新的。它介绍了使用霍夫曼编码进行数据压缩的方法。
- 附录 I 是全新的。它给出了大 O、大 Omega 和大 Theta 表示法的精确数学定义。

灵活的章节顺序

可采用灵活的章节顺序阅读本版，如下图所示。

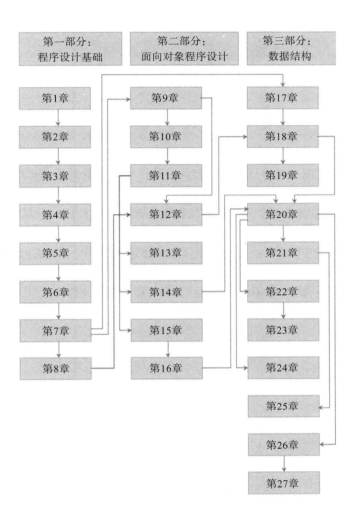

补充说明

由于中文版未获得英文版 Revel 版本（互动式数字教材）的授权，因此大量视频和动画内容无法通过纸质版本有效呈现。我们在书中提供了部分互动内容的访问地址，包括 CodeAnimation、LiveExample 和编程练习等。读者可通过以下二维码获得完整的互动内容链接列表。

作者简介
Introduction to C++ Programming and Data Structures, Fifth Edition

梁勇博士于 1991 年在俄克拉何马大学获得计算机科学博士学位,并于 1986 年和 1983 年在复旦大学分别获得计算机科学硕士和学士学位。在加入阿姆斯特朗州立大学(现已与佐治亚南方大学合并)之前,他曾任普渡大学计算机科学系副教授,在那里他曾两次获得卓越研究奖。

梁勇博士目前是佐治亚南方大学计算机科学系教授。他的研究领域是理论计算机科学。他曾在 *SIAM Journal on Computing*、*Discrete Applied Mathematics*、*Acta Informatica* 和 *Information Processing Letters* 等期刊上发表论文。他撰写了三十余本著作,其中广受欢迎的计算机科学教材在世界各地得到广泛使用。

2005 年,梁勇博士被 Sun Microsystems 公司(现为甲骨文公司)评选为 Java Champion。他还曾在多个国家做过关于程序设计的讲座。

目 录

译者序
前言
作者简介

第 17 章 递归 ································ 1
17.1 简介 ····································· 1
17.2 案例研究：计算阶乘 ················ 2
17.3 案例研究：斐波那契数 ············· 5
17.4 使用递归解决问题 ··················· 7
17.5 递归辅助函数 ························· 9
    17.5.1 选择排序 ···················· 10
    17.5.2 二分查找 ···················· 12
17.6 汉诺塔 ································· 13
17.7 八皇后问题 ·························· 16
17.8 递归与迭代 ·························· 19
17.9 尾递归 ································· 19
关键术语 ······································· 21
章节总结 ······································· 21
编程练习 ······································· 21

第 18 章 开发高效算法 ················ 30
18.1 简介 ··································· 30
18.2 使用大 O 表示法衡量算法效率 ····· 30
18.3 示例：确定大 O ····················· 32
18.4 分析算法时间复杂度 ·············· 34
    18.4.1 分析二分查找 ············· 35
    18.4.2 分析选择排序 ············· 35
    18.4.3 分析汉诺塔问题 ·········· 35
    18.4.4 常见的递归关系 ·········· 36
    18.4.5 比较常见的增长函数 ···· 36
18.5 使用动态规划求斐波那契数 ····· 37
18.6 使用欧几里得算法求最大
     公约数 ································· 39
18.7 寻找质数的高效算法 ·············· 43

18.8 使用分治法寻找最近点对 ········· 51
18.9 使用回溯法解决八皇后问题 ······ 53
18.10 案例研究：寻找凸包 ············· 56
    18.10.1 礼品包装算法 ············ 57
    18.10.2 Graham 算法 ············· 58
18.11 字符串匹配 ·························· 59
    18.11.1 Boyer-Moore 算法 ······· 61
    18.11.2 Knuth-Morris-Pratt 算法 ····· 64
关键术语 ······································· 67
章节总结 ······································· 68
编程练习 ······································· 68

第 19 章 排序 ······························ 74
19.1 简介 ··································· 74
19.2 插入排序 ····························· 74
19.3 冒泡排序 ····························· 77
19.4 归并排序 ····························· 79
19.5 快速排序 ····························· 82
19.6 堆排序 ································ 86
    19.6.1 存储堆 ······················ 86
    19.6.2 添加新节点 ················ 87
    19.6.3 删除根 ······················ 89
    19.6.4 Heap 类 ····················· 92
    19.6.5 使用 Heap 类进行排序 ····· 94
    19.6.6 堆排序的时间复杂度 ···· 95
19.7 桶排序和基数排序 ················· 96
19.8 外部排序 ····························· 97
    19.8.1 实现第一阶段 ············· 99
    19.8.2 实现第二阶段 ············· 100
    19.8.3 合成两个阶段 ············· 102
    19.8.4 外部排序复杂度 ·········· 107
关键术语 ······································· 107
章节总结 ······································· 107
编程练习 ······································· 107

第 20 章　链表、队列和优先级队列 … 109
- 20.1　简介 … 109
- 20.2　节点 … 109
- 20.3　LinkedList 类 … 112
- 20.4　实现 LinkedList … 114
  - 20.4.1　实现 addFirst(T element) … 115
  - 20.4.2　实现 addLast(T element) … 116
  - 20.4.3　实现 add(int index, T element) … 118
  - 20.4.4　实现 removeFirst() … 119
  - 20.4.5　实现 removeLast() … 120
  - 20.4.6　实现 removeAt(int index) … 122
  - 20.4.7　LinkedList 的源代码 … 123
  - 20.4.8　LinkedList 的时间复杂度 … 129
- 20.5　迭代器 … 130
- 20.6　C++11 foreach 循环 … 133
- 20.7　链表的变体 … 135
- 20.8　队列 … 135
- 20.9　优先级队列 … 138
- 关键术语 … 141
- 章节总结 … 141
- 编程练习 … 141

第 21 章　二叉查找树 … 144
- 21.1　简介 … 144
- 21.2　二叉查找树基础知识 … 144
- 21.3　表示二叉查找树 … 145
- 21.4　访问二叉查找树中的节点 … 146
- 21.5　查找元素 … 146
- 21.6　将元素插入二叉查找树 … 146
- 21.7　树的遍历 … 148
- 21.8　BST 类 … 150
- 21.9　删除二叉查找树中的元素 … 160
- 21.10　BST 的迭代器 … 165
- 21.11　案例研究：数据压缩 … 167
- 关键术语 … 172
- 章节总结 … 172
- 编程练习 … 173

第 22 章　STL 容器 … 174
- 22.1　简介 … 174
- 22.2　STL 基础 … 174
- 22.3　STL 迭代器 … 179
  - 22.3.1　迭代器的类型 … 181
  - 22.3.2　迭代器运算符 … 182
  - 22.3.3　预定义迭代器 … 184
  - 22.3.4　istream_iterator 和 ostream_iterator … 185
- 22.4　C++11 自动类型推断 … 187
- 22.5　序列容器 … 187
  - 22.5.1　序列容器：vector … 188
  - 22.5.2　序列容器：deque … 189
  - 22.5.3　序列容器：list … 191
- 22.6　关联容器 … 194
  - 22.6.1　关联容器：set 和 multiset … 195
  - 22.6.2　关联容器：map 和 multimap … 196
- 22.7　容器适配器 … 198
  - 22.7.1　容器适配器：stack … 198
  - 22.7.2　容器适配器：queue … 200
  - 22.7.3　容器适配器：priority_queue … 201
- 关键术语 … 202
- 章节总结 … 203
- 编程练习 … 203

第 23 章　STL 算法 … 207
- 23.1　简介 … 207
- 23.2　算法类型 … 208
- 23.3　copy 函数 … 209
- 23.4　fill 和 fill_n … 211
- 23.5　将函数作为参数传递 … 212
- 23.6　generate 和 generate_n … 215
- 23.7　remove、remove_if、remove_copy 和 remove_copy_if … 216

23.8	replace、replace_if、replace_copy 和 replace_copy_if … 220		24.4.3	双重散列 … 258
23.9	find、find_if、find_end 和 find_first_of … 223		24.5	用独立链处理冲突 … 259
			24.6	负载因子和再散列 … 259
23.10	search 和 search_n … 227		24.7	用散列实现映射 … 260
23.11	sort 和 binary_search … 229		24.8	用散列实现集合 … 268

23.8 ~ 23.23 (左栏)

- 23.8 replace、replace_if、replace_copy 和 replace_copy_if … 220
- 23.9 find、find_if、find_end 和 find_first_of … 223
- 23.10 search 和 search_n … 227
- 23.11 sort 和 binary_search … 229
- 23.12 adjacent_find、merge 和 inplace_merge … 231
- 23.13 reverse 和 reverse_copy … 233
- 23.14 rotate 和 rotate_copy … 234
- 23.15 swap、iter_swap 和 swap_ranges … 236
- 23.16 count 和 count_if … 237
- 23.17 max_element 和 min_element … 238
- 23.18 random_shuffle … 239
- 23.19 for_each 和 transform … 240
- 23.20 includes、set_union、set_difference、set_intersection 和 set_symmetric_difference … 242
- 23.21 accumulate、adjacent_difference、inner_product 和 partial_sum … 244
- 23.22 lambda 表达式 … 247
- 23.23 新的 C++11 STL 算法 … 250
- 关键术语 … 251
- 章节总结 … 251
- 编程练习 … 252

第 24 章 散列 … 254

- 24.1 简介 … 254
- 24.2 散列是什么 … 254
- 24.3 散列函数和散列码 … 255
  - 24.3.1 基元类型的散列码 … 255
  - 24.3.2 字符串的散列码 … 256
  - 24.3.3 压缩散列码 … 256
- 24.4 用开放寻址处理冲突 … 257
  - 24.4.1 线性探测 … 257
  - 24.4.2 平方探测 … 258
  - 24.4.3 双重散列 … 258
- 24.5 用独立链处理冲突 … 259
- 24.6 负载因子和再散列 … 259
- 24.7 用散列实现映射 … 260
- 24.8 用散列实现集合 … 268
- 关键术语 … 275
- 章节总结 … 275
- 编程练习 … 275

第 25 章 AVL 树 … 277

- 25.1 简介 … 277
- 25.2 再平衡树 … 278
- 25.3 设计 AVL 树类 … 279
- 25.4 重写 insert 函数 … 280
- 25.5 实现旋转 … 281
- 25.6 实现 remove 函数 … 282
- 25.7 AVLTree 类 … 282
- 25.8 测试 AVLTree 类 … 288
- 25.9 AVL 树的时间复杂度分析 … 291
- 关键术语 … 292
- 章节总结 … 292
- 编程练习 … 292

第 26 章 图及其应用 … 294

- 26.1 简介 … 294
- 26.2 基本图术语 … 295
- 26.3 图的表示 … 296
  - 26.3.1 顶点的表示 … 296
  - 26.3.2 边的表示（用于输入）：边数组 … 297
  - 26.3.3 边的表示（用于输入）：Edge 对象 … 298
  - 26.3.4 边的表示：邻接矩阵 … 299
  - 26.3.5 边的表示：邻接列表 … 299
- 26.4 Graph 类 … 301
- 26.5 图遍历 … 311
- 26.6 深度优先搜索 … 314
  - 26.6.1 深度优先搜索算法 … 314
  - 26.6.2 深度优先搜索的实现 … 315
  - 26.6.3 DFS 的应用 … 318

26.7 广度优先搜索 ………………… 318
   26.7.1 广度优先搜索算法 ………… 318
   26.7.2 广度优先搜索的实现 ……… 319
   26.7.3 BFS 的应用 ………………… 321
26.8 案例研究：九枚硬币翻转问题 …… 322
关键术语 …………………………… 328
章节总结 …………………………… 328
编程练习 …………………………… 329

第 27 章 加权图及其应用 …………… 332

27.1 简介 …………………………… 332
27.2 加权图的表示 ………………… 333
   27.2.1 加权边的表示：边数组 …… 333
   27.2.2 加权相邻列表 ……………… 333
27.3 `WeightedGraph` 类 …………… 335
27.4 最小生成树 …………………… 341
   27.4.1 最小生成树算法 …………… 342
27.5 寻找最短路径 ………………… 350

27.6 案例研究：加权九枚硬币
     翻转问题 ……………………… 359
关键术语 …………………………… 363
章节总结 …………………………… 363
编程练习 …………………………… 363

附录 A C++ 关键字 ……………… 367
附录 B ASCII 字符集 …………… 368
附录 C 运算符优先级表 ………… 369
附录 D 数字系统 ………………… 371
附录 E 按位运算 ………………… 375
附录 F 使用命令行参数 ………… 376
附录 G 枚举类型 ………………… 379
附录 H 正则表达式 ……………… 383
附录 I   大 O、大 Omega 和大 Theta
       表示法 ……………………… 390

第 17 章

Introduction to C++ Programming and Data Structures, Fifth Edition

递　　归

学习目标

1. 描述什么是递归函数以及使用递归的好处（17.1 节）。
2. 为递归数学函数开发递归程序（17.2 ～ 17.3 节）。
3. 解释如何在调用栈中处理递归函数调用（17.2 ～ 17.3 节）。
4. 递归地思考（17.4 节）。
5. 使用重载的辅助函数来派生递归函数（17.5 节）。
6. 使用递归解决选择排序（17.5.1 节）。
7. 使用递归解决二分查找（17.5.2 节）。
8. 使用递归解决汉诺塔问题（17.6 节）。
9. 使用递归解决八皇后问题（17.7 节）。
10. 理解递归和迭代之间的关系和区别（17.8 节）。
11. 了解尾递归函数以及为什么它们是可取的（17.9 节）。

17.1　简介

要点提示：递归是一种技术，它可以为固有的递归问题提供优雅的解决方案。

假设你希望打印字符串的所有排列。例如，对于字符串 abc，其排列为 abc、acb、bac、bca、cab 和 cba。你是如何解决这个问题的？有几种方法可以做到。一个直观且高效的解决方案是使用递归。

如图 17.1 所示，H 树在超大规模集成电路（VLSI）设计中用作时钟分布网络，用于以相等的传播延迟将定时信号路由到芯片的所有部分。如何编写显示 H 树的程序？一种好方法是使用递归。

提示：你可以输入一个新的序号，然后按 Enter 键来显示一个新的 H 树。

图 17.1　H 形动画——使用递归显示 H 树

使用递归就是使用递归函数编程，递归函数是调用自身的函数。递归是一种有用的编程技术。在某些情况下，它使你能够为其他困难问题开发一个自然、直接、简单的解决方案。本章介绍递归程序设计的概念和技术，并举例说明如何"递归思考"。

17.2 案例研究：计算阶乘

要点提示：递归函数是一个调用自身的函数。

许多数学函数都是使用递归定义的。我们从一个演示递归的简单示例开始。

数字 n 的阶乘可以递归地定义如下：

```
0! = 1;
n! = n × (n - 1)!; n > 0
```

对于给定的 n，怎么求 n!？我们很容易求得 1!，因为已知 0! 是 1，1! 是 1×0!。假设已知 (n-1)!，可以立即使用 n×(n-1)! 得到 n!。因此，计算 n! 被简化为计算 (n-1)!。当计算 (n-1)! 时，可以递归地应用相同的思想，直到 n 减少到 0。

设 `factorial(n)` 是计算 n! 的函数。如果用 n=0 调用函数，会立即返回结果。函数知道如何解决最简单的情况，即**基本情况**或**停止条件**。如果用 n>0 调用函数，函数会将问题简化为计算 n-1 的阶乘的子问题。这个子问题本质上与原问题相同，但比原问题更简单或更小。因为子问题具有与原问题相同的属性，可以用不同的参数调用函数，这被称为**递归调用**。

计算 `factorial(n)` 的递归算法可以如下简单描述：

```
if (n == 0)
    return 1;
else
    return n * factorial(n - 1);
```

一个递归调用可能会导致更多的递归调用，因为函数将子问题划分为新的子问题。为了终止递归函数，必须最终将问题简化为停止情况。此时，函数将向调用者返回一个结果。然后，调用者执行计算并将结果返回给它自己的调用者。这个过程一直持续到将结果传递回原始调用者。原来的问题现在可以通过将 n 乘以 `factorial(n-1)` 的结果来解决。

LiveExample 17.1 是一个完整的程序，它提示用户输入一个非负整数，并显示该数字的阶乘。

CodeAnimation[⊖] 17.1 的互动程序请访问 https://liangcpp.pearsoncmg.com/codeanimation5ecpp/ComputeFactorial.html，LiveExample 17.1 的互动程序请访问 https://liangcpp.pearsoncmg.com/LiveRunCpp5e/faces/LiveExample.xhtml?header=off&programName=ComputeFactorial&fileType=.cpp&programHeight=490&resultHeight=180。

CodeAnimation 17.1　ComputeFactorial.cpp

```
1  #include <iostream>
2  using namespace std;
```

⊖ CodeAnimation 和 LiveExample 代码一致，在后文中不再给出 CodeAnimation 的截图，有需要的读者可自行访问互动程序网址。——编辑注

```cpp
 3
 4  // Return the factorial for a specified index
 5  long long factorial(int);
 6
 7  int main()
 8  {
 9    // Prompt the user to enter an integer
10    cout << "Enter a non-negative integer: ";
11    int n;
12    cin >> n;
13
14    // Display factorial
15    cout << "Factorial of " << n << " is " << factorial(n);
16
17    return 0;
18  }
19
20  // Return the factorial for a specified index
21  long long factorial(int n)
22  {
23    if (n == 0) // Base case
24      return 1;
25    else
26      return n * factorial(n - 1); // Recursive call
27  }
```

LiveExample 17.1   ComputeFactorial.cpp

Source Code Editor:

```cpp
#include <iostream>
using namespace std;

// Return the factorial for a specified index
long long factorial(int);

int main()
{
  // Prompt the user to enter an integer
  cout << "Enter a non-negative integer: ";
  int n;
  cin >> n;

  // Display factorial
  cout << "Factorial of " << n << " is " << factorial(n);

  return 0;
}

// Return the factorial for a specified index
long long factorial(int n)
{
  if (n == 0) // Base case
    return 1;
  else
    return n * factorial(n - 1); // Recursive call
} // You can view the Code Animation for this program from
// https://liangcpp.pearsoncmg.com/codeanimation5ecpp/ComputeFactorial.html
```

```
Enter input data for the program (Sample data provided below. You may modify it.)
9
```
[Automatic Check] [Compile/Run] [Reset] [Answer]   Choose a Compiler: VC++

Execution Result:
```
command>cl ComputeFactorial.cpp
Microsoft C++ Compiler 2019
Compiled successful (cl is the VC++ compile/link command)

command>ComputeFactorial
Enter a non-negative integer: 9
Factorial of 9 is 362880

command>
```

factorial 函数（第 21 ~ 27 行）本质上是将阶乘的递归数学定义直接转换为 C++ 代码。对 factorial 的调用是递归的，因为它调用自己。传递给 factorial 的参数递减，直到达到 0 的基本情况。

现在我们知道了如何写递归函数。那递归是如何工作的呢？图 17.2 展示了递归调用的执行过程，从 n=4 开始。

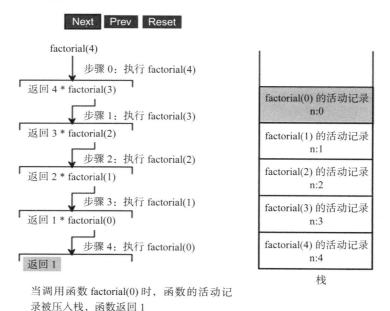

图 17.2 递归调用堆动画——调用 factorial(4) 派生对 factorial 的递归调用

注意：如果递归不能最终收敛到基本情况，或者没有指定基本情况，则可能发生**无限递归**。例如，假设你错误地将 factorial 函数写成如下形式：

```cpp
long long factorial(int n)
{
    return n * factorial(n - 1);
}
```

该函数将无限运行并导致栈溢出。

教学提示：使用循环实现 `factorial` 函数更简单、更高效。然而，递归 `factorial` 函数是演示递归概念的一个很好的例子。

注意：到目前为止讨论的示例显示了一个调用自身的递归函数，这被称为直接递归。也可以创建间接递归。当函数 A 调用函数 B，而函数 B 又调用函数 A 时，就发生这种情况。递归中甚至可以包含更多的函数。例如，函数 A 调用函数 B，函数 B 调用函数 C，函数 C 调用函数 A。

17.3 案例研究：斐波那契数

要点提示：在某些情况下，递归能够帮助我们创建一个直观、直接、简单的问题解决方案。

上一节中的 `factorial` 函数可以在不使用递归的情况下轻松重写。但在某些情况下，使用递归可以为程序提供一个自然、直接、简单的解决方案，否则很难解决问题。考虑著名的斐波那契数列问题，如下所示：

```
数列: 0 1 1 2 3 5 8 13 21 34 55 89 ...
索引: 0 1 2 3 4 5 6 7  8  9 10 11
```

斐波那契数列以 0 和 1 开始，后面的每个数字都是数列中该数前两个数字的和。该数列可以递归定义如下：

```
fib(0) = 0;
fib(1) = 1;
fib(index) = fib(index - 2) + fib(index - 1); index >= 2
```

斐波那契数列是以中世纪数学家莱昂纳多·斐波那契的名字命名的，他发现了斐波那契数列并用来模拟兔子种群的增长。它可以应用于数值优化和其他各种领域。

如何找到给定 `index` 的 `fib(index)`？找到 `fib(2)` 很容易，因为知道 `fib(0)` 和 `fib(1)`。假设知道 `fib(index-2)` 和 `fib(index-1)`，那么可以立即获得 `fib(index)`。因此，计算 `fib(index)` 的问题简化为计算 `fib(index-2)` 和 `fib(index-1)`。当计算 `fib(index-2)` 和 `fib(index-1)` 时，递归地应用这个思想，直到 `index` 减少到 0 或 1。

基本情况是 `index=0` 或 `index=1`。如果用 `index=0` 或 `index=1` 调用函数，它会立即返回结果。如果用 `index>=2` 调用函数，它会将问题分为两个子问题，分别使用递归调用计算 `fib(index-1)` 和 `fib(index-2)`。计算 `fib(index)` 的递归算法可以简单地描述为

```
if (index == 0)
  return 0;
else if (index == 1)
  return 1;
else
  return fib(index - 1) + fib(index - 2);
```

LiveExample17.2 是一个完整的程序，它提示用户输入索引并计算该索引的斐波那契数。LiveExample 17.2 的互动程序请访问 https://liangcpp.pearsoncmg.com/LiveRunCpp5e/faces/LiveExample.xhtml?header=off&programName=ComputeFibonacci&programHeight=530&resultHeight=180。

LiveExample 17.2　ComputeFibonacci.cpp

Source Code Editor:
```cpp
#include <iostream>
using namespace std;

// The function for finding the Fibonacci number
int fib(int);

int main()
{
  // Prompt the user to enter an integer
  cout << "Enter an index for the Fibonacci number: ";
  int index;
  cin >> index;

  // Display factorial
  cout << "Fibonacci number at index " << index << " is "
    << fib(index) << endl;

  return 0;
}

// The function for finding the Fibonacci number
int fib(int index)
{
  if (index == 0) // Base case
    return 0;
  else if (index == 1) // Base case
    return 1;
  else // Reduction and recursive calls
    return fib(index - 1) + fib(index - 2);
}
```

Enter input data for the program (Sample data provided below. You may modify it.)

9

[Automatic Check] [Compile/Run] [Reset] [Answer]　　Choose a Compiler: VC++ ▽

Execution Result:
```
command>cl ComputeFibonacci.cpp
Microsoft C++ Compiler 2019
Compiled successful (cl is the VC++ compile/link command)

command>ComputeFibonacci
Enter an index for the Fibonacci number: 9
Fibonacci number at index 9 is 34

command>
```

该程序没有显示计算机在幕后所做的大量工作。然而，图17.3显示了用于计算fib(4)的连续递归调用。原始函数fib(4)进行两次递归调用，即fib(3)和fib(2)，然后返回fib(3)+fib(2)。但这些函数的调用顺序是什么？在C++中，二元运算符+的操作数可以按任意顺序求值。假设它是从左到右计算的。图17.3中的标签显示了调用函数的顺序。

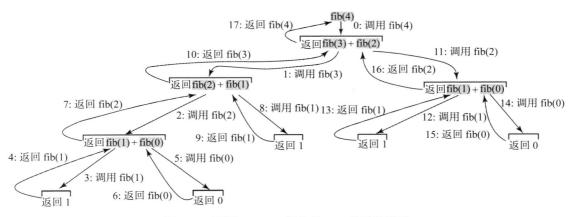

图 17.3 调用 fib(4) 派生对 fib 的递归调用

如图 17.3 所示，存在许多重复的递归调用。例如，fib(2) 被调用两次，fib(1) 被调用三次，而 fib(0) 被调用两次。一般来说，计算 fib(index) 所需的递归调用是计算 fib(index-1) 所需的两倍。当尝试更大的索引值时，调用次数会显著增加，如表 17.1 所示。

表 17.1 fib(index) 的递归调用次数

索引	2	3	4	10	20	30	40	50
调用次数	3	5	9	177	21 891	2 692 537	331 160 281	2 075 316 483

教学提示：fib 函数的递归实现非常简单明了，但效率不高，因为它需要更多的时间和空间来运行递归函数。有关使用循环的高效解决方案，请参阅编程练习 17.2。递归 fib 函数是演示如何编写递归函数的一个很好的例子，尽管它并不实用。

17.4 使用递归解决问题

要点提示：如果你递归地思考，可以使用递归来解决很多问题。

前面介绍了两个经典的递归示例。所有递归函数都具有以下特征：

- 该函数使用 if-else 或 switch 语句实现，会导致不同的情况。
- 一个或多个基本情况（最简单的情况）用于停止递归。
- 每次递归调用都会缩小原始问题，使其越来越接近基本情况，直到它变成基本情况。

通常，使用递归解决问题时，需要将其分解为子问题。如果子问题类似于原始问题，则可以应用相同的方法递归地求解子问题。这个子问题在性质上几乎与原始问题相同，只是规模较小。

递归无处不在。递归地思考很有趣。考虑喝咖啡的情况。可以如下所示递归地描述该过程：

```
void drinkCoffee(Cup& cup)
{
  if (!cup.isEmpty())
  {
    cup.takeOneSip(); // Take one sip
    drinkCoffee(cup);
  }
}
```

假设 cup 是一杯咖啡的对象,它具有实例函数 isEmpty() 和 takeOneSip()。我们可以将问题分解为两个子问题:一个是喝一口咖啡,另一个是把杯子里剩下的咖啡都喝了。第二个问题与原始问题相同,但规模较小。这个问题的基本情况是 cup 空了。

让我们考虑一个简单的问题,将一条消息打印 n 次。可以将问题分解为两个子问题:一个是打印一次消息,另一个是将消息打印 n-1 次。第二个问题与原始问题相同,只是规模较小。这个问题的基本情况是 n=0。使用递归可以解决此问题,如下所示:

```cpp
void nPrintln(const string& message, int times)
{
  if (times >= 1)
  {
    cout << message << endl;
    nPrintln(message, times - 1);
  } // The base case is times == 0
}
```

注意,LiveExample17.2 中的 fib 函数会向其调用者返回一个值,但 nPrintln 函数是 void,并不会向调用者返回值。

如果递归地思考,那么在基础篇中提出的许多问题都可以使用递归来解决。考虑 LiveExample 5.15 中的回文问题。回想一下,如果一个字符串从左到右和从右到左读起来都是一样的,那么它就是回文。例如,mom 和 dad 是回文,但 uncle 和 aunt 不是。检查字符串是否为回文的问题可以分为两个子问题:

- 检查字符串的第一个字符和最后一个字符是否相同。
- 忽略这两个边字符,检查子字符串的其余部分是否为回文。

第二个子问题与原始问题相同,但规模较小。有两种基本情况:(1)两个结束字符不相同;(2)字符串大小为 0 或 1。在情况 1 中,字符串不是回文;在情况 2 中,字符串是回文。这个问题的递归函数可以在 LiveExample17.3 中实现。

LiveExample 17.3 的互动程序请访问 https://liangcpp.pearsoncmg.com/LiveRunCpp5e/faces/LiveExample.xhtml?header=off&programName=RecursivePalindrome&fileType=.cpp&programHeight=480&resultHeight=180。

LiveExample 17.3   RecursivePalindrome.cpp

```cpp
#include <iostream>
#include <string>
using namespace std;

bool isPalindrome(const string& s)
{
  if (s.size() <= 1) // Base case
    return true;
  else if (s[0] != s[s.size() - 1]) // Base case
    return false;
  else
    return isPalindrome(s.substr(1, s.size() - 2));
}

int main()
{
  cout << "Enter a string: ";
```

```
18      string s;
19      getline(cin, s);
20
21      if (isPalindrome(s))
22        cout << s << " is a palindrome" << endl;
23      else
24        cout << s << " is not a palindrome" << endl;
25
26      return 0;
27    }
```

Enter input data for the program (Sample data provided below. You may modify it.)

abccba

[Automatic Check] [Compile/Run] [Reset] [Answer]   Choose a Compiler: VC++

Execution Result:

```
command>cl RecursivePalindrome.cpp
Microsoft C++ Compiler 2019
Compiled successful (cl is the VC++ compile/link command)

command>RecursivePalindrome
Enter a string: abccba
abccba is a palindrome

command>
```

isPalindrome 函数检查字符串的大小是否小于或等于 1（第 7 行）。如果是，则该字符串为回文。该函数检查字符串的第一个元素和最后一个元素是否相同（第 9 行）。如果不相同，则该字符串不是回文。否则，使用 s.substr(1, s.size()-2) 获得 s 的子字符串，并使用新字符串递归调用 isPalindrome（第 12 行）。

17.5 递归辅助函数

要点提示： 有时，通过为类似于原始问题的问题定义递归函数，可以找到原始问题的解决方案。这个新函数称为递归辅助函数。原始问题可以通过调用递归辅助函数来解决。

前面的递归函数 isPalindrome 效率不高，因为它为每个递归调用创建一个新字符串。为了避免创建新字符串，可以使用 low 索引和 high 索引来指示子字符串的范围。这两个索引必须传递给递归函数。由于原始函数是 isPalindrome(const string& s)，因此必须创建一个新的**递归辅助函数** isPalindrome(const string& s,int low, int high) 来接收有关字符串的附加信息，如 LiveExample 17.4 所示。

LiveExample 17.4 的互动程序请访问 https://liangcpp.pearsoncmg.com/LiveRunCpp5e/faces/LiveExample.xhtml?header=off&programName=RecursivePalindromeUsingHelperFunction&fileType=.cpp&programHeight=560&resultHeight=180。

LiveExample 17.4 RecursivePalindromeUsingHelperFunction.cpp

Source Code Editor:
```
1  #include <iostream>
2  #include <string>
```

```cpp
 3  using namespace std;
 4
 5  bool isPalindrome(const string& s, int low, int high)
 6  {
 7    if (high <= low) // Base case
 8      return true;
 9    else if (s[low] != s[high]) // Base case
10      return false;
11    else
12      return isPalindrome(s, low + 1, high - 1);
13  }
14
15  bool isPalindrome(const string& s)
16  {
17    return isPalindrome(s, 0, s.size() - 1);
18  }
19
20  int main()
21  {
22    cout << "Enter a string: ";
23    string s;
24    getline(cin, s);
25
26    if (isPalindrome(s))
27      cout << s << " is a palindrome" << endl;
28    else
29      cout << s << " is not a palindrome" << endl;
30
31    return 0;
32  }
```

Enter input data for the program (Sample data provided below. You may modify it.)

abccba

Execution Result:

```
command>cl RecursivePalindromeUsingHelperFunction.cpp
Microsoft C++ Compiler 2019
Compiled successful (cl is the VC++ compile/link command)

command>RecursivePalindromeUsingHelperFunction
Enter a string: abccba
abccba is a palindrome

command>
```

这里定义了两个重载的 isPalindrome 函数。函数 isPalindrome(const string& s)（第 15 行）检查字符串是否为回文，第二个函数 isPalindrome(const string& s, int low, int high)（第 5 行）检查子字符串 s(low..high) 是否为回文。第一个函数将 low=0 和 high=s.size()-1 的字符串 s 传递给第二个函数。第二个函数可以递归调用，以检查不断收缩的子字符串是否为回文。在递归程序设计中，定义接收附加参数的第二个函数是一种常见的设计技术。这样的函数被称为**递归辅助函数**。

辅助函数对于设计涉及字符串和数组的问题的递归解决方案非常有用。下面再举两个例子。

17.5.1 选择排序

选择排序在 7.10 节中介绍过。现在我们为字符串中的字符引入递归选择排序。选择排序的变体如下所示。它会找到列表中最大的元素并将其放在最后。然后，它会找到剩余的最大元素，并将其放在最后一个元素的前面，以此类推，直到列表只包含一个元素。该问题可分为两个子问题：
- 找到列表中最大的元素，并将其与最后一个元素交换。
- 忽略最后一个元素，并递归地对剩余的较小列表进行排序。

基本情况是列表只包含一个元素。

LiveExample17.5 给出了递归排序函数。

LiveExample 17.5 的互动程序请访问 https://liangcpp.pearsoncmg.com/LiveRunCpp5e/faces/LiveExample.xhtml?header=off&programName=RecursiveSelectionSort&programHeight=780&resultHeight=160。

LiveExample 17.5　RecursiveSelectionSort.cpp

```cpp
#include <iostream>
#include <string>
using namespace std;

void sort(string& s)
{
  sort(s, s.size() - 1);
}

void sort(string& s, int high)
{
  if (high > 0)
  {
    // Find the largest element and its index
    int indexOfMax = 0;
    char max = s[0];
    for (int i = 1; i <= high; i++)
    {
      if (s[i] > max)
      {
        max = s[i];
        indexOfMax = i;
      }
    }

    // Swap the largest with the last element in the list
    s[indexOfMax] = s[high];
    s[high] = max;

    // Sort the remaining list
    sort(s, high - 1);
  }
}

int main()
{
  cout << "Enter a string: ";
  string s;
  getline(cin, s);

  sort(s);
```

```
43    cout << "The sorted string is " << s << endl;
44
45    return 0;
46  }
```

Enter input data for the program (Sample data provided below. You may modify it.)

abccba

[Automatic Check] [Compile/Run] [Reset] [Answer]   Choose a Compiler: [VC++ ▼]

Execution Result:

```
command>cl RecursiveSelectionSort.cpp
Microsoft C++ Compiler 2019
Compiled successful (cl is the VC++ compile/link command)

command>RecursiveSelectionSort
Enter a string: abccba
The sorted string is aabbcc

command>
```

第 5～33 行定义了两个重载 sort 函数。函数 sort(string& s) 对 s[0..s.size() - 1] 中的字符进行排序，第二个函数 sort(string& s, int high) 对 s[0..high] 中的字符进行排序。可以递归调用辅助函数来对不断收缩的子字符串排序。

17.5.2 二分查找

二分查找在 7.9.2 节中介绍过。要使二分查找法奏效，数组中的元素必须已经排序。二分查找首先将键与数组中间的元素进行比较。考虑以下情况：

- 情况 1：如果键小于中间元素，则递归在数组前半部分查找键。
- 情况 2：如果键等于中间元素，则查找以匹配结束。
- 情况 3：如果键大于中间元素，则递归在数组后半部分查找键。

情况 1 和情况 3 将查找已经减小的列表。情况 2 是找到匹配项时的基本情况。另一个基本情况是查找完成但是没有找到匹配项。LiveExample 17.6 为使用递归的二分查找问题提供了一个清晰、简单的解决方案。

LiveExample 17.6 的互动程序请访问 https://liangcpp.pearsoncmg.com/LiveRunCpp5e/faces/LiveExample.xhtml?header=off&programName=RecursiveBinarySearch&fileType=.cpp&programHeight=630&resultHeight=190。

LiveExample 17.6   RecursiveBinarySearch.cpp

Source Code Editor:

```
1  #include <iostream>
2  using namespace std;
3
4  int binarySearch(const int list[], int key, int low, int high)
5  {
6    if (low > high)  // The list has been exhausted without a match
7      return -low - 1; // key not found, return the insertion point
8
9    int mid = (low + high) / 2;
10   if (key < list[mid])
11     return binarySearch(list, key, low, mid - 1);
12   else if (key == list[mid])
13     return mid;
```

```
14      else
15        return binarySearch(list, key, mid + 1, high);
16    }
17
18    int binarySearch(const int list[], int key, int size)
19    {
20      int low = 0;
21      int high = size - 1;
22      return binarySearch(list, key, low, high);
23    }
24
25    int main()
26    {
27      int list[] = {2, 4, 7, 10, 11, 45, 50, 59, 60, 66, 69, 70, 79};
28      int i = binarySearch(list, 2, 13); // Returns 0
29      int j = binarySearch(list, 11, 13); // Returns 4
30      int k = binarySearch(list, 12, 13); // Returns -1
31
32      cout << "binarySearch(list, 2, 13) returns " << i << endl;
33      cout << "binarySearch(list, 11, 13) returns " << j << endl;
34      cout << "binarySearch(list, 12, 13) returns " << k << endl;
35
36      return 0;
37    }
```

Execution Result:

```
command>cl RecursiveBinarySearch.cpp
Microsoft C++ Compiler 2019
Compiled successful (cl is the VC++ compile/link command)

command>RecursiveBinarySearch
binarySearch(list, 2, 13) returns 0
binarySearch(list, 11, 13) returns 4
binarySearch(list, 12, 13) returns -6

command>
```

第 18 行中的 binarySearch 函数在整个列表中查找键。第 4 行中的辅助函数 binarySearch 在由索引从 low 到 high 确定的列表中查找键。

第 18 行中的 binarySearch 函数将 low=0 和 high=size-1 的初始数组传递给辅助函数 binarySearch。辅助函数被递归调用，以便在不断缩小的子数组中找到键。

17.6 汉诺塔

要点提示：用递归很容易解决经典的汉诺塔问题，但用其他方法很难解决。

汉诺塔问题是一个经典的递归的例子。使用递归很容易地解决它，但用其他方法很难解决这个问题。

该问题涉及将指定数量的不同大小的盘子从一个塔移动到另一个塔，同时遵守以下规则：
- 有 n 个盘子，分别标记为 1，2，3，⋯，n，还有标记为 A、B 和 C 的三个塔。
- 任何时候都不能在较小盘子的上方放置比它大的盘子。
- 所有盘子最初都放置在塔 A 上。
- 一次只能移动一个盘子，并且必须是塔顶部的盘子。

目标是在 C 的帮助下将所有盘子从 A 移动到 B。例如，如果有三个盘子，则将所有盘子从 A 移到 B，如图 17.4 所示。

如果是三个盘子，则可以手动查找解决方案。但是，对于数量更多的盘子（即使是 4 个），问题将相当复杂。幸运的是，这个问题具有递归性质，因而可以得到一个简单的递归解决方案。

提示：单击 Start 按钮将所有的盘子从塔 A 移动到塔 B。单击 Reset 按钮将重置到初始状态。

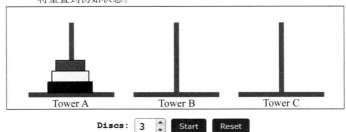

图 17.4 汉诺塔动画。汉诺塔问题的目标是在不违反规则的情况下将盘子从塔 A 移动到塔 B

这个问题的基本情况是 n=1。如果 n==1，可以简单地将盘子从 A 移动到 B。当 n>1 时，可以将原始问题拆分为三个子问题并依次求解。

1. 在塔 B 的帮助下，递归地将前 n-1 个盘子从 A 移动到 C，如图 17.5 中的步骤 1 所示。
2. 将盘子 n 从 A 移动到 B，如图 17.5 中的步骤 2 所示。
3. 在塔 A 的帮助下，递归地将 n-1 个盘子从 C 移动到 B，如图 17.5 中的步骤 3 所示。

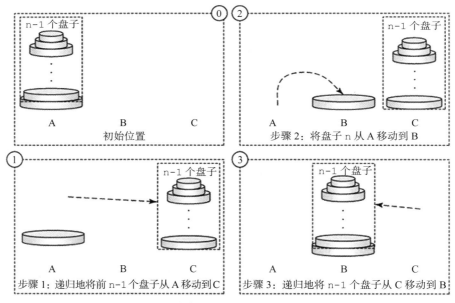

图 17.5 汉诺塔问题可以分解为三个子问题

下面的函数在 auxTower 的帮助下将 *n* 个盘子从 fromTower 移动到 toTower：

```
void moveDisks(int n, char fromTower, char toTower, char auxTower)
```

该函数的算法如下：

```
if (n == 1) // 停止条件
    将盘子 1 从 fromTower 移动到 toTower;
else
```

```
    {
      moveDisks(n - 1, fromTower, auxTower, toTower);
      将盘子 n 从 fromTower 移动到 toTower;
      moveDisks(n - 1, auxTower, toTower, fromTower);
    }
```

LiveExample 17.7 提示用户输入盘子数量,并调用递归函数 `moveDisks` 来显示移动盘子的解决方案。

LiveExample 17.7 的互动程序请访问 https://liangcpp.pearsoncmg.com/LiveRunCpp5e/faces/LiveExample.xhtml?header=off&programName=TowerOfHanoi&fileType=.cpp&programHeight=580&resultHeight=300。

LiveExample 17.7 TowerOfHanoi.cpp

Source Code Editor:

```cpp
#include <iostream>
using namespace std;

// The function for finding the solution to move n disks
// from fromTower to toTower with auxTower
void moveDisks(int n, char fromTower,
    char toTower, char auxTower)
{
  if (n == 1) // Stopping condition
    cout << "Move disk " << n << " from " <<
      fromTower << " to " << toTower << endl;
  else
  {
    moveDisks(n - 1, fromTower, auxTower, toTower);
    cout << "Move disk " << n << " from " <<
      fromTower << " to " << toTower << endl;
    moveDisks(n - 1, auxTower, toTower, fromTower);
  }
}

int main()
{
  // Read number of disks, n
  cout << "Enter number of disks: ";
  int n;
  cin >> n;

  // Find the solution recursively
  cout << "The moves are: " << endl;
  moveDisks(n, 'A', 'B', 'C');

  return 0;
}
```

Enter input data for the program (Sample data provided below. You may modify it.)

```
3
```

Automatic Check | Compile/Run | Reset | Answer   Choose a Compiler: VC++

Execution Result:

```
command>cl TowerOfHanoi.cpp
Microsoft C++ Compiler 2019
Compiled successful (cl is the VC++ compile/link command)
```

```
command>TowerOfHanoi
Enter number of disks: 3
The moves are:
Move disk 1 from A to B
Move disk 2 from A to C
Move disk 1 from B to C
Move disk 3 from A to B
Move disk 1 from C to A
Move disk 2 from C to B
Move disk 1 from A to B

command>
```

这个问题本质上是递归的。使用递归可以找到一个自然、简单的解决方案。如果不使用递归，则很难解决这个问题。

考虑跟踪 n=3 的程序。连续的递归调用如图 17.6 所示。正如所看到的，编写程序比跟踪递归调用更容易。系统使用栈在后台跟踪调用。在某种程度上，递归提供了一个向用户隐藏迭代和其他细节的抽象级别。

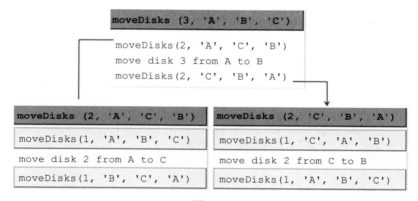

图 17.6

17.7 八皇后问题

要点提示：八皇后问题可以使用递归来解决。

本节给出了八皇后问题的递归解法。任务是在棋盘上的每一行放置一个皇后，要满足任意两个皇后都不能互相攻击。可以使用二维数组表示棋盘。但由于每行只能有一个皇后，因此使用一维数组表示皇后在该行中的位置就足够了。因此，我们如下声明 queens 数组：

```
int queens[8];
```

将 j 赋值给 queen[i] 表示皇后被放置在第 i 行和第 j 列中，如图 17.7 所示。将 -1 赋值给 queen[i] 表示第 i 行还没有放置皇后。

LiveExample17.8 是一个为八皇后问题找到解决方案的程序。

LiveExample 17.8 的互动程序请访问 https://liangcpp.pearsoncmg.com/LiveRunCpp5e/faces/LiveExample.xhtml?header=off&programName=EightQueen&fileType=.cpp&programHeight=900&resultHeight=430。

◀)) 提示：这个动画展示了八皇后算法的查找过程。单击 Next 按钮查找下一个皇后的新位置，或者回溯查找前一行中的新位置。

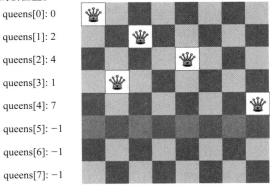

图 17.7 八皇后动画。queens[i] 表示第 i 行中皇后的位置

LiveExample 17.8   EightQueen.cpp

Source Code Editor:

```cpp
#include <iostream>
using namespace std;

const int NUMBER_OF_QUEENS = 8; // Constant: eight queens
int queens[NUMBER_OF_QUEENS];

// Check whether a queen can be placed at row i and column j
bool isValid(int row, int column)
{
  for (int i = 1; i <= row; i++)
    if (queens[row - i] == column       // Check column
      || queens[row - i] == column - i  // Check upper left diagonal
      || queens[row - i] == column + i) // Check upper right diagonal
      return false; // There is a conflict
  return true; // No conflict
}

// Display the chessboard with eight queens
void printResult()
{
  cout << "\n---------------------------------\n";
  for (int row = 0; row < NUMBER_OF_QUEENS; row++)
  {
    for (int column = 0; column < NUMBER_OF_QUEENS; column++)
      printf(column == queens[row] ? "| Q " : "|   ");
    cout << "|\n---------------------------------\n";
  }
}

// Search to place a queen at the specified row
bool search(int row)
{
  if (row == NUMBER_OF_QUEENS) // Stopping condition
    return true; // A solution found to place 8 queens in 8 rows

  for (int column = 0; column < NUMBER_OF_QUEENS; column++)
  {
    queens[row] = column; // Place a queen at (row, column)
```

```
39        if (isValid(row, column) && search(row + 1))
40          return true; // Found, thus return true to exit for loop
41      }
42
43      // No solution for a queen placed at any column of this row
44      return false;
45    }
46
47    int main()
48    {
49      search(0); // Start search from row 0. Note row indices are 0 to 7
50      printResult(); // Display result
51
52      return 0;
53    }
```

[Automatic Check] [Compile/Run] [Reset] [Answer]　　　　　　Choose a Compiler: VC++ ∨

Execution Result:

```
command>cl EightQueen.cpp
Microsoft C++ Compiler 2019
Compiled successful (cl is the VC++ compile/link command)

command>EightQueen

---------------------------------
| Q |   |   |   |   |   |   |   |
---------------------------------
|   |   |   |   | Q |   |   |   |
---------------------------------
|   |   |   |   |   |   |   | Q |
---------------------------------
|   |   |   |   |   | Q |   |   |
---------------------------------
|   |   | Q |   |   |   |   |   |
---------------------------------
|   |   |   |   |   |   | Q |   |
---------------------------------
|   | Q |   |   |   |   |   |   |
---------------------------------
|   |   |   | Q |   |   |   |   |
---------------------------------
command>
```

该程序调用 search(0)（第 49 行）在第 0 行开始查找解决方案，该查找递归地调用 search(1)，search(2)，…，search(7)（第 39 行）。

如果所有行都已填充（第 39～40 行），递归 search(row) 函数将返回 true。该函数在 for 循环中检查皇后是否可以放置在第 0，1，2，…，7 列（第 36 行）。将皇后放置在列中（第 38 行）。如果放置有效，则通过调用 search(row+1)（第 39 行）递归查找下一行。如果查找成功，则返回 true（第 40 行）退出 for 循环。在这种情况下，不需要查找行中的下一列。如果没有找到将皇后放在此行任何列上的解决方案，则函数返回 false（第 44 行）。

假设 row 为 3，调用 search(row)，如图 17.8 所示。函数尝试按以下顺序在第 0，1，2，…列中填充一个皇后。对于每个试验，都会调用 isValid(row, column) 函数（第 39 行）来检查将皇后放在指定位置是否会导致与放在该行之前的皇后发生冲突。它确保没有皇后被放在同一列（第11行），没有皇后被放置在左上对角线（第12行），也没有皇后被置于右上对角线（第13行），如图17.8所示。如果 isValid(row, column) 返回 false，就检查下一列。如果 isValid(row, column) 返回 true，则递归调用 search(row+1)。如果 search(row+1) 返回 false，则检查上一行的下一列。

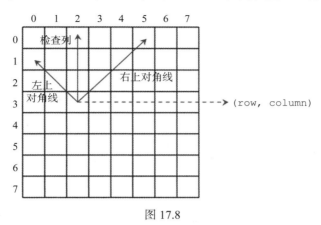

图 17.8

17.8 递归与迭代

要点提示：递归是程序控制的另一种形式。它本质上是没有循环控制的重复。

递归是程序控制的另一种形式。它本质上是没有循环控制的重复。使用循环时，可以指定循环体。循环体的重复由循环控制结构控制。在递归中，函数本身被重复调用，须用选择语句控制是否递归调用函数。

递归产生大量开销。每次程序调用函数时，系统都必须为所有函数的局部变量和参数分配空间。这可能会消耗大量内存，并且需要额外的时间来管理额外的空间。

任何能递归解决的问题都能通过迭代以非递归方式解决。递归有一些缺点：使用了太多的时间和太多的内存。那么，为什么要用递归呢？因为在某些情况下，使用递归可以为难以解决的固有递归问题指定一个清晰、简单的解决方案。汉诺塔问题就是这样一个例子，不使用递归很难解决。

使用递归还是迭代取决于你试图解决的问题的性质和你对它的理解。经验法则是使用这两种方法中最能开发出自然反映问题的直观解决方案的方法。如果迭代解决方案是显而易见的，那么就用它。它通常会比递归效率更高。

注意：递归程序可能会用光内存，导致栈溢出运行时错误。

提示：如果关注程序的性能，则应避免使用递归，因为它比迭代花费更多的时间和内存。

17.9 尾递归

要点提示：尾递归函数对于减少栈空间是有效的。

如果递归调用返回时没有要执行的操作，则递归函数称为**尾递归**，如图17.9a所示。但图17.9b中的函数 B 不是尾递归，因为函数调用返回后有要执行的操作。

例如，LiveExample 17.4 中的递归 isPalindrome 函数（第5～13行）是尾递归的，因为在第12行递归调用 isPalindrome 后没有等待执行的操作。但 LiveExample 17.1 中的递归 factorial 函数（第21～27行）不是尾递归的，因为每次递归调用返回时都要执行一个操作，即乘法。

```
Recursive Function A
  ...
  ...
  ...
  ...
  Invoke function A recursively
```
a) 尾递归

```
Recursive Function B
  ...
  ...
  Invoke function B recursively
  ...
  ...
```
b) 非尾递归

图 17.9 尾递归函数在递归调用后没有要执行的操作

尾递归是可取的，因为函数在最后一次递归调用结束时结束。因此，不需要将中间调用存储在栈中。现代编译器对尾递归进行优化以减少栈空间。

非尾递归函数通常可以通过使用辅助参数转换为尾递归函数。这些参数用来存放结果。其思想是将等待执行的操作合并到辅助参数中，使递归调用不再有等待执行的操作。可以用辅助参数定义一个新的辅助递归函数。此函数可能会重载具有相同名称但不同签名的原始函数。例如，LiveExample 17.1 中的 `factorial` 函数可以在 LiveExample 17.9 中以尾递归的方式编写。此程序使用调用栈来显示尾递归函数的高效。

LiveExample 17.9 的互动程序请访问 https://liangcpp.pearsoncmg.com/LiveRunCpp5e/faces/LiveExample.xhtml?header=off&programName=ComputeFactorialTailRecursion&fileType=.cpp&programHeight=650&resultHeight=180。

LiveExample 17.9　ComputeFactorialTailRecursion.cpp

Source Code Editor:

```cpp
#include <iostream>
using namespace std;

// Return the factorial for a specified index
long long factorial(int);

// Auxiliary tail-recursive function for factorial
long long factorial(int n, long long result);

int main()
{
  // Prompt the user to enter an integer
  cout << "Please enter a non-negative integer: ";
  int n;
  cin >> n;

  // Display factorial
  cout << "Factorial of " << n << " is " << factorial(n);

  return 0;
}

// Return the factorial for a specified number
long long factorial(int n)
{
  return factorial(n, 1); // Call auxiliary function
}

// Auxiliary tail-recursive function for factorial
```

```
30  long long factorial(int n, long long result)
31  {
32    if (n == 0)
33      return result;
34    else
35      return factorial(n - 1, n * result); // Recursive call
36  } // You can view the Code Animation for this program from
37  //https://liangcpp.pearsoncmg.com/codeanimation5ecpp/ComputeFactorialTailRecursion.html
```

Enter input data for the program (Sample data provided below. You may modify it.)

9

[Automatic Check] [Compile/Run] [Reset] [Answer]   Choose a Compiler: VC++

Execution Result:

```
command>cl ComputeFactorialTailRecursion.cpp
Microsoft C++ Compiler 2019
Compiled successful (cl is the VC++ compile/link command)

command>ComputeFactorialTailRecursion
Please enter a non-negative integer: 9
Factorial of 9 is 362880

command>
```

第一个 factorial 函数只是调用第二个辅助函数（第 26 行）。第二个函数包含一个辅助参数 result，用来存储 n 的阶乘结果。该函数在第 35 行递归调用。返回调用后没有等待执行的操作。最终结果在第 33 行返回，这也是第 26 行调用 factorial(n,1) 的返回值。

关键术语

base case（基本情况）
infinite recursion（无限递归）
recursive function（递归函数）
recursive helper function（递归辅助函数）
stopping condition（停止条件）
tail recursion（尾递归）

章节总结

1. 递归函数是指直接或间接调用自身的函数。要终止递归函数，必须有一个或多个基本情况。
2. 递归是程序控制的另一种形式。它本质上是没有循环控制的重复。它可以用来为原本难以解决的递归问题编写简单、清晰的解决方案。
3. 有时，为了递归调用，需要修改原始函数以接收额外的参数。可以为此目的定义递归辅助函数。
4. 递归产生大量开销。每次程序调用函数时，系统都必须为所有函数的局部变量和参数分配空间。这可能会消耗大量内存，并且需要额外的时间来管理额外的空间。
5. 如果递归调用返回时没有要执行的操作，则递归函数称为尾递归。一些编译器会优化尾递归以减少栈空间。

编程练习

互动程序请访问 https://liangcpp.pearsoncmg.com/CheckExerciseCpp/faces/CheckExercise5e.xhtml?chapter=17&programName=Exercise17_01。

注：题目的难度等级分为容易（无星）、中等（*）、难（**）以及非常难（***）。

17.2 ～ 17.3 节

17.1 （计算阶乘）使用迭代重写 LiveExample 17.1 中的 factorial 函数。

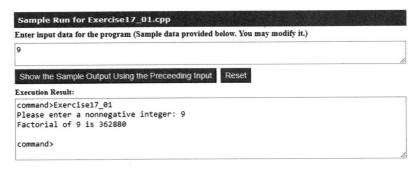

*17.2 （斐波那契数）使用迭代重写 LiveExample17.2 中的 fib 函数。

提示：要在不递归的情况下计算 fib(n)，需要首先获得 fib(n-2) 和 fib(n-1)。设 f0 和 f1 表示前两个斐波那契数。那么当前的斐波那契数是 f0+f1。该算法如下：

```
f0 = 0; // For fib(0)
f1 = 1; // For fib(1)
for (int i = 2; i <= n; i++)
{
  currentFib = f0 + f1;
  f0 = f1;
  f1 = currentFib;
}
// After the loop, currentFib is fib(n)
```

以下是此算法的动画。

提示：计算斐波那契数。单击 Step 按钮查找数列中的下一个斐波那契数。动画在索引 11 处结束。单击 Reset 按钮重新开始。

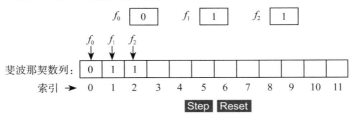

图 17.10 斐波那契数动画

编写一个测试程序，提示用户输入索引并显示其斐波那契数。

17.3 （使用递归计算最大公约数）gcd(m, n) 也可以递归定义如下：
- 如果 m%n 是 0，则 gcd(m, n) 是 n。
- 否则，gcd(m, n) 就是 gcd(n, m%n)。

编写一个递归函数来查找最大公约数。编写一个测试程序，提示用户输入两个整数并显示其最大公约数。

```
Sample Run for Exercise17_03.cpp
Enter input data for the program (Sample data provided below. You may modify it.)
59 57

Show the Sample Output Using the Preceeding Input   Reset
Execution Result:
command>Exercise17_03
Enter the first number: 59
Enter the second number: 57
The GCD of 59 and 57 is 1

command>
```

17.4 （求和级数）编写一个递归函数来计算以下级数：

$$m(i) = 1 + \frac{1}{2} + \frac{1}{3} + \cdots + \frac{1}{i}$$

编写一个测试程序，当 i=1, 2, …, 10 时显示 m(i)。

```
Sample Run for Exercise17_04.cpp
Execution Result:
command>Exercise17_04
i    m(i)
1    1.0000
2    1.5000
3    1.8333
4    2.0833
5    2.2833
6    2.4500
7    2.5929
8    2.7179
9    2.8290
10   2.9290

command>
```

17.5 （求和级数）编写一个递归函数来计算以下级数：

$$m(i) = \frac{1}{3} + \frac{2}{5} + \frac{3}{7} + \frac{4}{9} + \frac{5}{11} + \frac{6}{13} + \cdots + \frac{i}{2i+1}$$

编写一个测试程序，当 i=1, 2, …, 10 时显示 m(i)。

```
Sample Run for Exercise17_05.cpp
Execution Result:
command>Exercise17_05
m(1): 0.333333
m(2): 0.733333
m(3): 1.1619
m(4): 1.60635
m(5): 2.06089
m(6): 2.52243
m(7): 2.9891
m(8): 3.45969
m(9): 3.93337
m(10): 4.40956

command>
```

**17.6 （求和级数）编写一个递归函数来计算以下级数：

$$m(i) = \frac{1}{2} + \frac{2}{3} + \cdots + \frac{i}{i+1}$$

编写一个测试程序，当 i=1，2，…，10 时显示 m(i)。

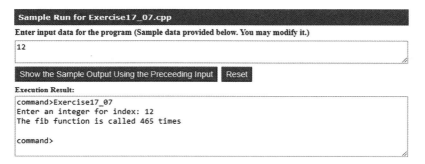

17.7 （斐波那契数列）修改 LiveExample 17.2，使程序找到调用 fib 函数的次数。（提示：使用全局变量，并在每次调用函数时将其递增。）

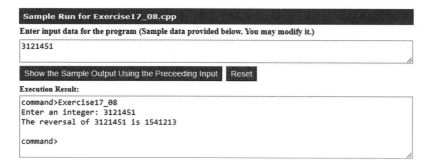

17.4 节

*17.8 （逆序输出整数中的各位数字）使用以下函数头编写一个递归函数，在控制台上逆序显示 int 值：

      void reverseDisplay(int value)

例如，reverseDisplay(12345) 显示 54321。编写一个测试程序，提示用户输入一个整数并逆序显示这个数。

**17.9 （逆序输出字符串中的字符）使用以下函数头编写一个递归函数，在控制台上逆序显示字符串：

      void reverseDisplay(const string& s)

例如，reverseDisplay("abcd") 显示 dcba。编写一个测试程序，提示用户输入字符串并逆序显示该字符串。

```
Sample Run for Exercise17_09.cpp
Enter input data for the program (Sample data provided below. You may modify it.)
abcdef

Execution Result:
command>Exercise17_09
Enter a string: abcdef
The reversal of abcdef is fedcba

command>
```

*17.10 （字符串中指定字符的出现次数）使用以下函数头编写一个递归函数，查找字符串中指定字母的出现次数。

```
int count(const string& s, char a)
```

例如，count("Welcome", 'e') 返回 2。编写一个测试程序，提示用户输入一个字符串和一个字符，并显示该字符在字符串中的出现次数。

```
Sample Run for Exercise17_10.cpp
Enter input data for the program (Sample data provided below. You may modify it.)
Welcome
e

Execution Result:
command>Exercise17_10
Enter a string: Welcome
Enter a character: e
e occurs 2 times in Welcome

command>
```

**17.11 （使用递归对整数中的各位数字求和）使用以下函数头编写一个递归函数，计算整数中各位数字的和：

```
int sumDigits(int n)
```

例如，sumDigits(234) 返回 2+3+4=9。编写一个测试程序，提示用户输入一个整数并显示其各位数字的和。

```
Sample Run for Exercise17_11.cpp
Enter input data for the program (Sample data provided below. You may modify it.)
1231879

Execution Result:
command>Exercise17_11
Ente an integer: 1231879
The sum of digits in 1231879 is 31

command>
```

17.5 节

**17.12 （逆序输出字符串中的字符）重写编程练习 17.9，使用辅助函数将子字符串索引 high 传递给函数。辅助函数头如下所示：

```
void reverseDisplay(const string& s, int high)
```

Sample Run for Exercise17_12.cpp
Enter input data for the program (Sample data provided below. You may modify it.)

abvdefg

Execution Result:
```
command>Exercise17_12
Enter a string: abvdefg
The reversal of abvdefg is gfedvba

command>
```

**17.13 （查找数组中最大的数）编写一个递归函数，返回数组中的最大整数。编写一个测试程序，提示用户输入一个由 8 个整数组成的列表，并显示最大整数。

Sample Run for Exercise17_13.cpp
Enter input data for the program (Sample data provided below. You may modify it.)

12 123 45 223 46 1212 623 87

Execution Result:
```
command>Exercise17_13
Enter 8 integers: 12 123 45 223 46 1212 623 87
The largest number is 1212

command>
```

17.14 （统计字符串中大写字母的数量）编写一个递归函数，返回字符串中的大写字母数量。需要定义以下两个函数。第二个是递归辅助函数。

```
int getNumberOfUppercaseLetters(const string& s)
int getNumberOfUppercaseLetters(const string& s, int high)
```

编写一个测试程序，提示用户输入字符串并显示字符串中的大写字母数。

Sample Run for Exercise17_14.cpp
Enter input data for the program (Sample data provided below. You may modify it.)

Programming is fun

Execution Result:
```
command>Exercise17_14
Enter a string: Programming is fun
The number of uppercase letters in Programming is fun is 1

command>
```

*17.15 （字符串中指定字符的出现次数）重写编程练习 17.10，使用辅助函数将子字符串索引 high 传递给函数。需要定义以下两个函数。第二个是递归辅助函数。

```
int count(const string& s, char a)
int count(const string& s, char a, int high)
```

编写一个测试程序，提示用户输入一个字符串和一个字符，并显示该字符在字符串中的出现次数。

```
Sample Run for Exercise17_15.cpp
Enter input data for the program (Sample data provided below. You may modify it.)
Welcome
e

Show the Sample Output Using the Preceeding Input    Reset
Execution Result:
command>Exercise17_15
Enter a string: Welcome
Enter a character: e
e appears 1 time in Welcome

command>
```

17.6 节

*17.16 （汉诺塔）修改 LiveExample 17.7，使程序得到将 n 个盘子从塔 A 移动到塔 B 所需的移动次数。（提示：使用全局变量，每次调用函数时将其递增。）

```
Sample Run for Exercise17_16.cpp
Enter input data for the program (Sample data provided below. You may modify it.)
2

Show the Sample Output Using the Preceeding Input    Reset
Execution Result:
command>Exercise17_16
Enter number of disks: 2
The moves are:
Move disk 1 from A to C
Move disk 2 from A to B
Move disk 1 from C to B
The number of the moves is 3

command>
```

综合题

***17.17 （字符串排列）编写一个递归函数来打印字符串的所有排列。例如，对于字符串 abc，排列为

```
abc
acb
bac
bca
cab
cba
```

（提示：定义以下两个函数。第二个是辅助函数。）

```
void displayPermuation(const string& s)
void displayPermuation(const string& s1, const string& s2)
```

第一个函数只调用 displayPermuation("", s)。第二个函数使用循环将一个字符从 s2 移动到 s1，并用新的 s1 和 s2 递归调用它。基本情况是 s2 为空，并将 s1 打印到控制台。编写一个测试程序，提示用户输入字符串并显示其所有排列。

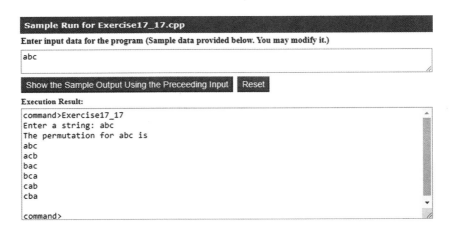

***17.18 （游戏：多个八皇后解决方案）使用递归重写 LiveExample 17.8。

*17.19 （十进制数转换成二进制数）编写一个递归函数，将十进制数转换为字符串形式的二进制数。函数头为：

```
string decimalToBinary(int value)
```

编写一个测试程序，提示用户输入一个十进制数并显示其等效的二进制数。

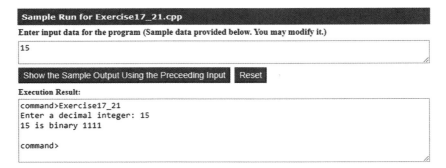

*17.20 （十进制数转换成十六进制数）编写一个递归函数，将十进制数转换为字符串形式的十六进制数。函数头为：

```
string decimalToHex(int value)
```

编写一个测试程序，提示用户输入十进制数并显示其等效的十六进制数。

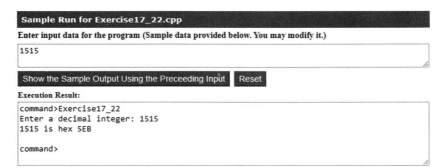

*17.21 （二进制数转换成十进制数）编写一个递归函数，将字符串形式的二进制数转换为十进制整数。

函数头为：

```
int binaryToDecimal(const string& binaryString)
```

编写一个测试程序，提示用户输入一个二进制字符串并显示其等效的十进制数。

```
Sample Run for Exercise17_23.cpp
Enter input data for the program (Sample data provided below. You may modify it.)
10101011111

[Show the Sample Output Using the Preceeding Input]  [Reset]
Execution Result:
command>Exercise17_23
Enter a binary number: 10101011111
10101011111 is decimal 1375

command>
```

*17.22 （十六进制数转换成十进制数）编写一个递归函数，将字符串形式的十六进制数转换为十进制整数。函数头为：

```
int hexToDecimal(const string& hexString)
```

编写一个测试程序，提示用户输入一个十六进制字符串并显示其等效的十进制数。

```
Sample Run for Exercise17_24.cpp
Enter input data for the program (Sample data provided below. You may modify it.)
1ABD2

[Show the Sample Output Using the Preceeding Input]  [Reset]
Execution Result:
command>Exercise17_24
Enter a hex number: 1ABD2
1ABD2 is decimal 109522

command>
```

第 18 章

Introduction to C++ Programming and Data Structures, Fifth Edition

开发高效算法

学习目标

1. 使用大 O 表示法评估算法效率（18.2 节）。
2. 解释增长率以及为什么常数和非支配项在评估时可以忽略（18.2 节）。
3. 确定各种类型算法的复杂度（18.3 节）。
4. 分析二分查找算法（18.4.1 节）。
5. 分析选择排序算法（18.4.2 节）。
6. 分析汉诺塔算法（18.4.3 节）。
7. 描述和比较常见的增长函数（常数、对数、对数 – 线性、二次、三次和指数）（18.4.4 ～ 18.4.5 节）。
8. 使用动态规划设计求斐波那契数的高效算法（18.5 节）。
9. 使用欧几里得算法求最大公约数（18.6 节）。
10. 使用埃拉托斯梯尼筛选法来寻找质数（18.7 节）。
11. 使用分治法设计寻找最近点对的高效算法（18.8 节）。
12. 使用回溯方法解决八皇后问题（18.9 节）。
13. 设计高效的算法来寻找一组点的凸包（18.10 节）。
14. 使用 Boyer-Moore 和 KMP 算法设计高效的字符串匹配算法（18.11 节）。

18.1 简介

要点提示：算法设计是为解决问题而开发一个数学处理过程。算法分析是预测一个算法的性能。

前面章节介绍了过程程序设计、面向对象程序设计和递归程序设计。本章将使用各种示例介绍开发高效算法的常见算法技术（动态规划、分治和回溯）。在介绍高效算法之前，我们将讨论如何衡量算法效率的问题。

18.2 使用大 O 表示法衡量算法效率

要点提示：大 O 表示法获得基于输入规模衡量算法时间复杂度的函数。可以忽略函数中的乘法常数和非支配项。

假设两个算法执行相同的任务，例如查找（线性查找与二分查找）。那么哪一个更好呢？为了回答这个问题，用户可以实现这些算法并运行程序来获得执行时间。但这种方法存在两个问题：

- 首先，许多任务可能同时在计算机上运行。特定程序的执行时间依赖于系统负载。
- 其次，执行时间会依赖于具体的输入。例如，在线性查找与二分查找中，如果要查找的元素恰好是列表中的第一个，则线性查找会比二分查找更快地找到该元素。

通过衡量算法的执行时间来比较算法是非常困难的。为了克服这些困难,人们开发了一种独立于计算机和特定输入的理论方法来分析算法。这种方法估算输入规模改变带来的影响。通过这种方式,可以看到算法的执行时间随着输入规模的增加而增加的速度,因此可以通过检查两种算法的**增长率**来比较它们。

考虑线性查找问题。线性查找算法按顺序将键与数组中的元素进行比较,直到找到键或数组查找完为止。如果键不在数组中,则对于大小为 n 的数组,进行 n 次比较。如果键在数组中,则平均需要 $n/2$ 次比较。算法的执行时间与数组的大小成正比。如果将数组的大小增加一倍,则比较次数将增加一倍。该算法以线性速率增长,增长率是 n 的数量级。计算机科学家使用大 O 表示法来表示"数量级"。使用该表示法,线性查找算法的复杂度为 $O(n)$。算法的时间复杂度(也称为运行时间)是使用大 O 表示法衡量的算法运行所需的时间。

注意:算法的时间复杂度(也称为运行时间)是用大 O 表示法衡量的算法运行所花费的时间。

对于相同的输入规模,算法的执行时间可能会随输入而变化。导致执行时间最短的输入被称为**最佳情况输入**,而导致执行时间最长的输入是**最坏情况输入**。最佳和最坏情况分析是在最佳和最坏输入情况下分析算法。最佳和最坏情况的分析并不具有代表性,但最坏情况分析非常有用。可以确保算法永远不会比最坏的情况慢。**平均情况分析**试图确定相同规模的所有可能输入之间的平均时间量。平均情况分析是理想的,但很难实现,因为对于许多问题,很难确定各种输入实例的相对概率和分布。由于最坏情况分析更容易做到,因此分析通常针对最坏情况进行。

如果用户几乎总是在查找列表中存在的内容,那么线性查找算法在最坏的情况下需要进行 n 次比较,在平均情况下需要进行 $n/2$ 次比较。使用大 O 表示法,这两种情况都需要 $O(n)$ 时间。乘法常数(1/2)可以省略。算法分析的重点是增长率,乘法常数对增长率没有影响。如表 18.1 所示,$n/2$ 或 $100n$ 的增长率与 n 的增长率相同。因此 $O(n) = O(n/2) = O(100n)$。

表 18.1

n	$f(n)$		
	n	$n/2$	$100n$
100	100	50	10000
200	200	100	20000
	2	2	2	$f(200)/f(100)$

考虑在 n 个元素的数组中查找最大值的算法。要找到最大值,如果 n 为 2,则需要一次比较;如果 n 为 3,则需要两次比较。一般来说,需要 $n-1$ 次比较才能找到 n 个元素的列表中的最大值。算法分析适用于较大的输入规模。如果输入规模较小,那么评估算法的效率就没有意义。随着 n 的增大,表达式 $n-1$ 中的 n 主导了复杂度。大 O 表示法支持忽略非支配部分(例如,表达式 $n-1$ 中的 -1),并突出重要部分(例如,表达式 $n-1$ 中的 n)。因此,该算法的复杂度为 $O(n)$。

大 O 表示法评估算法的执行时间与输入规模相关。如果时间与输入规模无关,则该算法被称为 $O(1)$ **常数**时间。例如,在数组中按给定索引检索元素的函数需要常数时间,因为时间不会随着数组大小的增加而增加。

以下数学求和公式在算法分析中经常用到:

$$1+2+3+\cdots+(n-2)+(n-1) = \frac{n(n-1)}{2} = O(n^2)$$

$$1+2+3+\cdots+(n-1)+n = \frac{n(n+1)}{2} = O(n^2)$$

$$a^0 + a^1 + a^2 + a^3 + \cdots + a^{(n-1)} + a^n = \frac{a^{n+1}-1}{a-1} = O(a^n)$$

$$2^0 + 2^1 + 2^2 + 2^3 + \cdots + 2^{(n-1)} + 2^n = \frac{2^{n+1}-1}{2-1} = 2^{n+1}-1 = O(2^n)$$

注意：时间复杂度是用大 O 表示法对执行时间的衡量。类似地，也可以使用大 O 表示法衡量空间复杂度。空间复杂度衡量算法使用的内存空间量。本书中介绍的大多数算法的空间复杂度是 $O(n)$，也就是说，它们使用的内存空间量对输入规模表现出线性增长率。例如，线性查找的空间复杂度为 $O(n)$。

注意：我们用非专业术语介绍了大 O 表示法。附录 I 给出了大 O 表示法以及大 Omega 表示法和大 Theta 表示法的精确数学定义。

18.3 示例：确定大 O

要点提示：本节给出几个确定大 O 的例子。

示例 1 考虑以下循环的时间复杂度：

```cpp
for (int i = 1; i <= n; i++)
{
  k = k + 5;
}
```

执行

```cpp
k = k + 5;
```

的常数时间为 c。由于循环执行了 n 次，因此循环的时间复杂度为

$$T(n) = (\text{aconstant} c) * n = O(n)$$

我们通过理论分析预测了算法的性能。为了了解该算法的执行情况，我们在 LiveExample 18.1 中运行代码，以获得 $n=25000000$、50000000、100000000 和 200000000 的执行时间。

LiveExample 18.1 的互动程序请访问 https://liangcpp.pearsoncmg.com/LiveRunCpp5e/faces/LiveExample.xhtml?header=off&programName=PerformanceTest&fileType=.cpp&programHeight=460&resultHeight=210。

LiveExample 18.1   PerformanceTest.cpp

```cpp
#include <iostream>
#include <ctime> // for time function
using namespace std;

void getTime(int n)
{
  int startTime = time(0);
  double k = 0;
  for (int i = 1; i <= n; i++)
  {
    k = k + 5;
  }
  int endTime = time(0);
  cout << "Execution time for n = " << n
    << " is " << (endTime - startTime) << " seconds" << endl;
```

```
16     }
17
18  int main()
19  {
20      getTime(250000000);
21      getTime(500000000);
22      getTime(1000000000);
23      getTime(2000000000);
24
25      return 0;
26  }
```

Compile/Run　Reset　Answer　　　　　　　　　Choose a Compiler: VC++

Execution Result:

```
command>cl PerformanceTest.cpp
Microsoft C++ Compiler 2019
Compiled successful (cl is the VC++ compile/link command)

command>PerformanceTest
Execution time for n = 250000000 is 1 seconds
Execution time for n = 500000000 is 1 seconds
Execution time for n = 1000000000 is 3 seconds
Execution time for n = 2000000000 is 5 seconds

command>
```

我们的分析预测了这个循环的线性时间复杂度。如示例输出所示，当输入规模增加两倍时，运行时间也大约增加两倍。执行与预测一致。

示例 2 以下循环的时间复杂度是多少？

```
for (int i = 1; i <= n; i++)
{
    for (int j = 1; j <= n; j++)
    {
        k = k + i + j;
    }
}
```

执行

```
k = k + i + j;
```

的常数时间为 c，外层循环执行 n 次。对于外层循环中的每次迭代，内层循环执行 n 次。因此，循环的时间复杂度为

$$T(n) = (\text{aconstant} c) * n * n = O(n^2)$$

时间复杂度为 $O(n^2)$ 的算法称为**二次算法**。二次算法随着问题规模的增加而快速增长。如果将输入规模增加一倍，算法的时间是原来的四倍。带有嵌套循环的算法通常是二次的。

示例 3 考虑以下循环：

```
for (int i = 1; i <= n; i++)
{
    for (int j = 1; j <= i; j++)
    {
        k = k + i + j;
    }
}
```

外层循环执行 n 次。对于 i = 1, 2, …, 内层循环分别执行 1 次、2 次和 n 次。因此，循环的时间复杂度为

$$T(n) = c + 2c + 3c + 4c + \cdots + nc$$
$$= cn(n+1)/2$$
$$= (c/2)n^2 + (c/2)n$$
$$= O(n^2)$$

示例 4 考虑以下循环：

```
for (int i = 1; i <= n; i++)
{
  for (int j = 1; j <= 20; j++)
  {
    k = k + i + j;
  }
}
```

内层循环执行 20 次，外部循环执行 n 次。因此，循环的时间复杂度为

$$T(n) = 20 * c * n = O(n)$$

示例 5 考虑以下序列：

```
for (int j = 1; j <= 10; j++)
{
  k = k + 4;
for (int i = 1; i <= n; i++)
{
  for (int j = 1; j <= 20; j++)
  {
    k = k + i + j;
  }
}
```

第一个循环执行 10 次，第二个循环执行 20*n 次。因此，循环的时间复杂度为

$$T(n) = 10 * c + 20 * c * n = O(n)$$

示例 6 考虑为一个整数 n 计算 a^n。一个简单的算法是将 a 乘 n 次，如下所示：

```
result = 1;
for (int i = 1; i <= n; i++)
  result *= a;
```

该算法需要 $O(n)$ 时间。在不失一般性的情况下，假设 $n=2^k$。可以使用以下方案改进算法：

```
result = a;
for (int i = 1; i <= k; i++)
  result = result * result;
```

该算法需要 $O(\log n)$ 时间。对于任意的 n，可以修改算法并证明复杂度仍然是 $O(\log n)$。

注意：时间复杂度为 $O(\log n)$ 的算法称为**对数算法**。对数的底数是 2，但底数不影响对数增长率，因此可以省略。

18.4 分析算法时间复杂度

要点提示：本节分析几种著名算法的复杂度：二分查找、选择排序和汉诺塔问题。

18.4.1 分析二分查找

LiveExample 7.9 中给出的二分查找算法在已排序数组中查找键。算法中的每次迭代都包含固定数量的操作，用 c 表示。设 $T(n)$ 表示在 n 个元素的列表上进行二分查找的时间复杂度。在不失一般性的情况下，假设 n 是 2 的幂并且 $k=\log n$。由于二分查找在两次比较之后消除一半的输入，

$$T(n) = T\left(\frac{n}{2}\right) + c = T\left(\frac{n}{2^2}\right) + c + c = T\left(\frac{n}{2^k}\right) + kc$$
$$= T(1) + c\log n = 1 + (\log n)c$$
$$= O(\log n)$$

忽略常数和非支配项，二分查找算法的复杂度为 $O(\log n)$。对数算法随着问题规模的增加而缓慢增长。在二分查找的情况下，每次将数组大小增加一倍时，最多需要再进行一次比较。如果将任何对数时间算法的输入规模平方，则只会使执行时间增加一倍。所以对数时间算法的效率很高。

18.4.2 分析选择排序

LiveExample 7.10 中给出的选择排序算法可以找到列表中最小的元素，并将其与第一个元素交换。然后，它再找到剩余元素中的最小元素，并将其与剩余列表中的第一个元素交换，以此类推，直到剩余列表只包含一个需要排序的元素。第一次迭代的比较次数为 $n-1$，第二次迭代的比较次数为 $n-2$，以此类推。设 $T(n)$ 表示选择排序的复杂度，c 表示每次迭代中其他操作（如赋值和额外比较）的总数。这样，

$$T(n) = (n-1) + c + (n-2) + c + \cdots + 2 + c + 1 + c$$
$$= \frac{(n-1)(n-1+1)}{2} + c(n-1) = \frac{n^2}{2} - \frac{n}{2} + cn - c$$
$$= O(n^2)$$

因此，选择排序算法的复杂度为 $O(n^2)$。

18.4.3 分析汉诺塔问题

LiveExample 17.7 中给出的汉诺塔问题，在塔 C 的帮助下，递归地将 n 个盘子从塔 A 移动到塔 B，如下所示：

1. 在塔 B 的帮助下，将前 $n-1$ 个盘子从 A 移动到 C。
2. 将盘子 n 从 A 移动到 B。
3. 在塔 A 的帮助下，将 $n-1$ 个盘子从 C 移动到 B。

该算法的复杂度通过移动次数来衡量。设 $T(n)$ 表示算法将 n 个盘子从塔 A 移动到塔 B 的移动次数。因此，$T(1)$ 为 1，并且

$$T(n) = T(n-1) + 1 + T(n-1)$$
$$= 2T(n-1) + 1$$
$$= 2(2T(n-2) + 1) + 1$$
$$= 2(2(2T(n-3) + 1) + 1) + 1$$
$$= 2^{n-1}T(1) + 2^{n-2} + \cdots + 2 + 1$$
$$= 2^{n-1} + 2^{n-2} + \cdots + 2 + 1 = (2^n - 1) = O(2^n)$$

具有时间复杂度 $O(2^n)$ 的算法称为**指数算法**。随着输入规模的增加，指数算法的时间呈指数增长。对于大的输入规模，指数算法是不可行的。假设盘子每秒移动一次，移动 32 个盘子则需要 2^{32}/(365*24*60*60)=136 年时间，移动 64 个盘子需要 2^{64}/(365*24*60*60)=5850 亿年时间。

18.4.4 常见的递归关系

递归关系是分析算法复杂度的有用工具。如前面例子所示，二分查找、选择排序和汉诺塔的复杂度分别为 $T(n)=T\left(\dfrac{n}{2}\right)+c, T(n)=T(n-1)+O(n)$ 和 $T(n)=2T(n-1)+O(1)$。表 18.2 总结了常见的递归关系函数。

表 18.2 常见的递归函数

递归关系	结果	示例
$T(n) = T(n/2) + O(1)$	$T(n) = O(\log n)$	二分查找，欧几里得法求最大公约数（18.6 节）
$T(n) = T(n-1) + O(1)$	$T(n) = O(n)$	线性查找
$T(n) = 2T(n/2) + O(1)$	$T(n) = O(n)$	
$T(n) = 2T(n/2) + O(n)$	$T(n) = O(n \log n)$	归并排序（第 19 章）
$T(n) = T(n-1) + O(n)$	$T(n) = O(n^2)$	选择排序
$T(n) = 2T(n-1) + O(1)$	$T(n) = O(2^n)$	汉诺塔
$T(n) = T(n-1) + T(n-2) + O(1)$	$T(n) = O(2^n)$	递归的斐波那契算法

18.4.5 比较常见的增长函数

前面分析了几种算法的复杂度。表 18.3 列出了一些常见的增长函数，并显示了随着输入规模从 n=25 翻倍到 n=50 时，增长率是如何变化的。

表 18.3 增长率的变化

函数	名称	n=25	n=50	$f(50) / f(25)$
$O(1)$	常数时间	1	1	1
$O(\log n)$	对数时间	4.64	5.64	1.21
$O(n)$	线性时间	25	50	2
$O(n \log n)$	对数－线性时间	116	282	2.43
$O(n^2)$	二次时间	625	2 500	4
$O(n^3)$	三次时间	15 625	125 000	8
$O(2^n)$	指数时间	3.36×10^7	1.27×10^{15}	3.35×10^7

这些函数如下排序，如图 18.1 所示。

$$O(1) < O(\log n) < O(n) < O(n \log n) < O(n^2) < O(n^3) < O(2^n)$$

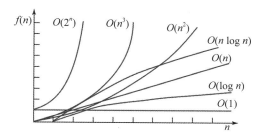

图 18.1 随着 n 的增大，函数的增长趋势

18.5 使用动态规划求斐波那契数

要点提示：本节分析并设计一种用动态规划求斐波那契数的高效算法。

17.3 节给出了一个求斐波那契数的递归函数。求斐波那契数的递归函数如下所示：

```cpp
// The function for finding the Fibonacci number
long fib(long index)
{
  if (index == 0) // Base case
    return 0;
  else if (index == 1) // Base case
    return 1;
  else // Reduction and recursive calls
    return fib(index - 1) + fib(index - 2);
}
```

我们现在可以证明这个算法的复杂度是 $O(2^n)$。方便起见，设 `index` 为 n。设 $T(n)$ 表示求 `fib(n)` 的算法的复杂度，c 表示将 `index` 与 0 和 1 进行比较的常数时间。因此

$$T(n) = T(n-1) + T(n-2) + c$$
$$\leq 2T(n-1) + c$$
$$\leq 2(2T(n-2) + c) + c$$
$$= 2^2 T(n-2) + 2c + c$$

类似于对汉诺塔问题的分析，我们可以证明 $T(n)$ 是 $O(2^n)$。但这种算法效率很低。有没有一种高效的算法可以求斐波那契数？递归 `fib` 函数的问题在于用相同的参数冗余调用函数。例如，为了计算 `fib(4)`，调用 `fib(3)` 和 `fib(2)`。为了计算 `fib(3)`，调用 `fib(2)` 和 `fib(1)`。我们注意到 `fib(2)` 是冗余调用的。可以通过避免重复调用相同参数的 `fib` 函数来改进它。注意到一个新的斐波那契数是通过将数列中的前两个数字相加而得的。如果用两个变量 `f0` 和 `f1` 来存储前面两个数字，则可以通过将 `f0` 与 `f1` 相加立即获得新的数字 `f2`。现在应该通过将 `f1` 赋值给 `f0`，并将 `f2` 赋值给 `f1` 来更新 `f0` 和 `f1`，如图 18.2 所示。

提示：计算斐波那契数。单击 Step 按钮求数列中的下一个斐波那契数。动画在索引 11 处结束。单击 Reset 按钮重新开始。

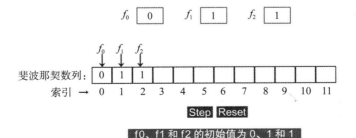

图 18.2 变量 `f0`、`f1` 和 `f2` 存储数列中三个连续的斐波那契数

新函数在 LiveExample 18.2 中实现。

LiveExample 18.2 的互动程序请访问 https://liangcpp.pearsoncmg.com/LiveRunCpp5e/faces/LiveExample.xhtml?header=off&programName=ImprovedFibonacci&programHeight=750&resultHeight=180。

LiveExample 18.2 ImprovedFibonacci.cpp

Source Code Editor:

```cpp
#include <iostream>
using namespace std;

// The function for finding the Fibonacci number
int fib(int);

int main()
{
  // Prompt the user to enter an integer
  cout << "Enter an index for the Fibonacci number: ";
  int index;
  cin >> index;

  // Display factorial
  cout << "Fibonacci number at index " << index << " is "
    << fib(index) << endl;

  return 0;
}

// The function for finding the Fibonacci number
int fib(int n)
{
  long f0 = 0; // For fib(0)
  long f1 = 1; // For fib(1)
  long f2 = 1; // For fib(2)

  if (n == 0)
    return f0;
  else if (n == 1)
    return f1;
  else if (n == 2)
    return f2;

  for (int i = 3; i <= n; i++)
  {
    f0 = f1;
    f1 = f2;
    f2 = f0 + f1;
  }

  return f2;
}
```

Enter input data for the program (Sample data provided below. You may modify it.)

```
19
```

[Automatic Check] [Compile/Run] [Reset] [Answer]   Choose a Compiler: VC++

Execution Result:

```
command>cl ImprovedFibonacci.cpp
Microsoft C++ Compiler 2019
Compiled successful (cl is the VC++ compile/link command)

command>ImprovedFibonacci
Enter an index for the Fibonacci number: 19
Fibonacci number at index 19 is 4181

command>
```

显然，这个新算法的复杂度是 $O(n)$。这是对递归 $O(2^n)$ 算法的巨大改进。

算法设计说明：这里介绍的计算斐波那契数的算法使用了一种称为**动态规划**的方法。动态规划是这样的过程：解决子问题，然后将子问题的解决方案组合起来获得整体解决方案。这自然会得到递归解决方案。但由于子问题重叠使递归效率变低。动态规划背后的关键思想是只解决每个子问题一次，存储子问题的结果供以后使用，从而避免子问题的冗余计算。

18.6 使用欧几里得算法求最大公约数

要点提示：本节介绍求两个整数的最大公约数的几种高效算法。

两个整数的最大公约数（GCD）是能被两个整数整除的最大除数。LiveExample 5.10 给出了一种求两个整数 m 和 n 的最大公约数的暴力算法。

算法设计说明：暴力指的是一种以最简单、最直接或最明显的方式解决问题的算法方法。结果是，这样的算法最终可能会比更聪明或更复杂的算法做更多的工作来解决给定的问题。另一方面，暴力算法通常比复杂的算法更容易实现，而且由于这种算法的简单性，有时它可能更高效。

暴力算法检查 k（对于 k=2，3，4，…）是否是 n1 和 n2 的公约数，直到 k 大于 n1 或 n2。该算法可以描述如下：

```
int gcd(int m, int n)
{
  int gcd = 1;
  for (int k = 2; k <= m && k <= n; k++)
  {
    if (m % k == 0 && n % k == 0)
      gcd = k;
  }
  return gcd;
}
```

假设 $m \geq n$，则该算法的复杂度明显为 $O(n)$。

有没有更好的算法来求最大公约数？与其从 1 开始向上搜索可能的除数，不如从 n 开始向下搜索更高效。一旦找到除数，该除数就是最大公约数。因此，可以使用以下循环来改进算法：

```
for (int k = n; k >= 1; k--)
{
  if (m % k == 0 && n % k == 0)
  {
    gcd = k;
    break;
  }
}
```

该算法比前面的算法好，但其最坏情况下的时间复杂度仍然是 $O(n)$。

数字 n 的除数不会大于 n/2，因此可以用以下循环进一步改进算法：

```
for (int k = n / 2; k >= 1; k--) {
  if (m % k == 0 && n % k == 0) {
    gcd = k;
    break;
  }
}
```

但这个算法是不正确的，因为 n 可以是 m 的除数。必须考虑这种情况。正确的算法如 LiveExample 18.3 所示。

LiveExample 18.3 的互动程序请访问 https://liangcpp.pearsoncmg.com/LiveRunCpp5e/faces/LiveExample.xhtml?header=off&programName=GCD&fileType=.cpp&programHeight=660&resultHeight=180。

LiveExample 18.3　GCD.cpp

Source Code Editor:

```cpp
#include <iostream>
using namespace std;

// Return the gcd of two integers
int gcd(int m, int n)
{
  int gcd = 1;

  if (m % n == 0) return n;

  for (int k = n / 2; k >= 1; k--)
  {
    if (m % k == 0 && n % k == 0)
    {
      gcd = k;
      break;
    }
  }

  return gcd;
}

int main()
{
  // Prompt the user to enter two integers
  cout << "Enter first integer: ";
  int n1;
  cin >> n1;

  cout << "Enter second integer: ";
  int n2;
  cin >> n2;

  cout << "The greatest common divisor for " << n1 <<
    " and " << n2 << " is " << gcd(n1, n2) << endl;

  return 0;
}
```

Enter input data for the program (Sample data provided below. You may modify it.)

45 75

[Automatic Check] [Compile/Run] [Reset] [Answer]　　Choose a Compiler: VC++

Execution Result:

```
command>cl GCD.cpp
Microsoft C++ Compiler 2019
Compiled successful (cl is the VC++ compile/link command)
```

```
command>GCD
Enter first integer: 45
Enter second integer: 75
The greatest common divisor for 45 and 75 is 15

command>
```

假设 $m \geq n$，for 循环最多执行 $n/2$ 次，时间比以前的算法减少了一半。该算法的时间复杂度仍然是 $O(n)$，但实际上，它比 LiveExample 5.10 中的算法快得多。

注意：大 O 表示法提供了算法效率的很好的理论评估。但具有相同时间复杂度的两种算法并不一定同样高效。如前一个例子所示，LiveExample 5.10 和 LiveExample 18.3 中的两种算法都具有相同的复杂度，但在实践中，LiveExample 18.3 的算法显然更好。

公元前 300 年左右，欧几里得发现了一种更高效的最大公约数算法。这是已知最古老的算法之一。它可以递归地定义如下：

设 gcd(m, n) 表示整数 m 和 n 的最大公约数：
- 如果 m%n 是 0，则 gcd(m, n) 是 n。
- 否则，gcd(m, n) 是 gcd(n, m%n)。

不难证明这个算法的正确性。假设 m%n=r，因此，m=qn+r，其中 q 是 m/n 的商。任何能整除 m 和 n 的数也必须整除 r。因此 gcd(m, n) 与 gcd(n, r) 相同，其中 r=m%n。该算法可以如 LiveExample 18.4 所示实现。

LiveExample 18.4 的互动程序请访问 https://liangcpp.pearsoncmg.com/LiveRunCpp5e/faces/LiveExample.xhtml?header=off&programName=GCDEuclid&fileType=.cpp&programHeight=480&resultHeight=180。

LiveExample 18.4   GCDEuclid.cpp

```cpp
#include <iostream>
using namespace std;

// Return the gcd of two integers
int gcd(int m, int n)
{
  if (m % n == 0)
    return n;
  else
    return gcd (n, m % n );
}

int main()
{
  // Prompt the user to enter two integers
  cout << "Enter first integer: ";
  int n1;
  cin >> n1;

  cout << "Enter second integer: ";
  int n2;
  cin >> n2;

  cout << "The greatest common divisor for " << n1 <<
    " and " << n2 << " is " << gcd(n1, n2) << endl;
```

```
    27        return 0;
    28    }
```

Enter input data for the program (Sample data provided below. You may modify it.)
```
45 75
```

[Automatic Check] [Compile/Run] [Reset] [Answer]   Choose a Compiler: [VC++ ▼]

Execution Result:
```
command>cl GCDEuclid.cpp
Microsoft C++ Compiler 2019
Compiled successful (cl is the VC++ compile/link command)

command>GCDEuclid
Enter first integer: 45
Enter second integer: 75
The greatest common divisor for 45 and 75 is 15

command>
```

在**最佳情况**下，当 m%n 为 0 时，算法只需一步即可求得最大公约数。很难分析一般情况。但我们可以证明最坏情况下的时间复杂度是 $O(\log n)$。

假设 $m \geq n$，我们可以证明 m%n<m/2，如下所示：

如果 n<=m/2，则 m%n<m/2，因为 m 除以 n 的余数总是小于 n。

如果 n>m/2，则 m%n=m-n<m/2。因此，m%n<m/2。

欧几里得的算法递归地调用 gcd 函数。它首先调用 gcd(m, n)，然后调用 gcd(n, m%n) 和 gcd(m%n, n%(m%n))，以此类推，如下所示：

```
  gcd(m, n)
= gcd(n, m % n)
= gcd(m % n, n % (m % n))
= ...
```

由于 m%n<m/2 和 n%(m%n)<n/2，传递给 gcd 函数的参数在每两次迭代后减少一半。在调用 gcd 两次之后，第二个参数小于 $n/2$。在调用 gcd 四次之后，第二个参数小于 $n/4$。在调用 gcd 六次之后，第二个参数小于 $n/2^3$。设 k 为调用 gcd 函数的次数。调用 gcd k 次后，第二个参数小于 $\dfrac{n}{2^{(k/2)}}$，这个数大于或等于 1。即，

$$\dfrac{n}{2^{(k/2)}} \geq 1 \rightarrow n \geq 2^{(k/2)} \rightarrow \log n \geq k/2 \rightarrow k \leq 2\log n$$

因此，$k \leq 2 \log n$，所以 gcd 函数的时间复杂度为 $O(\log n)$。

最坏的情况发生在两个数字产生最多除法的情况下。事实证明，两个连续的斐波那契数将产生最多的除法。回想一下，斐波那契数列以 0 和 1 开头，后面的每个数字都是数列中前两个数字的总和，例如：

$$0 \; 1 \; 1 \; 2 \; 3 \; 5 \; 8 \; 13 \; 21 \; 34 \; 55 \; 89 \; \cdots$$

该数列可以递归定义为

```
fib(0) = 0;
fib(1) = 1;
fib(index) = fib(index - 2) + fib(index - 1); index >= 2
```

对于两个连续的斐波那契数 fib(index) 和 fib(index-1),

```
gcd(fib(index), fib(index - 1))
= gcd(fib(index - 1), fib(index - 2))
= gcd(fib(index - 2), fib(index - 3))
= gcd(fib(index - 3), fib(index - 4))
= ...
= gcd(fib(2), fib(1))
= 1
```

例如,

```
gcd(21, 13)
= gcd(13, 8)
= gcd(8, 5)
= gcd(5, 3)
= gcd(3, 2)
= gcd(2, 1)
= 1
```

因此,调用 gcd 函数的次数与调用索引的次数相同。我们可以证明 index ⩽ 1.44log n,其中 n=fib(index−1)。这是一个比 index ⩽ 2 log n 更严格的界限。

表 18.4 总结了三种求最大公约数的算法的复杂度。

表 18.4 最大公约数算法的比较

算法	复杂度	描述
LiveExample 5.10	$O(n)$	暴力法,检查所有可能的除数
LiveExample 18.3	$O(n)$	检查所有可能除数的一半
LiveExample 18.4	$O(\log n)$	欧几里得算法

18.7 寻找质数的高效算法

要点提示:本节介绍寻找质数的几种高效算法。

第一个发现 100000000 位以上的质数的个人或团体会获得 150000 美元的奖金。你能设计一个寻找质数的快速算法吗?

如果一个大于 1 的整数,它的除数只有 1 或它自身,则它是个**质数**。例如,2、3、5 和 7 是质数,但 4、6、8 和 9 不是。

如何确定一个数字 n 是质数? LiveExample 5.16 给出了一个求质数的暴力算法。通过暴力算法可以找到质数。算法检查 2,3,4,5,⋯,n−1 是否可以整除 n。如果不能,则 n 是质数。该算法需要 $O(n)$ 时间来检查 n 是否为质数。我们注意到,只需要检查 2,3,4,5,⋯,n/2 是否能整除 n 即可。如果不能,则 n 是质数。该算法虽略有改善,但复杂度仍为 $O(n)$。

事实上,我们可以证明,如果 n 不是质数,那么 n 必有一个大于 1 且小于或等于 $\sqrt{n}$ 的因数。这里给出证明:由于 n 不是质数,因此存在两个数 p 和 q,使得 $n=pq$,其中 $1<p \leqslant q$。我们注意到 $n= \sqrt{n} \sqrt{n}$,所以 p 必须小于或等于 $\sqrt{n}$。因此,只需要检查 2,3,4,5,⋯,$\sqrt{n}$ 是否可以整除 n,如果不能,则 n 是质数。这将算法的时间复杂度显著地降低到 $O(\sqrt{n})$。

现在考虑寻找 n 以下所有质数的算法。一个直接的实现方法是检查 i 是否是质数,i=2,3,4,5,⋯,n,LiveExample 18.5 给出了该程序。

LiveExample 18.5 的互动程序请访问 https://liangcpp.pearsoncmg.com/LiveRunCpp5e/faces/LiveExample.xhtml?header=off&programName=PrimeNumbers&programHeight=930&resultHe

ight=470。

LiveExample 18.5   PrimeNumbers.cpp

Source Code Editor:

```cpp
#include <iostream>
#include <cmath>
using namespace std;

int main()
{
  cout << "Find all prime numbers <= n, enter n: ";
  int n;
  cin >> n;

  const int NUMBER_PER_LINE = 10; // Display 10 per line
  int count = 0; // Count the number of prime numbers
  int number = 2; // A number to be tested for primeness

  cout << "The prime numbers are:" << endl;

  // Repeatedly find prime numbers
  while (number <= n)
  {
    // Assume the number is prime
    bool isPrime = true; // Is the current number prime?

    // Test if number is prime
    for (int divisor = 2; divisor <= sqrt(number * 1.0); divisor++)
    {
      if (number % divisor == 0)
      { // If true, number is not prime
        isPrime = false; // Set isPrime to false
        break; // Exit the for loop
      }
    }

    // Print the prime number and increase the count
    if (isPrime)
    {
      count++; // Increase the count

      if (count % NUMBER_PER_LINE == 0)
      {
        // Print the number and advance to the new line
        cout << number << endl;
      }
      else
        cout << number << " ";
    }

    // Check whether the next number is prime
    number++;
  }

  cout << "\n" << count << " number of primes <= " << n << endl;

  return 0;
}
```

Enter input data for the program (Sample data provided below. You may modify it.)

```
1000
```

Execution Result:

```
command>cl PrimeNumbers.cpp
Microsoft C++ Compiler 2019
Compiled successful (cl is the VC++ compile/link command)

command>PrimeNumbers
Find all prime numbers <= n, enter n: 1000
The prime numbers are:
2 3 5 7 11 13 17 19 23 29
31 37 41 43 47 53 59 61 67 71
73 79 83 89 97 101 103 107 109 113
127 131 137 139 149 151 157 163 167 173
179 181 191 193 197 199 211 223 227 229
233 239 241 251 257 263 269 271 277 281
283 293 307 311 313 317 331 337 347 349
353 359 367 373 379 383 389 397 401 409
419 421 431 433 439 443 449 457 461 463
467 479 487 491 499 503 509 521 523 541
547 557 563 569 571 577 587 593 599 601
607 613 617 619 631 641 643 647 653 659
661 673 677 683 691 701 709 719 727 733
739 743 751 757 761 769 773 787 797 809
811 821 823 827 829 839 853 857 859 863
877 881 883 887 907 911 919 929 937 941
947 953 967 971 977 983 991 997
168 number of primes <= 1000

command>
```

如果必须为 for 循环（第 24 行）的每次迭代计算 sqrt(number)，那程序效率很低。好的编译器应该对整个 for 循环只计算一次 sqrt(number)。为此，可以用以下两行替换第 24 行：

```cpp
int squareRoot = sqrt(number);
for (int divisor = 2; divisor <= squareRoot; divisor++)
```

实际上并不需要为每个 number 计算 sqrt(number)。只需要寻找完美平方数，比如 4、9、16、25、36、49 等。注意到对于 36 和 48 之间的所有数字，它们的 static_cast<int>(sqrt(number)) 值都是 6。有了这一点，就可以将第 18～31 行中的代码替换为以下代码：

```cpp
...
int squareRoot = 1;
// Repeatedly find prime numbers
while (number <= n)
{
  // Assume the number is prime
  boolean isPrime = true; // Is the current number prime?
  if (squareRoot * squareRoot < number) squareRoot++;
  // Test whether number is prime
  for (int divisor = 2; divisor <= squareRoot; divisor++)
    if (number % divisor == 0) // If true, number is not prime
    {
      isPrime = false; // Set isPrime to false
      break; // Exit the for loop
    }
  }
...
```

现在我们将注意力转移到分析这个程序的复杂度上。由于在 for 循环（第 24 ～ 31 行）中需要 $\sqrt{i}$ 个步骤来检查数字 i 是否为质数，因此该算法需要 $\sqrt{2}+\sqrt{3}+\sqrt{4}+\cdots+\sqrt{n}$ 个步骤来找到所有小于或等于 n 的质数。观察到

$$\sqrt{2}+\sqrt{3}+\sqrt{4}+\cdots+\sqrt{n}\leqslant n\sqrt{n}$$

因此，该算法的时间复杂度为 $O(n\sqrt{n})$。

为了确定 i 是否是质数，算法检查 2，3，4，5，…，$\sqrt{i}$ 是否能整除 i。该算法可以进一步改进，事实上，只需要检查从 2 到 $\sqrt{i}$ 的质数是否是 i 的可能除数。

我们可以证明，如果 i 不是质数，那么必然存在一个质数 p，使得 $i=pq$，且 $p\leqslant q$。这里给出证明。假设 i 不是质数；设 p 是 i 的最小因数，p 必须是质数；否则，p 有一个因数 k，且 $2\leqslant k<p$。k 也是 i 的因数，这与 p 是 i 的最小因数相矛盾。因此，如果 i 不是质数，你可以找到一个从 2 到 $\sqrt{i}$ 整除 i 的质数。这得出了一个更高效的算法来找到 n 以内的所有质数，如 LiveExample 18.6 所示。

LiveExample 18.6 的互动程序请访问 https://liangcpp.pearsoncmg.com/LiveRunCpp5e/faces/LiveExample.xhtml?header=off&programName=EfficientPrimeNumbers&fileType=.cpp&programHeight=1030&resultHeight=500。

LiveExample 18.6　EfficientPrimeNumbers.cpp

Source Code Editor:

```cpp
#include <iostream>
#include <cmath>
#include <vector>
using namespace std;

int main()
{
  cout << "Find all prime numbers <= n, enter n: ";
  int n;
  cin >> n;

  const int NUMBER_PER_LINE = 10; // Display 10 per line
  int count = 0; // Count the number of prime numbers
  int number = 2; // A number to be tested for primeness
  // A vector to hold prime numbers
  vector<int> primeVector;
  int squareRoot = 1; // Check whether number <= squareRoot

  cout << "The prime numbers are:" << endl;

  // Repeatedly find prime numbers
  while (number <= n)
  {
    // Assume the number is prime
    bool isPrime = true; // Is the current number prime?

    if (squareRoot * squareRoot < number) squareRoot++;

    // Test if number is prime
    for (int k = 0; k < primeVector.size()
              && primeVector[k] <= squareRoot; k++)
    {
      if (number % primeVector[k] == 0) // If true, not prime
      {
        isPrime = false; // Set isPrime to false
        break; // Exit the for loop
```

```
37            }
38          }
39
40          // Print the prime number and increase the count
41          if (isPrime)
42          {
43            count++; // Increase the count
44            primeVector.push_back(number); // Add a new prime to the list
45            if (count % NUMBER_PER_LINE == 0)
46            {
47              // Print the number and advance to the new line
48              cout << number << endl;
49            }
50            else
51              cout << number << " ";
52          }
53
54          // Check if the next number is prime
55          number++;
56        }
57
58        cout << "\n" << count << " number of primes <= " << n << endl;
59        return 0;
60      }
```

Enter input data for the program (Sample data provided below. You may modify it.)

1000

[Automatic Check] [Compile/Run] [Reset] [Answer]   Choose a Compiler: VC++

Execution Result:

```
command>cl EfficientPrimeNumbers.cpp
Microsoft C++ Compiler 2019
Compiled successful (cl is the VC++ compile/link command)

command>EfficientPrimeNumbers
Find all prime numbers <= n, enter n: 1000
The prime numbers are:
2 3 5 7 11 13 17 19 23 29
31 37 41 43 47 53 59 61 67 71
73 79 83 89 97 101 103 107 109 113
127 131 137 139 149 151 157 163 167 173
179 181 191 193 197 199 211 223 227 229
233 239 241 251 257 263 269 271 277 281
283 293 307 311 313 317 331 337 347 349
353 359 367 373 379 383 389 397 401 409
419 421 431 433 439 443 449 457 461 463
467 479 487 491 499 503 509 521 523 541
547 557 563 569 571 577 587 593 599 601
607 613 617 619 631 641 643 647 653 659
661 673 677 683 691 701 709 719 727 733
739 743 751 757 761 769 773 787 797 809
811 821 823 827 829 839 853 857 859 863
877 881 883 887 907 911 919 929 937 941
947 953 967 971 977 983 991 997
168 number of primes <= 1000

command>
```

设 $\pi(i)$ 表示小于或等于 i 的质数的个数。20 以内的质数是 2、3、5、7、11、13、17 和 19。因此,

$\pi(2)$ 是 1，$\pi(3)$ 是 2，$\pi(6)$ 是 3，$\pi(20)$ 是 8。已经证明过 $\pi(i)$ 是近似的 $\dfrac{i}{\log i}$。

对于每个数字 i，算法检查小于或等于 $\sqrt{i}$ 的质数是否能整除 i，小于或等于 $\sqrt{i}$ 的质数的个数是 $\dfrac{\sqrt{i}}{\log\sqrt{i}} = \dfrac{2\sqrt{i}}{\log i}$，因此，找到 n 以内的所有质数的复杂度为

$$\dfrac{2\sqrt{2}}{\log 2} + \dfrac{2\sqrt{3}}{\log 3} + \dfrac{2\sqrt{4}}{\log 4} + \dfrac{2\sqrt{5}}{\log 5} + \dfrac{2\sqrt{6}}{\log 6} + \dfrac{2\sqrt{7}}{\log 7} + \dfrac{2\sqrt{8}}{\log 8} + \cdots + \dfrac{2\sqrt{n}}{\log n}$$

因为对于 $i < n$ 且 $n \geq 16$，有 $\dfrac{\sqrt{i}}{\log i} < \dfrac{\sqrt{n}}{\log n}$，所以

$$\dfrac{2\sqrt{2}}{\log 2} + \dfrac{2\sqrt{3}}{\log 3} + \dfrac{2\sqrt{4}}{\log 4} + \dfrac{2\sqrt{5}}{\log 5} + \dfrac{2\sqrt{6}}{\log 6} + \dfrac{2\sqrt{7}}{\log 7} + \dfrac{2\sqrt{8}}{\log 8} + \cdots + \dfrac{2\sqrt{n}}{\log n} < \dfrac{2n\sqrt{n}}{\log n}$$

所以，算法的时间复杂度是 $O\left(\dfrac{n\sqrt{n}}{\log n}\right)$。

有比这更好的算法吗？让我们来研究一下著名的求质数的埃拉托斯梯尼算法。埃拉托斯梯尼（公元前 276 年—前 194 年）是一位希腊数学家，他设计了一种聪明的算法，被称为埃拉托斯梯尼筛选法，用于寻找所有小于或等于 n 的质数。他的算法使用一个由 n 个布尔值组成的 `primes` 数组。初始状态下，`primes` 中的所有元素都设置为 `true`。由于 2 的倍数不是质数，因此对于所有 $2 \leq i \leq n/2$，将 `primes[2*i]` 设置为 `false`，如图 18.3 所示。由于我们不关心 `primes[0]` 和 `primes[1]`，因此这些值在图中标记为 ×。

primes 数组

索引	0	1	2	3	4	5	6	7	8	9	10	11	12	13	14	15	16	17	18	19	20	21	22	23	24	25	26	27
初始状态	×	×	T	T	T	T	T	T	T	T	T	T	T	T	T	T	T	T	T	T	T	T	T	T	T	T	T	T
$k=2$	×	×	T	T	F	T	F	T	F	T	F	T	F	T	F	T	F	T	F	T	F	T	F	T	F	T	F	T
$k=3$	×	×	T	T	F	T	F	T	F	F	F	T	F	T	F	F	F	T	F	T	F	F	F	T	F	T	F	F
$k=5$	×	×	T	T	F	T	F	T	F	F	F	T	F	T	F	F	F	T	F	T	F	F	F	T	F	F	F	F

图 18.3　primes 中的值随每个质数 k 而变化

由于 3 的倍数不是质数，因此对于所有 $3 \leq i \leq n/3$，将 `primes[3*i]` 设置为 `false`。由于 5 的倍数不是质数，因此对于所有 $5 \leq i \geq n/5$，将 `primes[5*i]` 设置为 `false`。注意不需要考虑 4 的倍数，因为它们也是 2 的倍数，我们已经考虑过了。类似地，不需要考虑 6、8 和 9 的倍数。只需要考虑质数 k=2、3、5、7、11、…的倍数，并将 `primes` 中的相应元素设置为 `false`。然后，如果 `primes[i]` 仍然为 `true`，那么 i 就是一个质数。如图 18.3 所示，2、3、5、7、11、13、17、19、23 是质数。LiveExample18.7 给出了使用埃拉托斯特尼筛选法求质数的程序。

LiveExample 18.7 的互动程序请访问 https://liangcpp.pearsoncmg.com/LiveRunCpp5e/faces/LiveExample.xhtml?header=off&programName=SieveOfEratosthenes&fileType=.cpp&programHeight=830&resultHeight=450。

LiveExample 18.7   SieveOfEratosthenes.cpp

Source Code Editor:

```cpp
#include <iostream>
using namespace std;

int main()
{
  cout << "Find all prime numbers <= n, enter n: ";
  int n;
  cin >> n;

  bool *primes = new bool[n + 1]; // Prime number sieve

  // Initialize primes[i] to true
  for (int i = 0; i < n + 1; i++)
  {
    primes[i] = true;
  }

  for (int k = 2; k <= n / k; k++)
  {
    if (primes[k])
    {
      for (int i = k; i <= n / k; i++)
      {
        primes[k * i] = false; // k * i is not prime
      }
    }
  }

  const int NUMBER_PER_LINE = 10; // Display 10 per line
  int count = 0; // Count the number of prime numbers found so far
  // Print prime numbers
  for (int i = 2; i < n + 1; i++)
  {
    if (primes[i])
    {
      count++;
      if (count % 10 == 0)
        cout << i << endl;
      else
        cout << i << " ";
    }
  }

  cout << "\n" << count << " number of primes <= " << n << endl;

  delete [] primes;

  return 0;
}
```

Enter input data for the program (Sample data provided below. You may modify it.)

```
1000
```

Automatic Check | Compile/Run | Reset | Answer   Choose a Compiler: VC++

Execution Result:

```
command>cl SieveOfEratosthenes.cpp
Microsoft C++ Compiler 2019
```

```
Compiled successful (cl is the VC++ compile/link command)

command>SieveOfEratosthenes
Find all prime numbers <= n, enter n: 1000
2 3 5 7 11 13 17 19 23 29
31 37 41 43 47 53 59 61 67 71
73 79 83 89 97 101 103 107 109 113
127 131 137 139 149 151 157 163 167 173
179 181 191 193 197 199 211 223 227 229
233 239 241 251 257 263 269 271 277 281
283 293 307 311 313 317 331 337 347 349
353 359 367 373 379 383 389 397 401 409
419 421 431 433 439 443 449 457 461 463
467 479 487 491 499 503 509 521 523 541
547 557 563 569 571 577 587 593 599 601
607 613 617 619 631 641 643 647 653 659
661 673 677 683 691 701 709 719 727 733
739 743 751 757 761 769 773 787 797 809
811 821 823 827 829 839 853 857 859 863
877 881 883 887 907 911 919 929 937 941
947 953 967 971 977 983 991 997
168 number of primes <= 1000

command>
```

注意 k<=n/k（第 18 行）。否则，k*i 将大于 n（第 20 行）。这个算法的时间复杂度是多少？

对于每个质数 k（第 20 行），算法将 primes[k*i] 设置为 false（第 24 行）。在 for 循环中执行 n/k-k+1 次（第 22 行）。因此，找到 n 以内的所有质数的复杂度是

$$\frac{n}{2}-2+1+\frac{n}{3}-3+1+\frac{n}{5}-5+1+\frac{n}{7}-7+1+\frac{n}{11}-11+1+\cdots$$
$$=O\left(\frac{n}{2}+\frac{n}{3}+\frac{n}{5}+\frac{n}{7}+\frac{n}{11}+\cdots\right)<O(n\pi(n))$$
$$=O\left(\frac{n\sqrt{n}}{\log n}\right)$$

$O\left(\frac{n\sqrt{n}}{\log n}\right)$ 的上限非常宽松。实际的时间复杂度要比 $O\left(\frac{n\sqrt{n}}{\log n}\right)$ 好得多。埃拉托斯特尼筛选法适用于小的 n 值，这样 primes 数组可以放入内存。

表 18.5 总结了寻找 n 以内所有质数的三种算法的复杂度。

表 18.5 比较质数查找算法

算法	复杂度	描述
LiveExample 5.16	$O(n^2)$	暴力法，检查所有可能的因数
LiveExample 18.5	$O(n\sqrt{n})$	检查 $\sqrt{n}$ 以内的因数
LiveExample 18.6	$O\left(\frac{n\sqrt{n}}{\log n}\right)$	检查 $\sqrt{n}$ 以内的质数因数
LiveExample 18.7	$O\left(\frac{n\sqrt{n}}{\log n}\right)$	埃拉托斯特尼筛选法

18.8 使用分治法寻找最近点对

要点提示：本节介绍使用分治法寻找最近点对的高效算法。

教学提示：从本节开始，我们将提出有趣且富有挑战性的问题。是时候开始学习高级算法成为一名熟练的程序员了。我们建议你学习算法并在练习中实现。

给定一组点，寻找最近点对问题就是找到彼此最近的两个点。如图 18.4 最近点对动画所示，绘制一条线连接最近的一对点。

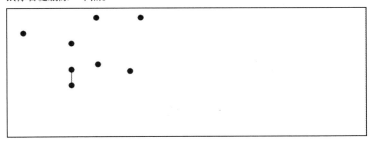

提示：动画绘制一条线连接最近的一对点。单击鼠标左键添加一个点，单击鼠标右键删除一个点。

图 18.4　最近点对动画

8.6 节给出了一种寻找最近点对的暴力算法。该算法计算所有点对之间的距离，并找到距离最小的点对。显然，该算法需要 $O(n^2)$ 时间。我们能设计一个更高效的算法吗？可以使用一种称为分治的方法在 $O(n\log n)$ 时间内高效地解决这个问题。

算法设计说明：分治法将问题划分为子问题，求解子问题，然后组合子问题的解以获得整个问题的解。与动态规划方法不同，分治法中的子问题不重叠。子问题像是规模更小的原始问题，因此可以应用递归来解决问题。事实上，所有递归问题的解决方案都遵循分治法。

Listing 18.1 描述了如何使用分治法解决最近点对问题。

Listing 18.1　查找最近点对的算法

步骤 1：按 x 坐标的升序对点进行排序。对于具有相同 x 坐标的点，按 y 坐标排序。形成点的排序列表 S。

步骤 2：使用排序列表中的中点将 S 划分为大小相等的两个子集 S_1 和 S_2，设列表中点在 S_1 中。递归查找 S_1 和 S_2 中的最近点对。设 d_1 和 d_2 分别表示各自子集中最近点对的距离。

步骤 3：找到 S_1 中的点和 S_2 中的点之间最近的点对，并将它们的距离表示为 d_3。最近点对是距离为 $\min(d_1, d_2, d_3)$ 的点对。

选择排序需要 $O(n^2)$ 时间。在第 19 章中，我们将介绍归并排序和堆排序。这些排序算法需要 $O(n\log n)$ 时间。步骤 1 可以在 $O(n\log n)$ 时间内完成。

步骤 3 可以在 $O(n)$ 时间内完成。设 $d=\min(d_1, d_2)$。我们已经知道，最近点对的距离不能大于 d。对于 S_1 中的点和 S_2 中的点，要形成 S 中的最近点对，如图 18.5a 所示，左边的点必须在条带 `stripL` 中，右边的点必须在条带 `stripR` 中。

对于 `stripL` 中的点 p，只需要考虑 $d \times 2d$ 矩形内的某个右点，如图 18.5b 所示。矩形外的任何右点都不能与 p 形成最近点对。由于 S_2 中的最近点对的距离大于或等于 d，所以矩形中最多只能有六个点。因此，对于 `stripL` 中的每个点，最多需要考虑 `stripR` 中的六个点。

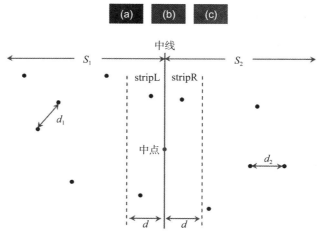

（a）中点将这些点划分为大小相等的两个集合 S_1 和 S_2。d_1 和 d_2 分别是 S_1 和 S_2 中最近点对的距离

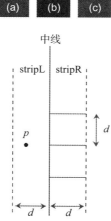

（b）$d = \min(d_1, d_2)$。对于 stripL 中的点 p，只需要考虑 stripR 中 $d \times 2d$ 矩形中的点

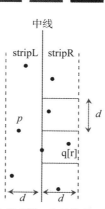

（c）stripL 和 stripR 中的点沿 y 坐标排列，以加快寻找 stripL 中的点与 stripR 中的点之间最近点对的过程

图 18.5　中点将这些点分成大小相等的两组

对于 stripL 中的每个点 p，如何定位 stripR 中相应的 $d \times 2d$ 矩形区域中的点？如果 stripL 和 stripR 中的点按其 y 坐标升序进行排序，则可以高效地做到这一点。设 pointsOrderedOnY 是按 y 坐标升序排序的点的列表。可以在算法中预先获得 pointsOrderedOnY。stripL 和 stripR 可以从步骤 3 中的 pointsOrderedOnY 中获得，如 Listing 18.2 所示。

Listing 18.2 获取 **stripL** 和 **stripR** 的算法

```
1  对于 pointsOrderedOnY 中的每个点 p
2    if (p 在 S1 中且 mid.x - p.x <= d)
3      p 附加到 stripL;
4    else if (p 在 S2 中且 p.x - mid.x <= d)
5      p 附加到 stripR;
```

设 stripL 和 stripR 中的点为 $\{p_0, p_1, \cdots, p_k\}$ 和 $\{q_0, q_1, \cdots, q_t\}$，如图 18.5c 所示。可以使用 Listing 18.3 中描述的算法找到 stripL 中的点和 stripR 中的点之间的最近点对。

Listing 18.3 步骤 3 中寻找最近点对的算法

```
1   d = min(d1, d2);
2   r = 0; // r 是 stripR 中的索引
3   for (stripL 中的每个点 p) {
4     // 跳过矩形区域下方的点
5     while (r < stripR.length && q[r].y <= p.y - d)
6       r++;
7   
8     令 r1 = r;
9     while (r1 < stripR.length && |q[r1].y - p.y| <= d) {
10      // 检查 (p, q[r1]) 是否是可能的最近点对
11      if (distance(p, q[r1]) < d) {
12        d = distance(p, q[r1]);
13        (p, q[r1]) 现在是最近的一对点;
14      }
15  
16      r1 = r1 + 1;
17    }
18  }
```

stripL 中的点按照 $p_0, p_1, \cdots, p_k$ 的顺序考虑。对于 stripL 中的点 p，stripR 中低于 p.y-d 的点将被跳过（第 5～6 行）。某个点一旦被跳过，就不再被考虑。while 循环（第 9～17 行）检查 (p, q[r1]) 是否是可能的最近点对。这样的 q[r1] 对最多有六个，因此在步骤 3 中找到最近点对的复杂度是 $O(n)$。

设 $T(n)$ 表示该算法的时间复杂度。那么

$$T(n) = 2T(n/2) + O(n) = O(n\log n)$$

因此，在 $O(n\log n)$ 时间内可以找到最近点对。此算法的完整实现留作练习（参见编程练习 18.9）。

18.9 使用回溯法解决八皇后问题

要点提示：本节用回溯方法解决八皇后问题。

17.7 节引入的八皇后问题要求找到在棋盘上的每一行都放置一个皇后且两个皇后不能互相攻击的解决方案。这个问题是用递归解决的。在本节中，我们将介绍一种常见的称为**回溯**

的算法设计技术来解决此问题。

算法设计说明：回溯法有许多可能的候选方案。如何找到解决方案？回溯法逐步搜索候选方案，一旦确定该候选方案不可能是高效的解决方案，就放弃该选项，然后寻找新的候选方案。

我们用一维数组来表示棋盘。

```
int queens[8];
```

将 j 赋值给 queen[i]，以表示一个皇后被放置在第 i 行和第 j 列中。将 -1 赋值给 queen[i]，以表示第 i 行还没有放置皇后。

从 k=0 的第 1 行开始搜索，其中 k 是所考虑的当前行的索引。对于 j=0,1,…,7 的顺序，该算法检查皇后是否可能被放置在该行的第 j 列中。搜索的实现方式如下：

- 如果成功，继续在下一行搜索皇后的位置。如果当前行是最后一行，则找到解决方案。
- 如果不成功，返回前一行，并继续在前一行的下一列中搜索新的位置。
- 如果算法回溯到第一行，并且无法在该行中为皇后找到新的位置，则无法找到解决方案。

LiveExample18.8 给出了显示八皇后问题解决方案的程序。

LiveExample 18.8 的互动程序请访问 https://liangcpp.pearsoncmg.com/LiveRunCpp5e/faces/LiveExample.xhtml?header=off&programName=EightQueenBacktracking&fileType=.cpp&programHeight=1380&resultHeight=450。

LiveExample 18.8　EightQueenBacktracking.cpp

Source Code Editor:

```cpp
#include <iostream>
using namespace std;

const int NUMBER_OF_QUEENS = 8; // Constant: eight queens
// queens are placed at (i, queens[i])
// -1 indicates that no queen is currently placed in the ith row
// Initially, place a queen at (0, 0) in the 0th row
int queens[NUMBER_OF_QUEENS] = {-1, -1, -1, -1, -1, -1, -1, -1};

// Check whether a queen can be placed at row i and column j
bool isValid(int row, int column)
{
  for (int i = 1; i <= row; i++)
    if (queens[row - i] == column          // Check column
      || queens[row - i] == column - i  // Check upper left diagonal
      || queens[row - i] == column + i) // Check upper right diagonal
      return false; // There is a conflict
  return true; // No conflict
}

// Display the chessboard with eight queens
void printResult()
{
  cout << "\n---------------------------------\n";
  for (int row = 0; row < NUMBER_OF_QUEENS; row++)
  {
    for (int column = 0; column < NUMBER_OF_QUEENS; column++)
      printf(column == queens[row] ? "| Q " : "|   ");
    cout << "|\n---------------------------------\n";
  }
}
```

```cpp
33    // Find a position to place a queen in row k
34    int findPosition(int k)
35    {
36      int start = queens[k] + 1; // Search for a new placement
37
38      for (int j = start; j < NUMBER_OF_QUEENS; j++)
39      {
40        if (isValid(k, j))
41          return j; // (k, j) is the place to put the queen now
42      }
43
44      return -1;
45    }
46
47    // Search for a solution
48    bool search()
49    {
50      // k - 1 indicates the number of queens placed so far
51      // We are looking for a position in the kth row to place a queen
52      int k = 0;
53      while (k >= 0 && k < NUMBER_OF_QUEENS)
54      {
55        // Find a position to place a queen in the kth row
56        int j = findPosition(k);
57        if (j < 0)
58        {
59          queens[k] = -1;
60          k--; // back track to the previous row
61        }
62        else
63        {
64          queens[k] = j;
65          k++;
66        }
67      }
68
69      if (k == -1)
70        return false; // No solution
71      else
72        return true; // A solution is found
73    }
74
75    int main()
76    {
77      search(); // Start search from row 0. Note row indices are 0 to 7
78      printResult(); // Display result
79
80      return 0;
81    }
```

Execution Result:

```
command>cl EightQueenBacktracking.cpp
Microsoft C++ Compiler 2019
Compiled successful (cl is the VC++ compile/link command)

command>EightQueenBacktracking

---------------------------------
| Q |   |   |   |   |   |   |   |
---------------------------------
|   |   |   |   | Q |   |   |   |
---------------------------------
```

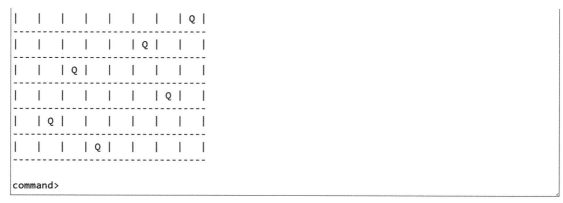

该程序调用 search()（第 77 行）搜索解决方案。最初，没有皇后被放置在任何一行（第 8 行）。现在从 k=0 的第 1 行开始搜索（第 52 行），并找到皇后的位置（第 56 行）。如果成功，将其放在该行（第 64 行）中，并考虑下一行（第 65 行）。如果不成功，则返回前一行（第 59 ~ 60 行）。

findPosition(k) 函数将皇后放置在第 k 行，从 queen[k]+1（第 36 行）开始搜索一个可能的位置。它依次检查皇后是否可以被放置在 start, start+1, …, 7 的位置（第 38 ~ 42 行）。如果可以，返回列索引（第 41 行）；否则，返回 -1（第 44 行）。

调用 isValid(row, column) 函数检查将皇后放在指定位置是否会导致与之前放置的皇后发生冲突（第 40 行）。确保没有皇后被放在同一列（第 14 行）、左上对角线（第 15 行）或右上对角线（16 行），如图 18.6 所示。

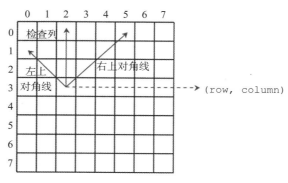

图 18.6　调用 isValid(row, column) 检查皇后是否可以放在 (row, column)

18.10　案例研究：寻找凸包

要点提示：本节介绍为一组点寻找凸包的高效几何算法。

给定一组点，**凸包**是包围所有这些点的最小凸多边形，如图 18.7a 所示。如果连接两个顶点的每条线都在多边形内，则多边形是凸的。例如，图 18.7a 中的顶点 v0、v1、v2、v3、v4 和 v5 形成了一个凸多边形，但图 18.7b 中不是，因为连接 v3 和 v1 的线不在多边形内部。

凸包在游戏编程、模式识别和图像处理中有许多应用。

人们已经开发了许多算法来寻找凸包。本节介绍两种流行的算法：礼品包装算法和 Graham 算法。

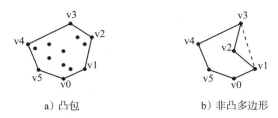

图 18.7 凸包是包含一组点的最小凸多边形

18.10.1 礼品包装算法

Listing 18.4 给出一种直观的方法，称为礼品包装算法。

Listing 18.4 凸包礼品包装算法

步骤 1：给定一个点集 S，S 中的点被标记为 S_0，S_1，$\cdots$，S_k。选择集合 S 中最右边的最低点 h_0。如图 18.8a 所示，h_0 就是这样一个点。把 h_0 添加到凸包 H。H 是一个初始为空的集合。设 t_0 为 h_0。

步骤 2：设 t_1 为 s_0。

对于 S 中的每个点 s，

如果 s 在从 t_0 到 t_1 的直线的右侧，则 t_1 赋值为 s。

（步骤 2 之后，从 t_0 到 t_1 的直线右侧没有任何点，如图 18.8b 所示。）

步骤 3：如果 t_1 是 h_0（见图 18.8d），则 H 中的点形成 S 的凸包。否则，将 t_1 添加到 H，设 t_0 为 t_1，然后转到步骤 2（见图 18.8c）。

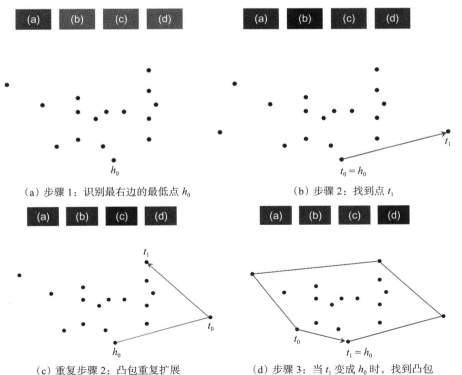

（a）步骤 1：识别最右边的最低点 h_0　　（b）步骤 2：找到点 t_1

（c）重复步骤 2：凸包重复扩展　　（d）步骤 3：当 t_1 变成 h_0 时，找到凸包

图 18.8 （a）h_0 是 S 中最右边的最低点，（b）步骤 2 找到点 t_1，（c）凸包重复扩展，（d）当 t_1 变成 h_0 时，找到凸包

凸包是逐步扩展的。其正确性由以下事实支持：在步骤 2 之后，没有点位于从 t_0 到 t_1 的直线的右侧。这样可以确保 S 中两点之间的每条线段都位于多边形内。

在步骤 1 中，找到最右边的最低点可以在 $O(n)$ 时间内完成。一个点是在直线的左侧、右侧还是在直线上，可以在 $O(1)$ 时间内决定（见编程练习 3.29）。因此，在步骤 2 中找到新的点 t_1 需要 $O(n)$ 时间。步骤 2 重复 h 次，其中 h 是凸包的边数。因此，该算法需要 $O(hn)$ 时间。在最坏的情况下，h 等于 n。这个算法的实现留作练习（见编程练习 18.11）。

18.10.2 Graham 算法

罗纳德·格雷厄姆（Ronald Graham）于 1972 年开发了一种更高效的寻找凸包的算法。它的工作原理如 Listing18.5 所示。

Listing18.5　凸包 Graham 算法

步骤 1：给定点集 S，选择集合 S 中最右边的最低点 p_0。如图 18.9a 所示，p_0 就是这样一个点。

步骤 2：对 S 中的点，以 p_0 为原点，按照沿 x 轴方向形成的角度排序，如图 18.9b 所示。如果两个点的角度相同，则丢弃最接近 p_0 的点。S 中的点现在被排序为 $p_0, p_1, p_2, \cdots, p_{n-1}$。

步骤 3：将 p_0、p_1 和 p_2 压入栈 H。

步骤 4：见如下代码。

```
i = 3;
while (i < n)
{
  令 t1 和 t2 为栈 H 中的顶部第一和第二个元素；
  if (pi 在从 t2 到 t1 的有向边的左侧)
  {
    pi 压入 H；
    i++; // 考虑 S 中的下一个点
  }
  else
    弹出栈 H 的栈顶元素
}
```

步骤 5：H 中的点构成凸包。

凸包是递增发现的。最初，p_0、p_1 和 p_2 形成一个凸包。考虑 p_3。p_3 在当前凸包之外，因为点是按其角度的递增顺序排序的。如果 p_3 严格地位于从 p_1 到 p_2 的直线的左侧（见图 18.9c），则将 p_3 压入 H。现在 p_0、p_1、p_2 和 p_3 形成凸包。如果 p_3 位于从 p_1 到 p_2 的直线的右侧（见图 18.9d），将 p_2 弹出 H，将 p_3 压入 H。现在 p_0、p_1 和 p_3 形成一个凸包，p_2 在这个凸包的内部。可以通过归纳法证明，步骤 5 中 H 中的所有点形成了输入集 S 中所有点的凸包。

在步骤 1 中，找到最右边的最低点可以在 $O(n)$ 时间内完成。可以使用三角函数来计算角度。不过可以对点进行排序，而无须实际计算它们的角度。我们注意到当且仅当 p_2 位于从 p_0 到 p_1 的直线的左侧时，p_2 会形成比 p_1 更大的角度。如编程练习 3.29 所示，一个点是否在一条直线的左侧，可以在 $O(1)$ 时间内确定。步骤 2 中的排序可以使用第 19 章中介绍的归并排序或堆排序算法在 $O(n\log n)$ 时间内完成。步骤 4 可以在 $O(n)$ 时间内完成。因此，该算法需要 $O(n\log n)$ 时间。

此算法的实现留作练习（参见编程练习 18.13）。

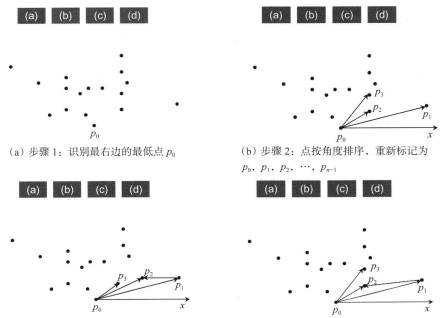

(a) 步骤 1：识别最右边的最低点 p_0

(b) 步骤 2：点按设定角度排序，重新标记为 $p_0, p_1, p_2, \cdots, p_{n-1}$

(c) 步骤 4：在这种假设情况下，p_3 严格地在从 p_1 到 p_2 这条直线的左侧。p_3 压入 H

(d) 步骤 4：在这种假设情况下，p_3 严格地在从 p_1 到 p_2 这条直线的右侧。弹出 p_2，并将 p_3 压入 H

图 18.9 （a）p_0 是 S 中最右边的最低点。（b）点按角度排序。（c）和（d）凸包是递增发现的

18.11 字符串匹配

要点提示：本节介绍用于字符串匹配的暴力、Boyer-Moore 和 Knuth-Morris-Pratt 算法。

字符串匹配是在字符串中查找子字符串的匹配项。字符串通常被称为文本，子字符串被称为模式。字符串匹配是计算机编程中的一项常见任务。`string` 类用 `find(pattern)` 函数来查找模式。人们已经做了大量的研究以寻找高效的字符串匹配算法。本节介绍三种算法：暴力算法、Boyer-Moore 算法和 Knuth-Morris-Pratt 算法。

暴力算法只是将模式与文本中的每个可能的子字符串进行比较。假设文本和模式的长度分别为 n 和 m。该算法如 Listing 18.6 所示。

Listing 18.6 暴力字符串匹配算法

```
for i 从 0 到 n - m
{
   检查模式是否匹配 text[i .. i + m]
}
```

这里，`text[i..j]` 表示从索引 i 到索引 j 的文本中的子字符串。

LiveExample 18.9 给出了暴力算法的实现。

LiveExample 18.9 的互动程序请访问 https://liangcpp.pearsoncmg.com/LiveRunCpp5e/faces/LiveExample.xhtml?header=off&programName=StringMatch&programHeight=750&resultHeight=200。

LiveExample 18.9   StringMatch.cpp

Source Code Editor:

```cpp
#include <iostream>
#include <string>
using namespace std;

// Test if pattern matches text starting at index i
bool isMatched(int i, string text, string pattern)
{
    for (unsigned k = 0; k < pattern.length(); k++)
    {
        if (pattern[k] != text[i + k])
            return false;
    }

    return true;
}

// Return the index of the first match. -1 otherwise.
int match(string text, string pattern)
{
    for (unsigned i = 0; i < text.length() - pattern.length() + 1; i++)
    {
        if (isMatched(i, text, pattern))
            return i;
    }

    return -1;
}

int main()
{
    cout << "Enter a string text: ";
    string text;
    getline(cin, text);
    cout << "Enter a string pattern: ";
    string pattern;
    getline(cin, pattern);

    int index = match(text, pattern);
    if (index >= 0)
        cout << "matched at index " << index << endl;
    else
        cout << "unmatched" <<endl;
}
```

Enter input data for the program (Sample data provided below. You may modify it.)

```
aaaaaaaaaab
aab
```

[Automatic Check] [Compile/Run] [Reset] [Answer]   Choose a Compiler: VC++

Execution Result:

```
command>cl StringMatch.cpp
Microsoft C++ Compiler 2019
Compiled successful (cl is the VC++ compile/link command)

command>StringMatch
Enter a string text: aaaaaaaaaab
Enter a string pattern: aaab
matched at index 7

command>
```

函数 match(text, pattern)（第 18～27 行）检查 pattern 是否与 text 中的子字符串匹配。函数 isMatched(i, text, pattern)（第 6～15 行）检查 pattern 是否与从索引 i 开始的 text[i, i+m] 匹配。

显然，该算法需要时间 $O(nm)$，因为检查 pattern 是否与 text[i, i+m] 匹配需要时间 $O(m)$。

18.11.1 Boyer-Moore 算法

暴力算法通过检查所有对齐来搜索文本中的模式匹配，这是没必要的。Boyer-Moore 算法通过从右到左将模式与文本中的子字符串进行比较来找到匹配。如果文本中的某个字符与模式中的字符不匹配，并且该字符不在模式的其余部分，则可以将模式一直滑动到经过该字符的位置。Boyer-Moore 算法如 Listing 18.7 所示。

Listing 18.7　Boyer-Moore 字符串匹配算法

```
i = m - 1;
while i <= n - 1
    将 pattern 与 text[i-(m-1)..i] 从右向左逐一进行比较，如图 18.10 所示。
    如果都匹配，则完成。否则，设 text[k] 是第一个与 pattern 中相应字符不匹配的字符。
    考虑两种情况：
    情况 1：如果 text[k] 不在剩余的 pattern 中，则滑动 pattern 经过 text[k]，如图 18.11
        所示。设置 i=k+m；
    情况 2：如果 text[k] 在 pattern 中，在与 text[k] 匹配的 pattern 中找到最后一个字符，
        如 pattern[j]，然后向右滑动 pattern，使 pattern[j] 与 text[k] 对齐，如图 18.12
        所示。设置 i=k+m-j-1。
```

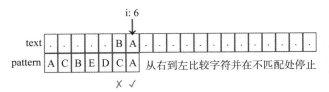

图 18.10　从右到左比较字符并在不匹配处停止，检查模式是否与子字符串匹配

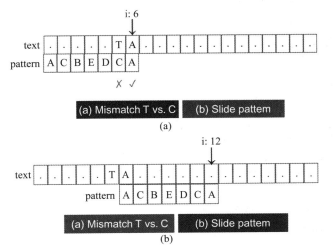

图 18.11　由于 T 不在 pattern 的剩余部分，滑动 pattern 经过 T，开始下一次检查

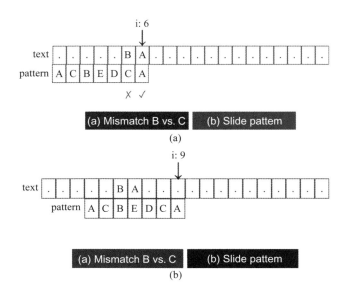

图 18.12 B 与 C 不匹配。B 首先从右到左匹配了 `pattern` 剩余部分的 `pattren[2]`, 滑动 `pattern` 使 `text` 中的 B 与 `pattern[2]` 对齐

LiveExample 18.10 给出了 Boyer-Moore 算法的一个实现。

LiveExample 18.10 的互动程序请访问 https://liangcpp.pearsoncmg.com/LiveRunCpp5e/faces/LiveExample.xhtml?header=off&programName=StringMatchBoyerMoore&programHeight=1080&resultHeight=200。

LiveExample 18.10　StringMatchBoyerMoore.cpp

```cpp
#include <iostream>
#include <string>
using namespace std;

// Return the index of the last element in pattern[0 .. j]
// that matches ch. -1 otherwise.
int findLastIndex(char ch, int j, string pattern)
{
  for (int k = j; k >= 0; k--)
  {
    if (ch == pattern[k])
      return k;
  }

  return -1;
}

// Return the index of the first match. -1 otherwise.
int match(string text, string pattern)
{
  int i = pattern.length() - 1;
  while (i < text.length())
  {
    int k = i;
    int j = pattern.length() - 1;
    while (j >= 0) {
      if (text[k] == pattern[j])
      {
        k--; j--;
      }
```

```cpp
31        else {
32          break;
33        }
34      }
35
36      if (j < 0)
37        return i - pattern.length() + 1; // A match found
38
39      int index = findLastIndex(text[k], j - 1, pattern);
40      if (index >= 0) // text[k] is in the remaining part of the pattern
41        i = k + pattern.length() - 1 - index;
42      else // text[k] is not in the remaining part of the pattern
43        i = k + pattern.length();
44    }
45
46    return -1;
47  }
48
49  int main()
50  {
51    cout << "Enter a string text: ";
52    string text;
53    getline(cin, text);
54    cout << "Enter a string pattern: ";
55    string pattern;
56    getline(cin, pattern);
57
58    int index = match(text, pattern);
59    if (index >= 0)
60      cout << "matched at index " << index << endl;
61    else
62      cout << "unmatched" <<endl;
63  }
```

Enter input data for the program (Sample data provided below. You may modify it.)

aaaaaaaaab
aab

[Automatic Check] [Compile/Run] [Reset] [Answer]　　　Choose a Compiler: VC++

Execution Result:

```
command>cl StringMatchBoyerMoore.cpp
Microsoft C++ Compiler 2019
Compiled successful (cl is the VC++ compile/link command)

command>StringMatchBoyerMoore
Enter a string text: aaaaaaaaab
Enter a string pattern: aaab
matched at index 7

command>
```

函数 match(text, pattern)（第 19～47 行）检查 pattern 是否与 text 中的子字符串匹配。i 表示子字符串的最后一个索引。它从 i=pattern.length()-1（第 21 行）开始，往回比较 text[i] 与 pattern[j]、text[i-1] 与 pattern[j-1]，以此类推（第 26～34 行）。如果 j<0，则找到匹配（第 36～37 行）。否则，使用 findLastIndex 函数查找 pattern[0..j-1] 中 text[k] 的最后一个匹配元素的索引。如果 index>=0，则设置 i 为 k+m-1-index（第 41 行），其中 m 是 pattern.length()。否则，设置 i 为 k+m（第 43 行）。

在最坏的情况下，Boyer-Moore 算法需要时间 $O(nm)$。最坏情况的示例如图 18.13 所示。

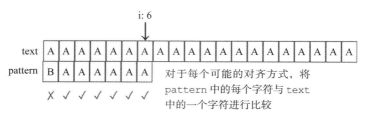

图 18.13 text 全部为 A，pattern 为 BAAAAAA

但平均而言，Boyer-Moore 算法是高效的，因为该算法通常能够跳过大部分文本。Boyer-Moore 算法有几种变体。本节介绍了一个简化版本。

18.11.2　Knuth-Morris-Pratt 算法

Knuth-Morris-Pratt（KMP）算法是高效的。它在最坏情况下的复杂度达到 $O(m+n)$。它是最优算法，因为在最坏情况下，文本和模式中的每个字符都必须至少检查一次。在暴力或 Boyer-Moore 算法中，一旦发现不匹配，该算法就重新开始搜索下一个可能的匹配，方法是将模式向右移动一个位置（暴力算法），也可能向右移动多个位置（Boyer-Moore 算法）。这样做将忽略不匹配之前字符的成功匹配。KMP 算法考虑成功的匹配，以在继续下一次搜索之前找到要在模式中移动的最大位置数。

为了找到模式中移动的最大位置数，我们首先将失败函数 fail[k] 定义为模式的最长前缀的长度，该前缀是 pattern[0..k] 的后缀。对给定的模式可以预先计算失败函数。失败函数实际上是一个包含 m 个元素的数组。假设模式为 ABCABCDABC。该模式的失败函数如图 18.14 所示：

k	0	1	2	3	4	5	6	7	8	9
pattern	A	B	C	A	B	C	D	A	B	C
fail	0	0	0	1	2	3	0	1	2	3

图 18.14　失败函数 fail[k] 是与 pattern[0..k] 中的后缀匹配的最长前缀的长度

例如，fail[5] 是 3，因为 ABC 是 ABCABC 后缀中最长的前缀。fail[7] 为 1，因为 A 是 ABCABCDA 后缀中最长的前缀。当比较文本和模式时，一旦在模式中的索引 k 处发现不匹配，可以移动模式将索引 fail[k-1] 处的模式与文本对齐，如图 18.15 所示。

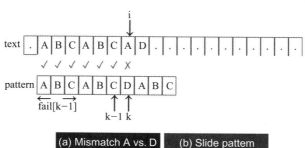

(a) pattern[0..k-1] 匹配 text[i-k..i-1]

图 18.15　在 text[i] 处不匹配时，模式向右移动，以将前缀中的第一个 fail[k-1] 元素与 text[i-1] 对齐

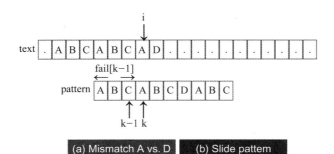

(b) 滑动 pattern，使 pattern[fail[k-1]] 与 text[i] 对齐

图 18.15 在 text[i] 处不匹配时，模式向右移动，以将前缀中的第一个 fail[k-1] 元素与 text[i-1] 对齐（续）

Listing 18.8 中给出了 KMP 算法。

Listing 18.8　KMP 字符串匹配算法

步骤 1：首先我们预计算失败函数。现在从 i=0 和 k=0 开始。

步骤 2：将 text[i] 与 pattern[k] 进行比较。考虑两种情况。

情况 1　text[i] 等于 pattern[k]：如果 k 是 m-1，则找到匹配项并返回 i-m+1。否则，将 i 和 k 增加 1。

情况 2　text[i] 不等于 pattern[k]：如果 k>0，移动与 pattern[k-1] 中后缀匹配的最长前缀，通过设置 k=fail[k-1] 使前缀中的最后一个字符与 text[i-1] 对齐，否则将 i 增加 1。

步骤 3：如果 i<n，则重复步骤 2。

现在让我们把注意力转向计算失败函数。这可以通过以下方式将模式与自身进行比较来实现：

步骤 1：失败函数是一个包含 m 个元素的数组。最初，将所有元素设置为 0。我们从 i=1 和 k=0 开始。

步骤 2：比较 pattern[i] 和 pattern[k]。考虑两种情况。

情况 1　pattern[i]==pattern[k]：fail[i]=k+1。将 i 和 k 增加 1。

情况 2　pattern[i]!=pattern[k]：如果 k>0，则设置 k=fail[k-1]，否则将 i 增加 1。

步骤 3：如果 i<m，重复步骤 2。注意，如果 pattern[i]==pattern[k]，则 k 表示 pattern[0..i-1] 中最长前缀的长度。

LiveExample18.11 给出了 KMP 算法的一个实现。

LiveExample 18.11 的互动程序请访问 https://liangcpp.pearsoncmg.com/LiveRunCpp5e/faces/LiveExample.xhtml?header=off&programName=StringMatchKMP&programHeight=1230&resultHeight=200。

函数 match(text, pattern)（第 30 ~ 56 行）检查 pattern 是否与 text 中的子字符串匹配。i 表示 text 中的当前位置，它总是向前移动。k 表示 pattern 中的当前位置。如果 text[i]==pattern[k]（第 37 行），则 i 和 k 都增加 1（第 43 ~ 44 行）。否则，如果 k>0，则将 fail[k-1] 设置为 k，从而滑动 pattern 使 pattern[k] 与 text[i] 对齐（第 49 行），否则将 i 增加 1（第 51 行）。

LiveExample 18.11　StringMatchKMP.cpp

Source Code Editor:

```cpp
#include <iostream>
#include <string>
#include <vector>
using namespace std;

// Compute failure function
vector<int> getFailure(string pattern)
{
  vector<int> fail(pattern.length());
  int i = 1;
  int k = 0;
  while (i < pattern.length())
  {
    if (pattern[i] == pattern[k])
    {
      fail[i] = k + 1;
      i++;
      k++;
    }
    else if (k > 0)
      k = fail[k - 1];
    else
      i++;
  }

  return fail;
}

// Return the index of the first match. -1 otherwise.
int match(string text, string pattern)
{
  vector<int> fail = getFailure(pattern);
  int i = 0; // Index on text
  int k = 0; // Index on pattern
  while (i < text.length())
  {
    if (text[i] == pattern[k])
    {
      if (k == pattern.length() - 1)
      {
        return i - pattern.length() + 1; // pattern matched
      }
      i++; // Compare the next pair of characters
      k++;
    }
    else
    {
      if (k > 0)
        k = fail[k - 1]; // Matching prefix position
      else
        i++; // No prefix
    }
  }

  return -1;
}

int main()
{
  cout << "Enter a string text: ";
```

```
61    string text;
62    getline(cin, text);
63    cout << "Enter a string pattern: ";
64    string pattern;
65    getline(cin, pattern);
66
67    int index = match(text, pattern);
68    if (index >= 0)
69      cout << "matched at index " << index << endl;
70    else
71      cout << "unmatched" <<endl;
72  }
```

Enter input data for the program (Sample data provided below. You may modify it.)
```
aaaaaaaaaab
aab
```

[Automatic Check] [Compile/Run] [Reset] [Answer]   Choose a Compiler: VC++

Execution Result:
```
command>cl StringMatchKMP.cpp
Microsoft C++ Compiler 2019
Compiled successful (cl is the VC++ compile/link command)

command>StringMatchKMP
Enter a string text: aaaaaaaaaab
Enter a string pattern: aab
matched at index 7

command>
```

getFailure(pattern) 函数（第 7 ~ 27 行）将 pattern 与 pattern 进行比较，以获得最大前缀 fail[k] 的长度，该前缀是 pattern[0..k] 中的后缀。它将数组 fail 初始化为 0（第 9 行），并将 i 和 k 分别设置为 1 和 0（第 10 ~ 11 行）。i 表示第一个 pattern 中的当前位置，它总是向前移动。k 表示可能的前缀的当前最大长度，该前缀也是 pattern[0..i] 中的后缀。如果 pattern[i]==pattern[k]，则将 fail[i] 设置为 k+1（第 16 行），并将 i 和 k 都增加 1（第 17 ~ 18 行）。否则，如果 k>0，则将 k 设置为 fail[k-1] 来滑动第二个 pattern，使第一个 pattern 中的 pattern[i] 与第二个 pattern 中的 pattern[k] 对齐（第 21 行），否则将 i 增加 1（第 23 行）。

分析运行时间，考虑三种情况：

情况 1：text[i] 等于 pattern[k]。i 向前移动了一个位置。

情况 2：text[i] 不等于 pattern[k]，k 为 0。i 向前移动了一个位置。

情况 3：text[i] 不等于 pattern[k] 并且 k>0。pattern 向前移动至少一个位置。

在任何情况下，要么 i 在文本上向前移动一个位置，要么模式向右移动至少一个位置。因此，匹配函数的 while 循环中的迭代次数最多为 $2n$。类似地，getFailure 函数中的迭代次数最多为 $2m$。因此，KMP 算法的运行时间为 $O(n+m)$。

关键术语

average-case analysis（平均情况分析）   best-case input（最佳情况输入）
backtracking approach（回溯法）   Big O notation（大 O 表示法）

brute force（暴力）
constant time（常数时间）
convex hull（凸包）
divide-and-conquer approach（分治法）
dynamic programming approach（动态规划法）
exponential algorithm（指数算法）
growth rate（增长率）
logarithmic time（对数时间）
quadratic time（二次时间）
worst-case input（最坏情况输入）

章节总结

1. 大 O 表示法是分析算法性能的一种理论方法。它评估算法执行时间随着输入规模的增加而增加的速度。因此，可以通过检查两种算法的增长率来比较它们。
2. 导致执行时间最短的输入称为最佳情况输入，导致执行时间最长的输入称为最坏情况输入。
3. 最佳情况和最坏情况并不具有代表性，但最坏情况分析非常有用。可以肯定的是，算法永远不会比最坏的情况慢。
4. 平均情况分析试图在相同规模的所有可能输入之间确定平均时间量。
5. 平均情况分析是理想的，但很难执行，因为对于许多问题，很难确定各种输入实例的相对概率和分布。
6. 如果时间与输入规模无关，则称该算法使用常数时间，用 $O(1)$ 表示。
7. 线性查找需要 $O(n)$ 时间。时间复杂度为 $O(n)$ 的算法称为线性算法。
8. 二分查找需要 $O(\log n)$ 时间。时间复杂度为 $O(\log n)$ 的算法称为对数算法。
9. 选择排序的最坏时间复杂度是 $O(n^2)$。
10. 时间复杂度为 $O(n^2)$ 的算法称为二次算法。
11. 汉诺塔问题的时间复杂度为 $O(2^n)$。
12. 时间复杂度为 $O(2^n)$ 的算法称为指数算法。
13. 给定索引的斐波那契数可以在 $O(n)$ 时间内被找到。
14. 欧几里得的最大公约数算法需要 $O(\log n)$ 时间。
15. 所有小于或等于 n 的质数都可以在 $O\left(\dfrac{n\sqrt{n}}{\log n}\right)$ 时间内找到。
16. 使用分治法可以在 $O(n\log n)$ 时间内找到最近点对。
17. 分治法将问题划分为子问题，求解子问题，然后将子问题的解组合起来，以获得整个问题的解。与动态规划法不同，分治法中的子问题不重叠。子问题类似于规模较小的原始问题，因此可以用递归来解决问题。
18. 八皇后问题可以用回溯法来解决。
19. 回溯法逐步搜索候选解决方案，一旦确定该候选方案不可能是有效解决方案，就放弃该选项，然后寻找新的候选方案。
20. 在 $O(n^2)$ 时间内使用礼品包装算法可以找到一组点的凸包，在 $O(n\log n)$ 时间内可以使用 Graham 算法找到一组点的凸包。
21. 暴力和 Boyer-Moore 字符串匹配算法需要 $O(nm)$ 时间，而 KMP 字符串匹配算法则需要 $O(n+m)$ 时间。

编程练习

互动程序请访问 https://liangcpp.pearsoncmg.com/CheckExerciseCpp/faces/CheckExercise5e.xhtml?chapter=18&programName=Exercise18_01。

*18.1 （最大连续升序的子字符串）编写一个程序，提示用户输入字符串并显示最大连续升序的子字符串。分析程序的时间复杂度。

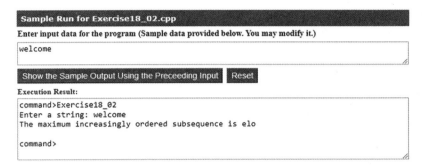

18.2 （最大升序的子序列）编写一个程序，提示用户输入字符串并显示最大升序的子序列。分析程序的时间复杂度。

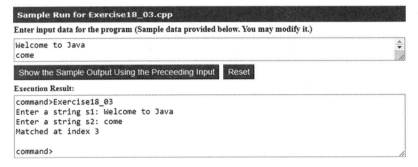

*18.3 （模式匹配）编写一个程序，提示用户输入两个字符串，并检查第二个字符串是否是第一个字符串中的子字符串。假设第二个字符串中的字符是不同的。算法需要在 $O(n)$ 时间内完成。

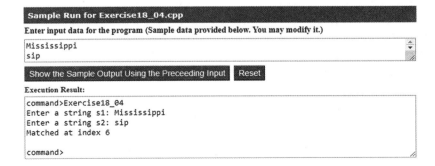

*18.4 （模式匹配）编写一个程序，提示用户输入两个字符串，并检查第二个字符串是否是第一个字符串中的子字符串。分析算法的时间复杂度。

*18.5 （相同数字子序列）编写一个时间复杂度为 $O(n)$ 的程序，提示用户输入以 0 结尾的整数序列，并找到数字相同的最长子序列。

```
Sample Run for Exercise18_05.cpp
Enter input data for the program (Sample data provided below. You may modify it.)
2 4 4 8 8 8 8 2 4 4 0

Show the Sample Output Using the Preceeding Input    Reset
Execution Result:
command>Exercise18_05
Enter a series of numbers ending with 0:
2 4 4 8 8 8 8 2 4 4 0
The longest same number sequence starts at index 4 with 4 values of 8

command>
```

*18.6 （最大公约数的执行时间）使用 LiveExample 18.3 和 18.4 中的算法，编写一个程序，获取查找从索引 40 到索引 45 的每两个连续斐波那契数的最大公约数的执行时间。程序打印如下表格：

	40	41	42	43	44	45
LiveExample 18.3						
LiveExample 18.4						

（提示：可以使用以下代码模板来获取执行时间。）

```
long startTime = time(0);
perform the task;
long endTime = time(0);
long executionTime = endTime - startTime;
```

**18.7 （质数的执行时间）使用 LiveExample 18.5 ～ 18.7 中的算法，编写一个程序，获取查找小于 8000000、10000000、12000000、14000000、16000000 和 18000000 的所有质数的执行时间。程序打印如下表格：

	8000000	10000000	12000000	14000000	16000000	18000000
LiveExample 18.5						
LiveExample 18.6						
LiveExample 18.7						

**18.8 （1000000000 以内的所有质数）编写一个程序，找出 1000000000 以内的所有质数。大约有 50847534 个这样的质数。程序应满足以下要求：
- 程序应该将质数存储在一个名为 Exercise18_8.dat 的二进制数据文件中。当找到新的质数时，该数字会追加到文件中。
- 要判断一个新数字是否为质数，程序应该将质数从文件加载到一个大小为 100000 的 long 类型的数组中。如果数组中没有数字是新数字的除数，继续从数据文件中读取接下来的 100000 个质数，直到找到除数或读取完文件中的所有数字。如果没有找到除数，新数就是质数。
- 由于此程序需要很长时间才能完成，因此应将其作为批处理作业在 UNIX 计算机上运行。如果机器关闭并重新启动，程序应该使用存储在二进制数据文件中的质数来恢复，而不是从头开始。

***18.9 （最近点对）18.8 节介绍了一种使用分治法寻找最近点对的算法。实现该算法并满足以下要求：
- 定义一个名为 Pair 的类，其中包含表示两点的数据字段 p1 和 p2，以及定义一个返回两点距离的名为 getDistance() 的函数。
- 实现以下函数：

```cpp
// Return the closest pair of points
Pair* getClosestPair(const vector<vector<double>>& points)
// Return the closest pair of points
Pair* getClosestPair(const vector<Point>& points)
// Return the distance of the closest pair of points
// in pointsOrderedOnX[low..high]. This is a recursive
// function. pointsOrderedOnX and pointsOrderedOnY are
// not changed in the subsequent recursive calls.
double distance(const vector<Point>& pointsOrderedOnX,
  int low, int high, const vector<Point>& pointsOrderedOnY)
// Return the distance between two points p1 and p2
double distance(const Point& p1, const Point& p2)
// Return the distance between points (x1, y1) and (x2, y2)
double distance(double x1, double y1, double x2, double y2)
```

18.10 （最后 10 个质数）编程练习 18.8 将质数存储在名为 Exercise18_8.dat 的文件中。编写一个高效的程序，读取文件中的最后 10 个数字。（提示：不要读取每个数字。跳过文件中最后 10 个数字之前的所有数字。）

**18.11 （几何：寻找凸包的礼品包装算法）18.10.1 节介绍了寻找一组点的凸包的礼品包装算法。使用以下函数实现算法：

```cpp
// Return the points that form a convex hull
vector<MyPoint> getConvexHull(const vector<MyPoint>& s)
```

MyPoint 类的定义如下：

```cpp
class MyPoint
{
public:
  double x, y;
  MyPoint(double x, double y)
  {
    this->x = x; this->y = y;
  }
};
```

编写一个测试程序，提示用户输入集合的大小和点，并显示形成凸包的点。注意，调试代码时，你会发现算法忽略了两种情况：当 t1=t0 时和当有一个点在 t0 到 t1 之间的直线上时。任何一种情况发生时，如果 t0 到 p 的距离大于 t0 到 t1 的距离，则用点 p 代替 t1。

```
Sample Run for Exercise18_11.cpp
Enter input data for the program (Sample data provided below. You may modify it.)
6
1 2.4 2.5 2 1.5 34.5 5.5 6 6 2.4 5.5 9

Show the Sample Output Using the Preceeding Input    Reset

Execution Result:
command>Exercise18_11
How many points are in the set? 6
Enter 6 points: 1 2.4 2.5 2 1.5 34.5 5.5 6 6 2.4 5.5 9
The convex hull is (2.5, 2.0) (6.0, 2.4) (5.5, 9.0) (1.5, 34.5) (1.0, 2.4)
command>
```

18.12 （质数的数量）编程练习 18.8 将质数存储在名为 Exercise18_8.dat 的文件中。编写一个程序，找出小于或等于 10、100、1000、10000、100000、1000000 和 10000000 的质数的数量。程序应该从 Exercise18_8.dat 中读取数据。注意，随着存储质数的增多，数据文件可能会持续增大。

**18.13 （几何：凸包）18.10.2 节介绍了为一组点寻找凸包的 Graham 算法。使用以下函数实现算法：

```cpp
// Return the points that form a convex hull
vector<MyPoint> getConvexHull(const vector<MyPoint>& s)
```

MyPoint的定义见编程练习9.4。编写一个测试程序，提示用户输入集合的大小和点，并显示形成凸包的点。

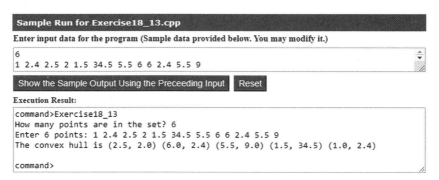

**18.14 （最大子方阵）在编程练习12.27中介绍了查找最大子方阵的问题。设计一个动态规划算法来解决这个问题，以便在$O(n^2)$时间内找到最大的子方阵。

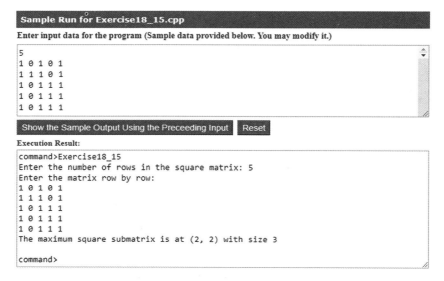

*18.15 （首次适合的箱子包装）箱子包装问题是将各种重量的物体装到箱子中。假设每个箱子最多可容纳10磅。该程序使用一种算法，将一个物体放入它所适合的第一个箱子中。程序提示用户输入物体的总数和每个物体的重量。程序显示包装物体所需的箱子总数以及每个箱子包含的物体。

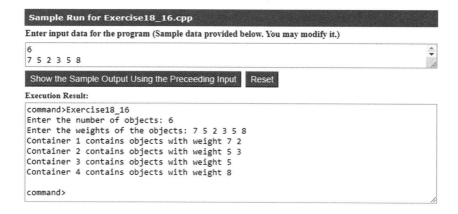

该程序是否得到一个最优解决方案，即找到包装物体的最小箱子数量？程序的时间复杂度是多少？

****18.16 （最优箱子包装）重写前面的程序来找到一个最优解决方案，即用最小数量的箱子包装所有物体。

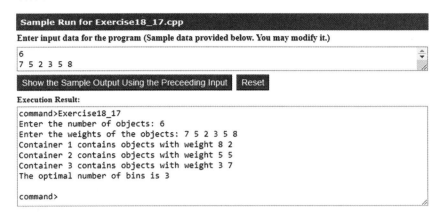

第 19 章
Introduction to C++ Programming and Data Structures, Fifth Edition

排　　序

学习目标
1. 研究和分析各种排序算法的时间复杂度（19.2～19.7 节）。
2. 设计、实现和分析插入排序（19.2 节）。
3. 设计、实现和分析冒泡排序（19.3 节）。
4. 设计、实现和分析归并排序（19.4 节）。
5. 设计、实现和分析快速排序（19.5 节）。
6. 设计和实现二元堆（19.6 节）。
7. 设计、实现和分析堆排序（19.6 节）。
8. 设计、实现和分析桶排序和基数排序（19.7 节）。
9. 设计、实现和分析支持大量数据文件的外部排序（19.8 节）

19.1　简介

要点提示：排序算法是研究算法设计和分析的好例子。

2007 年，当总统候选人巴拉克·奥巴马访问谷歌时，谷歌首席执行官埃里克·施密特问奥巴马对一百万个 32 位整数进行排序的最高效方法。奥巴马回答说，冒泡排序不是正确的方法。他说得对吗？我们将在本章中研究不同的排序算法，看看他的回答是否正确。

排序是计算机科学中的一个经典主题。研究排序算法有三个原因：
- 排序算法阐释了许多创造性的问题解决方法，这些方法可以应用于解决其他问题。
- 排序算法有利于练习使用选择语句、循环、函数和数组的基本编程技术。
- 排序算法是展示算法性能的优秀例子。

要排序的数据可能是整数、双精度数、字符串或其他类型。数据可以按升序或降序进行排序。为了简单起见，本章假设：
1. 要排序的数据是整数，
2. 数据存储在数组中，
3. 数据按升序排序。

现在有很多排序算法。我们已经学习了选择排序。本章介绍插入排序、冒泡排序、归并排序、快速排序、堆排序、桶排序、基数排序和外部排序。

19.2　插入排序

要点提示：插入排序算法通过在已排序的子列表中反复插入一个新元素，直到整个列表排序完毕，来实现对一列值的排序。

该算法可以描述如下：

```
for (int i = 1; i < listSize; i++)
{
  将 list[i] 插入到已排序的子列表 list[0..i-1] 中，从而对
  list[0..i] 进行排序。
}
```

为了将 list[i] 插入 list[0..i-1]，首先将 list[i] 保存到一个临时变量中，比如 currentElement。如果 list[i-1]>currentElement，则将 list[i-1] 移动到 list[i]，如果 list[i-2]>currentElement，则将 list[i-2] 移动到 list[i-1]，以此类推，直到 list[i-k]<=currentElement 或 k>i(我们经过了排序列表的第一个元素)。将 currentElement 赋值给 list[i-k+1]。例如，为了在图 19.1 中的步骤 4 中将 4 插入到 {2, 5, 9} 中，首先将 list[2]（9）移动到 list[3]，因为 9>4，并将 list[1]（5）移动到 list[2]，因为 5>4。最后，将 currentElement（4）移动到 list[1]。

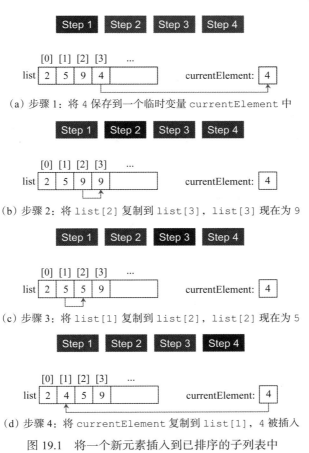

图 19.1 将一个新元素插入到已排序的子列表中

该算法可以如 LiveExample 19.1 这样扩展并实现。

LiveExample 19.1 的互动程序请访问 https://liangcpp.pearsoncmg.com/LiveRunCpp5e/faces/LiveExample.xhtml?header=off&programName=InsertionSort&fileType=.cpp&programHeight=540&resultHeight=160。

LiveExample 19.1　InsertionSort.cpp

Source Code Editor:

```cpp
#include <iostream>
using namespace std;

void insertionSort(int list[], int listSize)
{
  for (int i = 1; i < listSize; i++)
  {
    // Insert list[i] into a sorted sublist list[0..i-1] so that
    //   list[0..i] is sorted.
    int currentElement = list[i];
    int k;
    for (k = i - 1; k >= 0 && list[k] > currentElement; k--)
    {
      list[k + 1] = list[k];
    }

    // Insert the current element into list[k+1]
    list[k + 1] = currentElement;
  }
}

int main()
{
  const int SIZE = 9;
  int list[] = {1, 7, 3, 4, 9, 3, 3, 1, 2};
  insertionSort(list, SIZE);
  for (int i = 0; i < SIZE; i++)
    cout << list[i] << " ";

  return 0;
}
```

[Automatic Check] [Compile/Run] [Reset] [Answer]　　　　Choose a Compiler: VC++

Execution Result:

```
command>cl InsertionSort.cpp
Microsoft C++ Compiler 2019
Compiled successful (cl is the VC++ compile/link command)

command>InsertionSort
1 1 2 3 3 3 4 7 9

command>
```

　　函数 insertionSort 对 int 型元素的数组进行排序。该函数用嵌套的 for 循环实现。外层循环（带有循环控制变量 i）(第 6 行) 从 list[0] 到 list[i] 迭代，以获得排序的子列表。内层循环（带有循环控制变量 k）将 list[i] 插入到从 list[0] 到 list[i-1] 的子列表中。

　　要更好地理解此函数，用以下语句对其进行跟踪：

```cpp
int list[] = {1, 9, 4.5, 6.6, 5.7, -4.5};
insertionSort(list, 6);
```

　　这里给出的插入排序算法，通过不断将新元素插入到已排序的部分数组中，直到整个数组被排序完毕，来实现对元素列表的排序。在第 k 次迭代中，要将元素插入到大小为 k 的数组中，可能需要 k 次比较才能找到插入位置并且 k 次移动才能插入元素。设 $T(n)$ 表示插入

排序的复杂度，c 表示其他操作的总数，例如每次迭代中的赋值和额外比较。因此

$$\begin{aligned}T(n) &= (2+c) + (2\times 2 + c) + \cdots + (2\times(n-1) + c) \\ &= 2(1 + 2 + \cdots + n - 1) + c(n-1) \\ &= 2\frac{(n-1)n}{2} + cn - c = n^2 - n + cn - c \\ &= O(n^2)\end{aligned}$$

插入排序算法的复杂度为 $O(n^2)$。因此，选择排序和插入排序时间复杂度相同。

19.3 冒泡排序

要点提示：冒泡排序通过多次遍历数组来排序。每次遍历如果相邻元素不按顺序排列，则依次交换相邻元素。

冒泡排序算法会对数组进行多次遍历。每次遍历时连续地对相邻的两个数据进行比较。如果某对数据是降序排列的，则交换其值；否则，保持不变。这种技术称为**冒泡排序**或下沉排序，因为较小的值会逐渐"冒泡"到顶部，而较大的值则沉入底部。在第一次遍历之后，最后一个元素成为数组中最大的元素。在第二次遍历之后，倒数第二个元素成为数组中的第二大元素。此过程持续进行到所有元素排序完毕。

冒泡排序算法在 Listing 19.1 中给出。

Listing 19.1　冒泡排序算法

```
1  for (int k = 1; k < arraySize; k++)
2  {
3    // Perform the kth pass
4    for (int i = 0; i < arraySize - k; i++)
5    {
6      if (list[i] > list[i + 1])
7        swap list[i] with list[i + 1];
8    }
9  }
```

注意，如果在某次遍历中没有发生交换，则无须执行下次遍历，因为所有元素都已排序完毕。可以用这个属性改进 Listing 19.1 中的算法，如 Listing 19.2 所示。

Listing 19.2　改进的冒泡排序算法

```
1  bool needNextPass = true;
2  for (int k = 1; k < arraySize && needNextPass; k++)
3  {
4    // Array may be sorted and next pass not needed
5    needNextPass = false;
6    // Perform the kth pass
7    for (int i = 0; i < arraySize - k; i++)
8    {
9      if (list[i] > list[i + 1])
10     {
11       swap list[i] with list[i + 1];
12       needNextPass = true; // Next pass still needed
13     }
14   }
15 }
```

算法在 LiveExample 19.2 中实现。

LiveExample 19.2 的互动程序请访问 https://liangcpp.pearsoncmg.com/LiveRunCpp5e/faces/LiveExample.xhtml?header=off&programName=BubbleSort&fileType=.cpp&programHeight=660&resultHeight=160。

LiveExample 19.2　BubbleSort.cpp

```
#include <iostream>
using namespace std;

// The function for sorting the numbers
void bubbleSort(int list[], int arraySize)
{
  bool needNextPass = true;

  for (int k = 1; k < arraySize && needNextPass; k++)
  {
    // Array may be sorted and next pass not needed
    needNextPass = false;
    for (int i = 0; i < arraySize - k; i++)
    {
      if (list[i] > list[i + 1])
      {
        // Swap list[i] with list[i + 1]
        int temp = list[i];
        list[i] = list[i + 1];
        list[i + 1] = temp;

        needNextPass = true; // Next pass still needed
      }
    }
  }
}

int main()
{
  const int SIZE = 9;
  int list[] = {1, 7, 3, 4, 9, 3, 3, 1, 2};
  bubbleSort(list, SIZE);
  for (int i = 0; i < SIZE; i++)
    cout << list[i] << " ";

  return 0;
}
```

```
command>cl BubbleSort.cpp
Microsoft C++ Compiler 2019
Compiled successful (cl is the VC++ compile/link command)

command>BubbleSort
1 1 2 3 3 3 4 7 9

command>
```

在最佳情况下，冒泡排序算法只需要一次遍历就发现数组已经排序完毕——不需要下一次遍历。由于在第一次遍历中比较次数是 $n-1$，因此冒泡排序的最佳情况时间是 $O(n)$。

在最坏情况下，冒泡排序算法需要 $n-1$ 次遍历。第一次遍历进行 $n-1$ 次比较；第二次遍历进行 $n-2$ 次比较，以此类推，最后一次遍历进行 1 次比较。因此，比较总数为

$$(n-1)+(n-2)+\cdots+2+1$$
$$=\frac{(n-1)n}{2}=\frac{n^2}{2}-\frac{n}{2}=O(n^2)$$

所以，冒泡排序的最坏情况时间是 $O(n^2)$。

19.4 归并排序

要点提示：归并排序算法可以递归地描述如下：该算法将数组分为两半，在每一半上递归地应用归并排序。对两半进行排序后，将它们合并。

Listing 19.3 给出了**归并排序**算法。

Listing 19.3 归并排序算法

```
1    void mergeSort(int list[], int arraySize)
2    {
3      if (arraySize > 1)
4      {
5        在 list[0 ... arraySize / 2] 上归并排序；
6        在 list[arraySize /2 + 1 ... arraySize]) 上归并排序；
7        归并 list[0 ... arraySize / 2] 和
8          list[arraySize / 2 + 1 ... arraySize];
9      }
10   }
```

图 19.2 显示了 8 个元素（2 9 5 4 8 1 6 7）的数组的归并排序。原始数组分为（2 9 5 4）和（8 1 6 7）。在这两个子数组上递归地应用归并排序，将（2 9 5 4）拆分为（2 9）和（5 4），将（8 1 6 7）拆分为（8 1）和（6 7）。这个过程一直持续到子数组只包含一个元素。例如，数组（2 9）被拆分为子数组（2）和（9）。由于数组（2）包含单个元素，因此无法进一步拆分。现在将（2）与（9）归并为一个新的有序数组（2 9）；将（5）与（4）归并为新的有序数组（4 5）。将（2 9）与（4 5）归并为一个新的有序数组（2 4 5 9），最后将（2 4 5 9）与（1 6 7 8）归并为新的有序数组（1 2 4 5 6 7 8 9）。

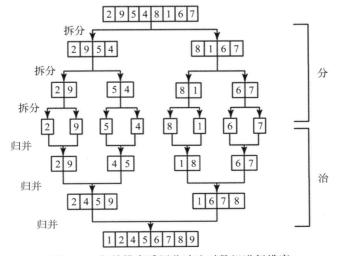

图 19.2 归并排序采用分治法对数组进行排序

递归调用不断将数组划分为子数组，直到每个子数组只包含一个元素。然后，该算法将这些小的子数组归并为较大的有序子数组，直到得到一个有序数组。

归并排序算法在 LiveExample 19.3 中实现。

LiveExample 19.3 的互动程序请访问 https://liangcpp.pearsoncmg.com/LiveRunCpp5e/faces/LiveExample.xhtml?header=off&programName=MergeSort&fileType=.cpp&programHeight=1310&resultHeight=160。

LiveExample 19.3　MergeSort.cpp

Source Code Editor:

```cpp
#include <iostream>
using namespace std;

// Function prototype
void arraycopy(int source[], int sourceStartIndex,
  int target[], int targetStartIndex, int length);

void merge(int list1[], int list1Size,
  int list2[], int list2Size, int temp[]);

// The function for sorting the numbers
void mergeSort(int list[], int arraySize)
{
  if (arraySize > 1)
  {
    // Merge sort the first half
    int* firstHalf = new int[arraySize / 2];
    arraycopy(list, 0, firstHalf, 0, arraySize / 2);
    mergeSort(firstHalf, arraySize / 2);

    // Merge sort the second half
    int secondHalfLength = arraySize - arraySize / 2;
    int* secondHalf = new int[secondHalfLength];
    arraycopy(list, arraySize / 2, secondHalf, 0, secondHalfLength);
    mergeSort(secondHalf, secondHalfLength);

    // Merge firstHalf with secondHalf
    merge(firstHalf, arraySize / 2, secondHalf, secondHalfLength,
      list);

    delete [] firstHalf;
    delete [] secondHalf;
  }
}

void merge(int list1[], int list1Size,
  int list2[], int list2Size, int temp[])
{
  int current1 = 0; // Current index in list1
  int current2 = 0; // Current index in list2
  int current3 = 0; // Current index in temp

  while (current1 < list1Size && current2 < list2Size)
  {
    if (list1[current1] < list2[current2])
      temp[current3++] = list1[current1++];
    else
      temp[current3++] = list2[current2++];
  }

  while (current1 < list1Size)
```

```cpp
52        temp[current3++] = list1[current1++];
53
54      while (current2 < list2Size)
55        temp[current3++] = list2[current2++];
56    }
57
58    void arraycopy(int source[], int sourceStartIndex,
59      int target[], int targetStartIndex, int length)
60    {
61      for (int i = 0; i < length; i++)
62      {
63        target[i + targetStartIndex] = source[i + sourceStartIndex];
64      }
65    }
66
67    int main()
68    {
69      const int SIZE = 9;
70      int list[] = {1, 7, 3, 4, 9, 3, 3, 1, 2};
71      mergeSort(list, SIZE);
72      for (int i = 0; i < SIZE; i++)
73        cout << list[i] << " ";
74
75      return 0;
76    }
```

Execution Result:

```
command>cl MergeSort.cpp
Microsoft C++ Compiler 2019
Compiled successful (cl is the VC++ compile/link command)

command>MergeSort
1 1 2 3 3 3 4 7 9

command>
```

函数mergeSort（第12～34行）创建一个新的数组firstHalf，它是list前半部分的副本（第18行）。该算法在firstHalf（第19行）上递归调用mergeSort。firstHalf的长度是arraySize/2，secondHalf的长度为arraySize-arraySize/2。创建新数组secondHalf是为了放置原始数组list的后半部分。该算法在secondHalf上递归调用mergeSort（第25行）。对firstHalf和secondHalf进行排序后，将它们归并到list中（第28～29行）。所以，数组list现在是有序的。

函数merge（第36～56行）归并两个已排序的数组。这个函数将list1和list2数组归并到一个新数组temp中。因此，temp的大小应该是list1Size+list2Size。current1和current2指向list1和list2中要考虑的当前元素（第39～41行）。该函数反复比较list1和list2的当前元素，并将较小的元素移动到temp中。如果较小的元素在list1中，current1将增加1（第46行）。如果较小的一个在list2中，则current2增加1（第48行）。最后，其中某一个列表的所有元素都被移动到temp中。如果list1中仍有未移动的元素，则将它们复制到temp中（第51～52行）。如果list2中仍有未移动的元素，将它们复制到temp中（第54～55行）。

函数mergeSort在拆分过程中创建两个临时数组（第17行和第23行），将数组的前半部分和后半部分复制到临时数组中(第 18 行和第 24 行),对临时数组进行排序(第 19 行和第 25 行),然后将其归并到原始数组中(第 28～29 行),如图 19.3a 所示。可以重写代码递归地对数组的前半部分和后半部分进行排序,而无须创建新的临时数组,然后将这两个数组归并为一个临时数组,并将其内容复制到原始数组中,如图 19.3b 所示。这留在编程练习 19.10 中完成。

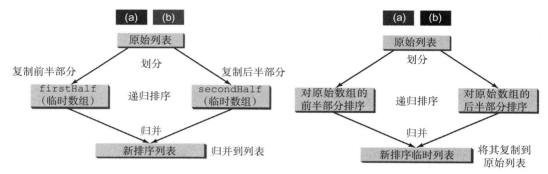

图 19.3 为支持归并排序创建临时数组

设 $T(n)$ 表示用归并排序对 n 个元素的数组进行排序所需的时间。在不失一般性的情况下,假设 n 是 2 的幂。归并排序算法将数组复制到两个子数组中,使用相同的算法递归排序,然后归并子数组。将 list 复制到 firstHalf 和 secondHalf 的时间为 n,归并两个已排序的子数组的时间为 $n-1$。所以

$$T(n) = T\left(\frac{n}{2}\right) + T\left(\frac{n}{2}\right) + n + n - 1$$

第一个 $T\left(\frac{n}{2}\right)$ 是排序前半部分数组的时间,第二个 $T\left(\frac{n}{2}\right)$ 是排序后半部分的时间。可以看出

$$T(n) = 2T\left(\frac{n}{2}\right) + 2n - 1 = O(n \log n)$$

归并排序的复杂度是 $O(n \log n)$。该算法优于选择排序、插入排序和冒泡排序。

19.5 快速排序

要点提示:快速排序的工作原理如下:算法在数组中选择一个称为轴心(pivot)的元素。它将数组分为两部分,使得第一部分中的所有元素都小于或等于轴心,而第二部分中的所有元素都大于轴心。然后将快速排序算法递归应用于第一部分和第二部分。

快速排序算法由 C.A.R.Hoare 于 1962 年开发,如 Listing19.4 所示。

Listing19.4 快速排序算法

```
1    void quickSort(int list[], int arraySize)
2    {
3        if (arraySize > 1)
4        {
5            选择一个轴心 (pivot);
6            将 list 拆分成 list1 和 list2,使得
```

```
 7         list1 中的所有元素 <= pivot,
 8         list2 中的所有元素 > pivot;
 9      在 list1 上快速排序;
10      在 list2 上快速排序;
11    }
12  }
```

每次划分将轴心放置在正确的位置。如下图所示，它将列表分为两个子列表。

```
   list   ──划分──▶  list1 pivot list2
```

轴心的选择会影响算法的性能。理想情况下，算法应该选择能将两部分平均划分的轴心。简单起见，假设选择数组中的第一个元素作为轴心。（编程练习 19.4 给出了选择轴心的替代策略。）

图 19.4 演示了如何使用快速排序对数组（5 2 9 3 8 4 0 1 6 7）进行排序。选择第一个元素 5 作为轴心。数组被分为两部分，如图 19.4a 所示。阴影显示的轴心被放置在数组中的正确位置。对两个部分数组（4 2 1 3 0）和（8 9 6 7），依次应用快速排序。轴心 4 将（4 2 1 3 0）只划分成一个部分数组（0 2 1 3），如图 19.4b 所示。对（0 2 1 3）应用快速排序。轴心 0 将其也只划分为一个部分数组（2 1 3），如图 19.4c 所示。对（2 1 3）应用快速排序。轴心 2 将其划分为（1）和（3），如图 19.4d 所示。对（1）应用快速排序。由于数组只包含一个元素，因此不需要进一步的划分。

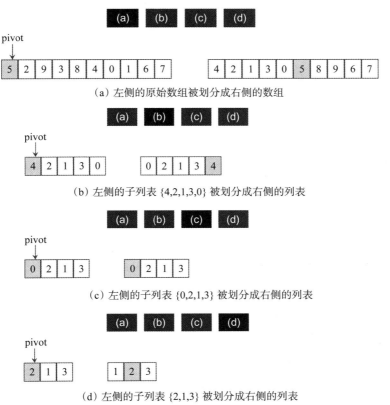

（a）左侧的原始数组被划分成右侧的数组

（b）左侧的子列表 {4,2,1,3,0} 被划分成右侧的列表

（c）左侧的子列表 {0,2,1,3} 被划分成右侧的列表

（d）左侧的子列表 {2,1,3} 被划分成右侧的列表

图 19.4   快速排序算法递归地应用于各个部分数组

快速排序算法在 LiveExample 19.4 中实现。类中有两个重载的 quickSort 函数。第一

个函数（第 5 行）对数组进行排序。第二个是辅助函数（第 6 行），对指定范围的部分数组进行排序。

LiveExample 19.4 的互动程序请访问 https://liangcpp.pearsoncmg.com/LiveRunCpp5e/faces/LiveExample.xhtml?header=off&programName=QuickSort&fileType=.cpp&programHeight=1290&resultHeight=160。

LiveExample 19.4　QuickSort.cpp

```cpp
#include <iostream>
using namespace std;

// Function prototypes
void quickSort(int list[], int arraySize);
void quickSort(int list[], int first, int last);
int partition(int list[], int first, int last);

void quickSort(int list[], int arraySize)
{
  quickSort(list, 0, arraySize - 1);
}

void quickSort(int list[], int first, int last)
{
  if (last > first)
  {
    int pivotIndex = partition(list, first, last);
    quickSort(list, first, pivotIndex - 1);
    quickSort(list, pivotIndex + 1, last);
  }
}

// Partition the array list[first..last]
int partition(int list[], int first, int last)
{
  int pivot = list[first]; // Choose the first element as the pivot
  int low = first + 1; // Index for forward search
  int high = last; // Index for backward search

  while (high > low)
  {
    // Search forward from left
    while (low <= high && list[low] <= pivot)
      low++;

    // Search backward from right
    while (low <= high && list[high] > pivot)
      high--;

    // Swap two elements in the list
    if (high > low)
    {
      int temp = list[high];
      list[high] = list[low];
      list[low] = temp;
    }
  }

  while (high > first && list[high] >= pivot)
    high--;

  // Swap pivot with list[high]
  if (pivot > list[high])
```

```
55      {
56        list[first] = list[high];
57        list[high] = pivot;
58        return high;
59      }
60      else
61      {
62        return first;
63      }
64    }
65
66    int main()
67    {
68      const int SIZE = 9;
69      int list[] = {1, 7, 3, 4, 9, 3, 3, 1, 2};
70      quickSort(list, SIZE);
71      for (int i = 0; i < SIZE; i++)
72        cout << list[i] << " ";
73
74      return 0;
75    }
```

Execution Result:
```
command>cl QuickSort.cpp
Microsoft C++ Compiler 2019
Compiled successful (cl is the VC++ compile/link command)

command>QuickSort
1 1 2 3 3 3 4 7 9

command>
```

函数 partition（第 25～64 行）用轴心对数组 list[first..last] 进行划分。选择部分数组中的第一个元素作为轴心（第 27 行）。开始 low 指向部分数组的第二个元素（第 28 行），high 指向部分数组的最后一个元素（第 29 行）。

函数从左向前搜索数组中大于轴心的第一个元素（第 34～35 行），然后从右向回搜索数组中小于或等于轴心的第一个元素（第 38～39 行）。交换这两个元素（第 42～47 行）。重复相同的搜索和交换操作，直到在 while 循环中搜索完所有元素（第 31～48 行）。

如果轴心已经移动，则函数返回将部分数组分为两部分的轴心的新索引（第 58 行）。否则，返回轴心的原始索引（第 62 行）。

要划分一个由 n 个元素组成的数组，最坏情况下需要进行 n 次比较和 n 次移动。因此，划分所需的时间是 $O(n)$。

最坏情况下，轴心每次都会将数组划分为一个子数组，而另一个子数组为空。大的子数组的大小比之前划分的子数组只小 1。该算法需要 $(n-1)+(n-2)+\cdots+2+1 = O(n^2)$ 时间。

最佳情况下，轴心每次都会将数组划分为大小大致相同的两部分。设 $T(n)$ 表示用快速排序对 n 个元素的数组进行排序所需的时间。因此，

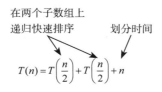

$$T(n) = T\left(\frac{n}{2}\right) + T\left(\frac{n}{2}\right) + n$$

与归并排序相似，$T(n) = O(n\log n)$。

平均而言，轴心不会每次将数组划分为两个相同大小的部分或一个空部分。从统计数据来看，这两个部分的大小非常接近。因此，平均时间为 $O(n\log n)$。精确的平均情况分析超出了本书的范围，因此不做介绍。

归并排序和快速排序都采用了分治法。对于归并排序，大部分工作是归并两个子列表，这发生在子列表排序之后。对于快速排序，大部分工作是将列表划分为两个子列表，这发生在子列表排序之前。最坏情况下，归并排序比快速排序效率高，但在平均情况下，两者的效率相仿。归并排序需要一个临时数组对两个子数组进行排序。快速排序不需要额外的数组空间。因此，快速排序比归并排序节省空间。

19.6 堆排序

要点提示：堆排序使用二元堆。它首先将所有元素添加到堆中，然后依次删除最大的元素，以获得排序列表。

堆排序使用二元堆，这是一个完整的二叉树。二叉树是一种层次结构。它要么是空的，要么由一个元素（称为根）和两个不同的二叉树（称为左子树和右子树）组成。路径的长度是路径中边的数量。节点的深度是从根到节点的路径长度。

二元堆是具有以下属性的二叉树。
- **形状属性**：它是一个完全二叉树。
- **堆属性**：每个节点都大于或等于其任意子节点。

如果一个二叉树的每一层都是满的，除了最后一层可能不满，且最后一层上的所有叶子都放在最左边，那么它是**完全二叉树**。例如，在图 19.5 中，a 和 b 中的二叉树是完全的，但 c 和 d 中的二叉树不是完全的。此外，a 中的二叉树是堆，但 b 中的二叉树不是堆，因为根（39）小于其右子节点（42）。注意，一个完全二叉树并不是一个满二叉树。在一个满二叉树中，每一层都是满的。

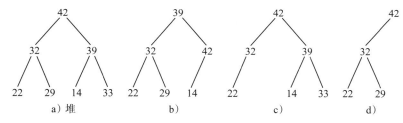

图 19.5　二元堆是一种特殊的完全二叉树

注意：堆在计算机科学中是一个有很多含义的术语。在本章中，堆的意思是二元堆。

注意：完全二叉树并不是满二叉树。满二叉树也被称为完美二叉树（perfect binary tree）。完美二叉树所有内部节点都有两个子节点，且所有叶子都有相同的深度或层级。

教学提示：堆可以高效实现插入键值和删除根。

19.6.1 存储堆

如果事先已知堆的大小，则可将堆存储在向量或数组中。图 19.6a 中的堆可以用图 19.6b 中的数组存储。根位于位置 0，它的两个子节点位于位置 1 和 2。对于位于位置 i 的节点，其左子节点位于位置 2i+1，其右子节点位于 2i+2，其父节点位于 (i-1)/2。例如，元素 39 的节点位于位置 4，因此其左子节点（元素 14）位于 9（2×4+1），其右子节点（元素 33）位于 10（2×4+2），其父节点（元素 42）位于 1（(4-1)/2）。

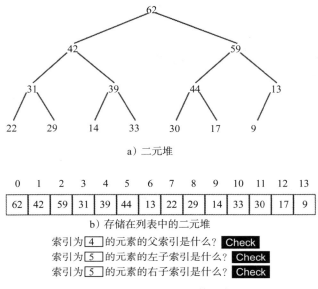

a) 二元堆

0	1	2	3	4	5	6	7	8	9	10	11	12	13
62	42	59	31	39	44	13	22	29	14	33	30	17	9

b) 存储在列表中的二元堆

索引为 4 的元素的父索引是什么？ Check
索引为 5 的元素的左子索引是什么？ Check
索引为 5 的元素的右子索引是什么？ Check

图 19.6　二元堆可以用数组实现

19.6.2　添加新节点

向堆中添加新节点，要先将其添加到堆的末尾，然后按如下方式重建树：

```
设最后一个节点为当前节点；
while (当前节点大于其父节点)
{
    将当前节点与其父节点交换；
    现在，当前节点向上一层；
}
```

假设堆最初是空的。按顺序添加数字 3、5、1、19、11 和 22 之后，堆如图 19.7 所示。

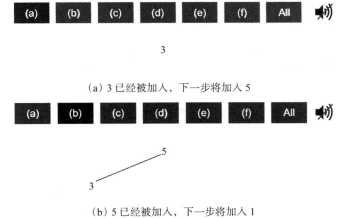

（a）3 已经被加入，下一步将加入 5

（b）5 已经被加入，下一步将加入 1

图 19.7　元素 3、5、1、19、11 和 22 被插入到堆中

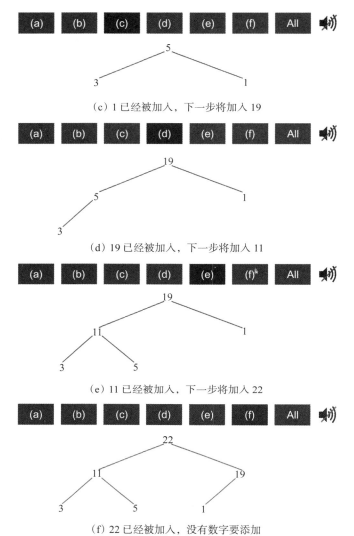

(c) 1 已经被加入，下一步将加入 19

(d) 19 已经被加入，下一步将加入 11

(e) 11 已经被加入，下一步将加入 22

(f) 22 已经被加入，没有数字要添加

图 19.7 元素 3、5、1、19、11 和 22 被插入到堆中（续）

现在要在堆中添加 88。将新节点 88 放置在树的末端，如图 19.8a 所示。将 88 与 19 交换，如图 19.8b 所示。用 22 交换 88，如图 19.8c 所示。

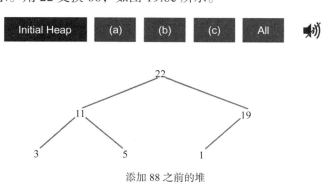

添加 88 之前的堆

图 19.8 添加新节点后重新生成堆

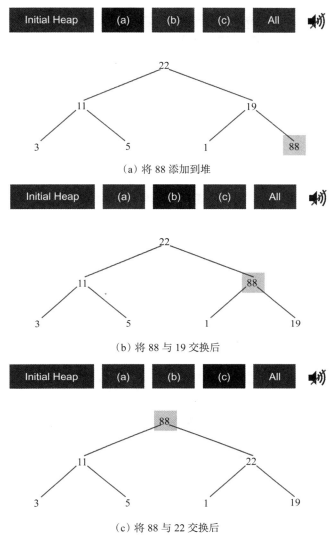

图 19.8 添加新节点后重新生成堆（续）

19.6.3 删除根

我们常常需要删除最大元素，它是堆的根。删除根之后，必须重建树以维护堆属性。用于重建树的算法可以描述如下：

```
移动最后一个节点替换根节点；
令根是当前节点；
while（当前节点有子节点，且当前节点小于其一个子节点）
{
    将当前节点与其较大的子节点交换；
    现在，当前节点向下一层；
}
```

图 19.9 显示了从图 19.6a 中删除根 62 后重建堆的过程。将最后一个节点 9 移动到根，如图 19.9a 所示。将 9 与 59 交换，如图 19.9b 所示；将 9 与 44 交换，如图 19.9c 所示；将 9

与 30 交换，如图 19.9d 所示。

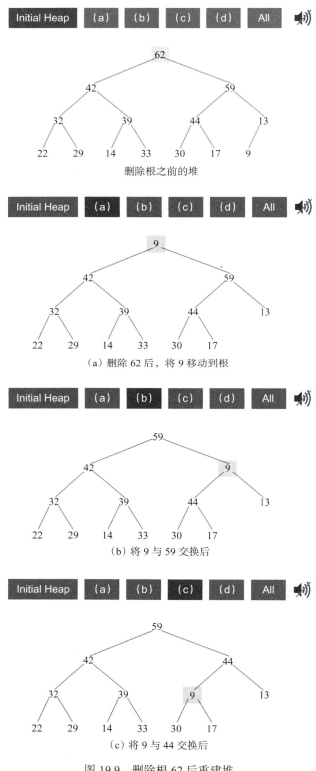

图 19.9 删除根 62 后重建堆

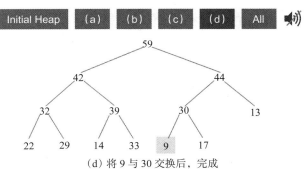

(d) 将 9 与 30 交换后，完成

图 19.9　删除根 62 后重建堆（续）

图 19.10 显示了从图 19.9d 中删除根 59 后重建堆的过程。将最后一个节点 17 移动到根，如图 19.10a 所示。将 17 与 44 交换，如图 19.10b 所示，然后将 17 与 30 交换，如图 19.10c 所示。

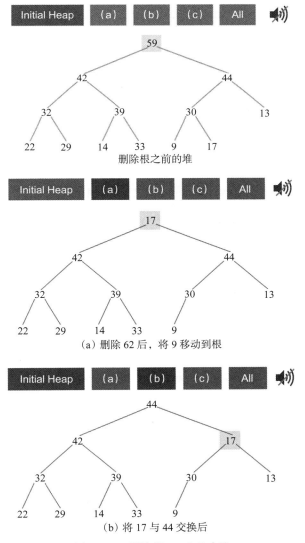

图 19.10　删除根 59 后重建堆

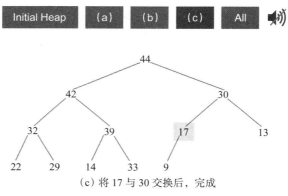

（c）将 17 与 30 交换后，完成

图 19.10　删除根 59 后重建堆（续）

19.6.4 Heap 类

现在我们已经准备好设计和实现 Heap 类了。类图如图 19.11 所示。它的实现在 Live-Example19.5 中给出。

图 19.11　Heap 类提供了操纵堆的操作

LiveExample 19.5 的互动程序请访问 https://liangcpp.pearsoncmg.com/LiveRunCpp5e/faces/LiveExample.xhtml?header=off&programName=Heap&fileType=.h&programHeight=1910&resultVisible=false。

LiveExample 19.5　Heap.h

Source Code Editor:

```cpp
#ifndef HEAP_H
#define HEAP_H
#include <vector>
#include <stdexcept>
using namespace std;

template<typename T>
class Heap
{
public:
    Heap();
    Heap(const T elements[], int arraySize);
    void add(const T& element);
    T remove();
    int getSize() const;

private:
    vector<T> v;
};
```

```cpp
20
21   template<typename T>
22   Heap<T>::Heap()
23   {
24   }
25
26   template<typename T>
27   Heap<T>::Heap(const T elements[], int arraySize)
28   {
29     for (int i = 0; i < arraySize; i++)
30     {
31       add(elements[i]);
32     }
33   }
34
35   // Insert element into the heap and maintain the heap property
36   template<typename T>
37   void Heap<T>::add(const T& element)
38   {
39     v.push_back(element); // Append element to the heap
40     int currentIndex = v.size() - 1; // The index of the last node
41
42     // Maintain the heap property
43     while (currentIndex > 0)
44     {
45       int parentIndex = (currentIndex - 1) / 2;
46       // Swap if the current element is greater than its parent
47       if (v[currentIndex] > v[parentIndex])
48       {
49         T temp = v[currentIndex];
50         v[currentIndex] = v[parentIndex];
51         v[parentIndex] = temp;
52       }
53       else
54         break; // the tree is a heap now
55
56       currentIndex = parentIndex;
57     }
58   }
59
60   // Remove the root from the heap
61   template<typename T>
62   T Heap<T>::remove()
63   {
64     if (v.size() == 0)
65       throw runtime_error("Heap is empty");
66
67     T removedElement = v[0];
68     v[0] = v[v.size() - 1]; // Copy the last to root
69     v.pop_back(); // Remove the last element
70
71     // Maintain the heap property
72     int currentIndex = 0;
73     while (currentIndex < v.size())
74     {
75       int leftChildIndex = 2 * currentIndex + 1;
76       int rightChildIndex = 2 * currentIndex + 2;
77
78       // Find the maximum between two children
79       if (leftChildIndex >= v.size()) break; // The tree is a heap
80       int maxIndex = leftChildIndex;
81       if (rightChildIndex < v.size())
82       {
83         if (v[maxIndex] < v[rightChildIndex])
84         {
```

```
 85            maxIndex = rightChildIndex;
 86        }
 87    }
 88
 89    // Swap if the current node is less than the maximum
 90    if (v[currentIndex] < v[maxIndex])
 91    {
 92       T temp = v[maxIndex];
 93       v[maxIndex] = v[currentIndex];
 94       v[currentIndex] = temp;
 95       currentIndex = maxIndex;
 96    }
 97    else
 98       break; // The tree is a heap
 99  }
100
101  return removedElement;
102 }
103
104 // Get the number of element in the heap
105 template<typename T>
106 int Heap<T>::getSize() const
107 {
108    return v.size();
109 }
110
111 #endif
```

堆在内部用向量表示（第 18 行）。可以将其更改为其他数据结构，但 Heap 类契约保持不变。

函数 add(T element)（第 36～58 行）将元素附加到树中，如果元素大于其父元素，则将其与其父元素交换。这个过程一直持续到新元素成为根元素或不大于其父元素。

函数 remove()（第 61～102 行）删除并返回根。为了维护堆属性，函数将最后一个元素移动到根位置，如果它小于较大的子元素，则将其与其较大的子节点交换。这个过程一直持续到最后一个元素变成叶子或者不小于它的子元素。

19.6.5 使用 Heap 类进行排序

要用堆对数组进行排序，首先用 Heap 类创建一个对象，用 add 函数将所有元素添加到堆中，然后用 remove 函数从堆中删除所有元素。元素按降序删除。LiveExample 19.6 给出了一个用堆对数组进行排序的程序。

LiveExample 19.6 的互动程序请访问 https://liangcpp.pearsoncmg.com/LiveRunCpp5e/faces/LiveExample.xhtml?header=off&programName=HeapSort&fileType=.cpp&programHeight=470&resultHeight=160。

LiveExample 19.6　HeapSort.cpp

Source Code Editor:

```
1  #include <iostream>
2  #include "Heap.h"
3  using namespace std;
4
```

```cpp
5  template <typename T>
6  void heapSort(T list[], int arraySize)
7  {
8    Heap<T> heap;
9  
10   for (int i = 0; i < arraySize; i++)
11     heap.add(list[i]);
12  
13   for (int i = 0; i < arraySize; i++)
14     list[arraySize - i - 1] = heap.remove();
15  }
16  
17  int main()
18  {
19    const int SIZE = 9;
20    int list[] = {1, 7, 3, 4, 9, 3, 3, 1, 2};
21    heapSort(list, SIZE);
22    for (int i = 0; i < SIZE; i++)
23      cout << list[i] << " ";
24  
25    return 0;
26  }
```

```
Execution Result:
command>cl HeapSort.cpp
Microsoft C++ Compiler 2019
Compiled successful (cl is the VC++ compile/link command)

command>HeapSort
1 1 2 3 3 3 4 7 9

command>
```

注意，堆中最大的元素首先被删除。因此，从堆中删除的元素以相反的顺序放置在数组中（第 13 ～ 14 行）。

19.6.6 堆排序的时间复杂度

让我们把注意力转向分析堆排序的时间复杂度。让 h 表示 n 个元素的**堆的高度**。非空树的高度是从根节点到其最远叶子的路径长度。包含单个节点的树的高度为 0。按照惯例，空树的高度是 -1。因为堆是一个完全二叉树，所以第一层有 1（2^0）个节点，第二层有 2（2^1）个节点。第 k 层具有 2^{k-1} 个节点，第 h 层有 2^{h-1} 个节点并且最后 $h+1$ 层有至少 1 个且至多 2^h 个节点。因此

$$1+2+\cdots+2^{h-1} < n \leqslant 1+2\cdots+2^{h-1}+2^h$$

即，

$$2^h - 1 < n \leqslant 2^{h+1} - 1$$
$$2^h < n+1 \leqslant 2^{h+1}$$
$$h < \log(n+1) \leqslant h+1$$

因此，$h < \log(n+1)$，且 $h \geqslant \log(n+1) - 1$，所以 $\log(n+1) - 1 \leqslant h < \log(n+1)$。因此，堆的高度是 $O(\log n)$。更确切地说，对于非空树，可以证明 $h=[\log n]$。

由于 add 函数跟踪从叶到根的路径，因此向堆中添加新元素最多需要 h 个步骤。因此，对于 n 个元素的数组，构建初始堆的总时间是 $O(n\log n)$。由于 remove 函数跟踪从根到叶的路径，因此从堆中删除根之后，重建堆最多需要 h 个步骤。由于 remove 函数被调用了 n 次，因此从堆中生成排序数组的总时间为 $O(n\log n)$。

归并排序和堆排序都需要 $O(n\log n)$ 时间。归并排序需要一个临时数组来归并两个子数组；堆排序不需要额外的数组空间。因此，堆排序比归并排序节省空间。

19.7 桶排序和基数排序

要点提示：桶排序和基数排序对整数排序是高效的。

到目前为止讨论的所有排序算法都是适用于任何键类型（例如，整数、字符串和任何可比较对象）的一般排序算法。这些算法通过比较键值对元素进行排序。一般排序算法的下限是 $O(n\log n)$，因此没有任何基于比较的排序算法的时间复杂度能比 $O(n\log n)$ 更低。但是，如果键值是小整数，则可以使用桶排序，而不必比较键值。

桶排序算法的工作原理如下。假设键值在 0 到 t 的范围内。我们需要 t+1 个标记为 $0,1,\cdots,$ t 的桶。如果一个元素的键值是 i，则该元素被放入桶 i 中。每个桶保存相同键值的元素。

可以用向量实现桶。显然，对列表进行排序需要 $O(n+t)$ 时间并使用 $O(n+t)$ 空间，其中 n 是列表大小。

注意，如果 t 太大，则不希望使用桶排序，而是使用基数排序。基数排序基于桶排序，但基数排序只使用十个桶。

值得注意的是，桶排序是稳定的，这意味着如果原始列表中的两个元素具有相同的键值，则它们在排序列表中的顺序不会改变。也就是说，如果元素 e_1 和元素 e_2 有相同的键值，并且在原始列表中 e_1 位于 e_2 前面，则在排序列表中 e_1 仍然位于 e_2 前面。

假设键值是正整数。基数排序的思想是根据键值的基数位置将其划分为多个子组。对基数位置上的键值从最低有效位置开始重复应用桶排序。

考虑对以下键值元素进行排序：

```
331, 454, 230, 34, 343, 45, 59, 453, 345, 231, 9
```

在最后一个基数位置应用桶排序，如下所示，元素被放入桶中：

从桶中取出元素后，元素按以下顺序排列：

```
230, 331, 231, 343, 453, 454, 34, 45, 345, 59, 9
```

在倒数第二个基数位置应用桶排序，如下所示，元素被放入桶中：

9			230 331 231 34	343 45 345	453 454 59				
bucket[0]	bucket[1]	bucket[2]	bucket[3]	bucket[4]	bucket[5]	bucket[6]	bucket[7]	bucket[8]	bucket[9]

从桶中取出元素后，元素按以下顺序排列：

```
9, 230, 331, 231, 34, 343, 45, 345, 453, 454, 59
```

（注意，9 是 009。）

在倒数第三个基数位置应用桶排序，元素被如下所示放入桶中：

9 34 45 59		230 231	331 343 345	453 454					
bucket[0]	bucket[1]	bucket[2]	bucket[3]	bucket[4]	bucket[5]	bucket[6]	bucket[7]	bucket[8]	bucket[9]

从桶中取出元素后，元素按以下顺序排列：

```
9, 34, 45, 59, 230, 231, 331, 343, 345, 453, 454
```

现在元素是有序的了。

通常，基数排序需要 $O(dn)$ 时间对具有整数键值的 n 个元素进行排序，其中 d 是所有键值中基数位置的最大数。

19.8 外部排序

要点提示：可以使用外部排序对大量数据进行排序。

前面讨论的所有排序算法都假设所有要排序的数据在内存（如数组）中一次可用。要对存储在外部文件中的数据进行排序，必须先将数据放入内存，然后在内部进行排序。

但是，如果文件太大，则无法同时将文件中的所有数据都放到内存中。本节讨论如何对大型外部文件中的数据进行排序。这被称为**外部排序**。

简单起见，假设 200 万个 int 值存储在一个名为 largedata.dat 的二进制文件中。此文件是用 LiveExample19.7 中的程序创建的。

LiveExample 19.7 的互动程序请访问 https://liangcpp.pearsoncmg.com/LiveRunCpp5e/faces/LiveExample.xhtml?header=off&programName=CreateLargeFile&fileType=.cpp&programHeight=590&resultHeight=150。

LiveExample 19.7   CreateLargeFile.cpp

```
5
6   int main()
7   {
8       fstream output;
9       output.open("largedata.dat", ios::out | ios::binary);
10
11      for (int i = 0; i < 2000000; i++)
12      {
13          int value = rand();
14          output.write(reinterpret_cast<char *>(&value), sizeof(value));
15      }
16
17      output.close();
18      cout << "File created" << endl;
19
20      fstream input;
21      input.open("largedata.dat", ios::in | ios::binary);
22      int value;
23
24      cout << "The first 10 numbers in the file are " << endl;
25      for (int i = 0; i < 10; i++)
26      {
27          input.read(reinterpret_cast<char *>(& value), sizeof(value));
28          cout << value << " ";
29      }
30
31      input.close();
32
33      return 0;
34  }
```

Compile/Run Reset   Choose a Compiler: VC++

Execution Result:

```
command>cl CreateLargeFile.cpp
Microsoft C++ Compiler 2019
Compiled successful (cl is the VC++ compile/link command)

command>CreateLargeFile
569193 131317 608695 776266 767910 624915 458599 5010 ... (omitted)

command>
```

使用归并排序的变体，可以分两个阶段对此文件进行排序：

第一阶段：重复地将数据从文件中取到数组中，使用内部排序算法对数组进行排序，并将数据从数组输出到临时文件中。该过程如图 19.12 所示。理想情况下，我们希望创建一个大数组，但其最大规模取决于操作系统为 JVM 分配的内存量。假设最大数组规模为 100000 个 int 值。在临时文件中，对每 100000 个 int 值进行排序。它们被表示为 S_1, S_2, ⋯, S_k，其中最后一个分段 S_k，可能包含的键值少于 100000 个。

第二阶段：将一对已排序的段（例如，S_1 和 S_2，S_3 和 S_4，等等）归并成一个较大的已排序段，并将新段保存到一个新的临时文件中。继续相同的过程，直到只产生一个排序的段。图 19.13 显示了如何归并八个分段。

注意：不用必须归并两个连续的段。例如，可以在第一个归并步骤中归并 S_1 和 S_5，S_2 和 S_6，S_3 和 S_7，S_4 和 S_8。这一观察有助于提高第二阶段的效率。

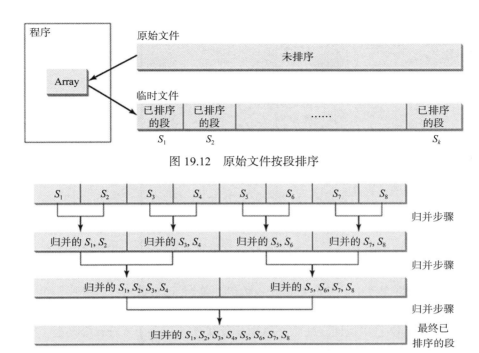

图 19.12 原始文件按段排序

图 19.13 迭代归并已排序的段

19.8.1 实现第一阶段

Listing19.5 给出了一个函数，它从文件中读取所有数据段，对数据段进行排序，并将排序后的数据段存储到一个新文件中。函数返回段的数量。

Listing 19.5　创建初始已排序段

```
1   // Sort original file into sorted segments
2   int initializeSegments(int segmentSize, string sourceFile, string f1)
3   {
4     int* list = new int[segmentSize];
5
6     fstream input;
7     input.open(sourceFile.c_str(), ios::in | ios::binary);
8     fstream output;
9     output.open(f1.c_str(), ios::out | ios::binary);
10
11    int numberOfSegments = 0;
12    while (!input.eof())
13    {
14      int i = 0;
15      for ( ; !input.eof() && i < segmentSize; i++)
16      {
17        input.read(reinterpret_cast<char*>
18          (&list[i]), sizeof(list[i]));
19      }
20
21      if (input.eof()) i--;
22      if (i <= 0)
23        break;
24      else
25        numberOfSegments++;
26
```

```
27          // Sort an array list[0..i-1]
28          quickSort(list, i);
29
30          // Write the array to f1.dat
31          for (int j = 0; j < i; j++)
32          {
33            output.write(reinterpret_cast<char*>
34              (&list[j]), sizeof(list[j]));
35          }
36        }
37
38        input.close();
39        output.close();
40        delete [] list;
41
42        return numberOfSegments;
43      }
```

该函数在第 4 行声明了一个指定段大小的数组，在第 7 行声明了原始文件的数据输入流，在第 9 行声明了临时文件的数据输出流。

第 15 ～ 19 行将文件中的一段数据读入数组。第 28 行对数组进行排序。第 31 ～ 35 行将数组中的数据写入临时文件。

段的数量在第 42 行中返回。注意，每个分段都有 segmentSize 个元素，但最后一个分段除外，它可能具有的元素数较少。

19.8.2 实现第二阶段

在每个归并步骤中，两个已排序的段被归并以形成一个新的段。新段的大小增加了一倍。在每个归并步骤之后，段的数量将减少一半。段太大，不能复制到内存中的数组中。为了实现归并步骤，首先将文件 f1.dat 中一半的段复制到临时文件 f2.dat 中。然后将 f1.dat 剩余的第一个段与 f2.dat 的第一个段归并到一个名为 f3.dat 的临时文件中，如图 19.14 所示。

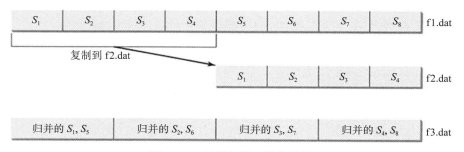

图 19.14　迭代归并已排序的段

注意：f1.dat 可能比 f2.dat 多一个段。如果是这样，在归并后将多出的最后一个段移到 f3.dat 中。

Listing 19.6 给出了将 f1.dat 中前半部分的段复制到 f2.dat 的函数。Listing19.7 给出了成对归并 f1.dat 和 f2.dat 中段的函数。Listing 19.8 给出了归并两个段的函数。

Listing19.6　复制前半部分的段

```
1   // Copy the first half of the number of segments from f1.dat to f2.dat
2   void copyHalfToF2(int numberOfSegments, int segmentSize,
3     fstream& f1, fstream& f2)
```

```
4    {
5      for (int i = 0; i < (numberOfSegments / 2) * segmentSize; i++)
6      {
7        int value;
8        f1.read(reinterpret_cast<char*>(& value), sizeof(value));
9        f2.write(reinterpret_cast<char*>(& value), sizeof(value));
10     }
11   }
```

Listing19.7 归并所有段

```
1    // Merge all segments
2    void mergeSegments(int numberOfSegments, int segmentSize,
3      fstream& f1, fstream& f2, fstream& f3)
4    {
5      for (int i = 0; i < numberOfSegments; i++)
6      {
7        mergeTwoSegments(segmentSize, f1, f2, f3);
8      }
9
10     // f1 may have one extra segment; copy it to f3
11     while (!f1.eof())
12     {
13       int value;
14       f1.read(reinterpret_cast<char*>(&value), sizeof(value));
15       if (f1.eof()) break;
16       f3.write(reinterpret_cast<char*>(&value), sizeof(value));
17     }
18   }
```

Listing19.8 归并两个段

```
1    // Merge two segments
2    void mergeTwoSegments(int segmentSize, fstream& f1, fstream& f2,
3      fstream& f3)
4    {
5      int intFromF1;
6      f1.read(reinterpret_cast<char*>(&intFromF1), sizeof(intFromF1));
7      int intFromF2;
8      f2.read(reinterpret_cast<char*>(&intFromF2), sizeof(intFromF2));
9      int f1Count = 1;
10       int f2Count = 1;
11
12       while (true)
13       {
14         if (intFromF1 < intFromF2)
15         {
16           f3.write(reinterpret_cast<char*>
17             (&intFromF1), sizeof(intFromF1));
18           if (f1.eof() || f1Count++ >= segmentSize)
19           {
20             if (f1.eof()) break;
21             f3.write(reinterpret_cast<char*>
22               (&intFromF2), sizeof(intFromF2));
23             break;
24           }
25           else
26           {
27             f1.read(reinterpret_cast<char*>
28               (& intFromF1), sizeof(intFromF1));
29           }
30         }
31         else
```

```cpp
32          {
33            f3.write(reinterpret_cast<char*>
34              (&intFromF2), sizeof(intFromF2));
35            if (f2.eof() || f2Count++ >= segmentSize)
36            {
37              if (f2.eof()) break;
38              f3.write(reinterpret_cast<char*>
39                (&intFromF1), sizeof(intFromF1));
40              break;
41            }
42            else
43            {
44              f2.read(reinterpret_cast<char*>
45                (&intFromF2), sizeof(intFromF2));
46            }
47          }
48        }
49
50        while (!f1.eof() && f1Count++ < segmentSize)
51        {
52          int value;
53          f1.read(reinterpret_cast<char*>
54            (&value), sizeof(value));
55          if (f1.eof()) break;
56          f3.write(reinterpret_cast<char*>
57            (&value), sizeof(value));
58        }
59
60        while (!f2.eof() && f2Count++ < segmentSize)
61        {
62          int value;
63          f2.read(reinterpret_cast<char*>(&value), sizeof(value));
64          if (f2.eof()) break;
65          f3.write(reinterpret_cast<char*>(&value), sizeof(value));
66        }
67      }
```

19.8.3 合成两个阶段

LiveExample19.8 给出了对 largedata.dat 中 int 值进行排序，并将排序后的数据存储在 sortedfile.dat 中的完整程序。

LiveExample 19.8 的互动程序请访问 https://liangcpp.pearsoncmg.com/LiveRunCpp5e/faces/LiveExample.xhtml?header=off&programName=SortLargeFile&fileType=.cpp&programHeight=4210&resultVisible=false。

LiveExample 19.8　ExternalSort.cpp

Source Code Editor:

```cpp
1  #include <iostream>
2  #include <fstream>
3  #include "QuickSort.h"
4  #include <string>
5  using namespace std;
6
7  // Function prototype
8  void sort(string sourcefile, string targetfile);
9  int initializeSegments(int segmentSize,
10   string sourcefile, string f1);
11 void mergeTwoSegments(int segmentSize, fstream &f1, fstream &f2,
12   fstream &f3);
13 void merge(int numberOfSegments, int segmentSize,
```

```cpp
14        string f1, string f2, string f3, string targetfile) ;
15   void copyHalfToF2(int numberOfSegments, int segmentSize,
16        fstream &f1, fstream &f2);
17   void mergeOneStep(int numberOfSegments, int segmentSize,
18        string f1, string f2, string f3);
19   void mergeSegments(int numberOfSegments, int segmentSize,
20        fstream &f1, fstream &f2, fstream &f3);
21   void copyFile(string f1, string targetfile);
22   void displayFile(string filename);
23
24   int main()
25   {
26     // Sort largedata.dat into sortedfile.dat
27     sort("largedata.dat", "sortedfile.dat");
28
29     // Display the first 100 numbers in sortedfile.dat
30     displayFile("sortedfile.dat");
31   }
32
33   /** Sort sourcefile into targetfile */
34   void sort(string sourcefile, string targetfile)
35   {
36     const int MAX_ARRAY_SIZE = 10000;
37
38     // Implement Phase 1: Create initial segments
39     int numberOfSegments =
40       initializeSegments(MAX_ARRAY_SIZE, sourcefile, "f1.dat");
41
42     // Implement Phase 2: Merge segments recursively
43     merge(numberOfSegments, MAX_ARRAY_SIZE,
44       "f1.dat", "f2.dat", "f3.dat", targetfile);
45   }
46
47   /* Sort original file into sorted segments */
48   int initializeSegments(int segmentSize, string sourceFile, string f1)
49   {
50     int *list = new int[segmentSize];
51
52     fstream input;
53     input.open(sourceFile.c_str(), ios::in | ios::binary);
54     fstream output;
55     output.open(f1.c_str(), ios::out | ios::binary);
56
57     int numberOfSegments = 0;
58     while (!input.eof())
59     {
60       int i = 0;
61       for ( ; !input.eof() && i < segmentSize; i++)
62       {
63         input.read(reinterpret_cast<char*>
64           (&list[i]), sizeof(list[i]));
65       }
66
67       if (input.eof()) i--;
68       if (i <= 0)
69         break;
70       else
71         numberOfSegments++;
72
73       // Sort an array list[0..i-1]
74       quickSort(list, i);
75
76       // Write the array to f1.dat
77       for (int j = 0; j < i; j++)
78       {
```

```cpp
 79           output.write(reinterpret_cast<char*>
 80               (&list[j]), sizeof(list[j]));
 81       }
 82     }
 83
 84     input.close();
 85     output.close();
 86     delete [] list;
 87
 88     return numberOfSegments;
 89   }
 90
 91   void merge(int numberOfSegments, int segmentSize,
 92     string f1, string f2, string f3, string targetfile)
 93   {
 94     if (numberOfSegments > 1)
 95     {
 96       mergeOneStep(numberOfSegments, segmentSize, f1, f2, f3);
 97       merge((numberOfSegments + 1) / 2, segmentSize * 2,
 98         f3, f1, f2, targetfile);
 99     }
100     else
101     { // rename f1 as the final sorted file
102       copyFile(f1, targetfile);
103       cout << "\nSorted into the file " << targetfile << endl;
104     }
105   }
106
107   void copyFile(string f1, string targetfile)
108   {
109     fstream input;
110     input.open(f1.c_str(), ios::in | ios::binary);
111
112     fstream output;
113     output.open(targetfile.c_str(), ios::out | ios::binary);
114     int i = 0;
115     while (!input.eof()) // Continue if not end of file
116     {
117       int value;
118       input.read(reinterpret_cast<char*>(&value), sizeof(value));
119       if (input.eof()) break;
120       output.write(reinterpret_cast<char*>(&value), sizeof(value));
121     }
122
123     input.close();
124     output.close();
125   }
126
127   void mergeOneStep(int numberOfSegments, int segmentSize, string f1,
128     string f2, string f3)
129   {
130     fstream f1Input;
131     f1Input.open(f1.c_str(), ios::in | ios::binary);
132
133     fstream f2Output;
134     f2Output.open(f2.c_str(), ios::out | ios::binary);
135
136     // Copy half number of segments from f1.dat to f2.dat
137     copyHalfToF2(numberOfSegments, segmentSize, f1Input, f2Output);
138     f2Output.close();
139
140     // Merge remaining segments in f1 with segments in f2 into f3
141     fstream f2Input;
142     f2Input.open(f2.c_str(), ios::in | ios::binary);
143     fstream f3Output;
```

```cpp
144     f3Output.open(f3.c_str(), ios::out | ios::binary);
145
146     mergeSegments(numberOfSegments / 2, segmentSize, f1Input, f2Input, f3Output);
147
148     f1Input.close();
149     f2Input.close();
150     f3Output.close();
151   }
152
153   /** Copy first half number of segments from f1.dat to f2.dat */
154   void copyHalfToF2(int numberOfSegments, int segmentSize, fstream &f1,
155     fstream &f2)
156   {
157     for (int i = 0; i < (numberOfSegments / 2) * segmentSize; i++)
158     {
159       int value;
160       f1.read(reinterpret_cast<char*>(&value), sizeof(value));
161       f2.write(reinterpret_cast<char*>(&value), sizeof(value));
162     }
163   }
164
165   /** Merge all segments */
166   void mergeSegments(int numberOfSegments, int segmentSize, fstream &f1,
167     fstream &f2, fstream &f3)
168   {
169     for (int i = 0; i < numberOfSegments; i++)
170     {
171       mergeTwoSegments(segmentSize, f1, f2, f3);
172     }
173
174     // f1 may have one extra segment, copy it to f3
175     while (!f1.eof())
176     {
177       int value;
178       f1.read(reinterpret_cast<char*>(&value), sizeof(value));
179       if (f1.eof()) break;
180       f3.write(reinterpret_cast<char*>(&value), sizeof(value));
181     }
182   }
183
184   /** Merge two segments */
185   void mergeTwoSegments(int segmentSize, fstream &f1, fstream &f2,
186     fstream &f3)
187   {
188     int intFromF1;
189     f1.read(reinterpret_cast<char*>(&intFromF1), sizeof(intFromF1));
190     int intFromF2;
191     f2.read(reinterpret_cast<char*>(&intFromF2), sizeof(intFromF2));
192     int f1Count = 1;
193     int f2Count = 1;
194
195     while (true)
196     {
197       if (intFromF1 < intFromF2)
198       {
199         f3.write(reinterpret_cast<char*>(&intFromF1), sizeof(intFromF1));
200         if (f1.eof() || f1Count++ >= segmentSize)
201         {
202           if (f1.eof()) break;
203           f3.write(reinterpret_cast<char*>(&intFromF2), sizeof(intFromF2));
204           break;
205         }
206         else
207         {
208           f1.read(reinterpret_cast<char*> (&intFromF1), sizeof(intFromF1));
```

```cpp
209        }
210      }
211      else
212      {
213        f3.write(reinterpret_cast<char*>(&intFromF2), sizeof(intFromF2));
214        if (f2.eof() || f2Count++ >= segmentSize)
215        {
216          if (f2.eof()) break;
217          f3.write(reinterpret_cast<char*>(&intFromF1), sizeof(intFromF1));
218          break;
219        }
220        else {
221          f2.read(reinterpret_cast<char*>(&intFromF2), sizeof(intFromF2));
222        }
223      }
224    }
225
226    while (!f1.eof() && f1Count++ < segmentSize) {
227      int value;
228      f1.read(reinterpret_cast<char*>(&value), sizeof(value));
229      if (f1.eof()) break;
230      f3.write(reinterpret_cast<char*>(&value), sizeof(value));
231    }
232
233    while (!f2.eof() && f2Count++ < segmentSize) {
234      int value;
235      f2.read(reinterpret_cast<char*>(&value), sizeof(value));
236      if (f2.eof()) break;
237      f3.write(reinterpret_cast<char*>(&value), sizeof(value));
238    }
239  }
240
241  /** Display the first 10 numbers in the specified file */
242  void displayFile(string filename)
243  {
244    fstream input(filename.c_str(), ios::in | ios::binary);
245    int value;
246    for (int i = 0; i < 100; i++)
247    {
248      input.read(reinterpret_cast<char*>(&value), sizeof(int));
249      cout << value << " ";
250    }
251
252    input.close();
253  }
```

运行此程序之前，首先运行 LiveExample 19.7，以创建 largedata.dat。调用 sort ("largedata.dat","sortedfile.dat")（第 27 行）从 largedata.dat 读取数据，并把排序后的数据写入 sortedfile.dat。调用 displayFile("sortedfile.dat")（第 30 行）显示指定文件中的前 100 个数。注意，这些文件是用二进制 I/O 创建的，不能用记事本等文本编辑器查看。

sort 函数首先从原始数组创建初始段，并将排序后的段存储在新文件 f1.dat 中（第 39～40 行），然后在 targetfile 中生成排序后的文件（第 43～44 行）。

归并函数（第 91～105 行）将 f1 中的段归并到 f3，用 f2 来辅助归并。merge 函数是在许多归并步骤中递归调用的。每个归并步骤都会将段数减少一半，并将已排序的段大小增加一倍。在完成一个归并步骤之后，下一个归并步骤递归地将 f3 中的新段归并到 f2，用 f1 辅助（第 97～98 行）。

下一个归并步骤的 numberOfSegments 是第 97 行中的 (numberOfSegments+1)/2。

例如，如果 `numberOfSegments` 为 5，则下一个归并步骤的 `numberOfSegments` 为 3，因为每两个段被归并，但还有一个未归并。

当 `numberOfSegments` 为 1 时，递归 `merge` 函数结束。在这种情况下，`f1` 包含已排序的数据。文件 `f1` 在第 102 行被复制到 `targetfile` 中。

19.8.4 外部排序复杂度

在外部排序中，主要成本是 I/O。假设 n 是在文件中要排序的元素数。在第一阶段，从原始文件中读取 n 个元素，并将其输出到临时文件中。因此，第一阶段的 I/O 为 $O(n)$。

在第二阶段中，在第一个归并步骤之前，已排序段的数量为 n/c，其中 c 为 `MAX_ARRAY_SIZE`。每个归并步骤都会将段数量减少一半。因此，在第一个归并步骤之后，段数是 $n/2c$；在第二个归并步骤之后，段数是 $\frac{n}{2^2c}$；在第三个归并步骤之后，段数是 $\frac{n}{2^3c}$。在 $\log\left(\frac{n}{c}\right)$ 个归并步骤之后，段数已经减少到 1。因此，归并步骤的总数为 $\log\left(\frac{n}{c}\right)$。

在每个归并步骤中，从文件 `f1` 读取一半数量的段，然后将其写入临时文件 `f2`。`f1` 中的剩余段与 `f2` 中的段归并。每个归并步骤中的 I/O 数为 $O(n)$。由于归并步骤的总数为 $\log\left(\frac{n}{c}\right)$，因此 I/O 的总数为

$$O(n) \times \log\left(\frac{n}{c}\right) = O(n \log n)$$

因此，外部排序的复杂度是 $O(n \log n)$。

关键术语

bubble sort（冒泡排序）
bucket sort（桶排序）
external sort（外部排序）
complete binary tree（完全二叉树）
heap（堆）

heap sort（堆排序）
height of a heap（堆的高度）
merge sort（归并排序）
quick sort（快速排序）
radix sort（基数排序）

章节总结

1. 选择排序、插入排序、冒泡排序和快速排序的最坏情况复杂度是 $O(n^2)$。
2. 归并排序的平均情况和最坏情况复杂度是 $O(n \log n)$。快速排序的平均时间也是 $O(n \log n)$。
3. 堆是设计高效算法（如排序）的有用数据结构。我们学习了如何定义和实现堆类，以及如何在堆中插入和删除元素。
4. 堆排序的时间复杂度是 $O(n \log n)$。
5. 桶排序和基数排序是整数键值的专用排序算法。这些算法使用桶而不是通过比较键值来对键值进行排序。它们比一般的排序算法更高效。
6. 归并排序的一种变体——称为外部排序——可以对外部文件中的大量数据进行排序。

编程练习

互动程序请访问 https://liangcpp.pearsoncmg.com/CheckExerciseCpp/faces/CheckExercise5e.xhtml?chapter=19&programName=Exercise19_01。

19.1 （泛型冒泡排序）为冒泡排序编写一个泛型函数。

19.2 （泛型归并排序）为归并排序编写一个泛型函数。

19.3 （泛型快速排序）用以下函数头编写快速排序的泛型函数：

```
template<typename T>
void quickSort(T list[], int arraySize);
```

编写一个测试程序，提示用户读取 10 个整数，然后用快速排序法排序并显示排序后的数字。

19.4 （改进快速排序）书中介绍的快速排序算法选择列表中的第一个元素作为轴心。用列表的第一个、中间和最后一个元素的中间值作为轴心来修改算法。

19.5 （泛型堆排序）编写一个测试程序，调用泛型堆排序函数对 `int` 值数组、`double` 值数组和字符串数组进行排序。

19.6 （检查顺序）编写以下函数，检查数组是按升序排列还是按降序排列。默认情况下，该函数检查升序。若要检查降序，将 `false` 传递给函数中的 `ascending` 参数。

```
// T is a generic type
template<typename T>
bool ordered(const T list[], int size, bool ascending = true)
```

*19.7 （基数排序）编写一个程序，随机生成 1000000 个整数，并用基数排序对其进行排序。

19.8 （排序的执行时间）编写一个程序，获取输入规模为 500000、1000000、1500000、2000000、2500000 和 3000000 的选择排序、插入排序、冒泡排序、归并排序、快速排序和堆排序的执行时间。程序打印如下表格：

```
Sample Run for Exercise19_08.cpp
Execution Result:

command>Exercise19_08
Array Selection Insertion  Bubble   Merge   Quick   Heap   Radix
 Size     Sort      Sort     Sort    Sort    Sort   Sort    Sort
10000       38        33      107       3       2     24      10
20000      142       121      463       4       2      7      13
30000      121        91     1073       6       2      7       3
40000      217       161     1924       9       3      9       5
50000      330       255     3038      11       5     13       7
60000      479       374     4403      18       6     14       6

command>
```

（提示：可以用以下代码模板获取执行时间。）

```
long startTime = time(0);
perform the task;
long endTime = time(0);
long executionTime = endTime - startTime;
```

19.9 （外部排序的执行时间）编写一个程序，获取 5000000、10000000、15000000、20000000、25000000 规模的整数的外部排序执行时间。程序打印如下表格：

```
Sample Run for Exercise19_09.cpp
Execution Result:

command>Exercise19_09
File size   5000000  10000000  15000000  20000000  25000000
Time            253      1251      4273     10932     30932

command>
```

*19.10 （修改归并排序）重写 `mergeSort` 函数，不创建新的临时数组，递归地对数组的前半部分和后半部分进行排序，然后将两者归并为一个临时数组，并将其内容复制到原始数组中，如图 19.3b 所示。

第 20 章

Introduction to C++ Programming and Data Structures, Fifth Edition

链表、队列和优先级队列

学习目标
1. 创建节点将元素存储在链表中（20.2 节）。
2. 通过指针访问链表中的节点（20.3 节）。
3. 定义 `LinkedList` 类存储和处理列表中的数据（20.4 节）。
4. 将元素添加到列表的头部（20.4.1 节）。
5. 将元素添加到列表的末尾（20.4.2 节）。
6. 在列表中插入元素（20.4.3 节）。
7. 删除列表的第一个元素（20.4.4 节）。
8. 删除列表的最后一个元素（20.4.5 节）。
9. 删除列表中指定位置的元素（20.4.6 节）。
10. 实现遍历各种类型容器中元素的迭代器（20.5 节）。
11. 使用 `foreach` 循环遍历容器中的元素（20.6 节）。
12. 探索链表的变体（20.7 节）。
13. 用链表实现 `Queue` 类（20.8 节）。
14. 用堆实现 `PriorityQueue` 类（20.9 节）。

20.1 简介

要点提示：本章的重点是设计和实现自定义数据结构。

12.4 节引入了一个泛型 `Stack` 栈类。栈中的元素存储在数组中。数组大小是固定的。如果数组太小，元素就不能存储在栈中；如果太大，会浪费很多空间。12.5 节提出了一种可能的解决方案。最初，栈使用一个小数组。当没有空间添加新元素时，栈会创建一个新数组，大小是旧数组的两倍，并将内容从旧数组复制到此新数组，然后丢弃旧数组。复制数组要耗费时间。

在 12.6 节中引入的向量类弥补了数组的缺点。向量本质上是一个灵活的数组。如果需要，它的大小会自动增长。但是，在向量的开头插入和删除元素效率很低。

本章介绍一种新的数据结构，称为**链表**。对于存储和管理数量变化的元素，链表效率很高。在列表的开头插入和删除元素尤其高效。本章还将讨论如何使用链表实现队列。

20.2 节点

要点提示：在链表中，每个元素都包含在一个称为节点的结构中。

当添加一个新元素时，会创建一个节点来包含它。所有节点都通过指针链接，如图 20.1 所示。

图 20.1 链表由链接在一起的任意数量的节点组成

可以用类定义节点。节点的类定义如下所示：

```
1   template<typename T>
2   class Node
3   {
4   public:
5     T element;  // Element contained in the node
6     Node* next; // Pointer to the next node
7
8     Node() // No-arg constructor
9     {
10      next = nullptr;
11    }
12
13    Node(T element) // Constructor
14    {
15      this->element = element;
16      next = nullptr;
17    }
18  };
```

Node 被定义为一个模板类，具有用于指定元素类型的类型参数 T。

按照惯例，名为 head 和 tail 的指针变量用于指向列表中的第一个和最后一个节点。如果列表为空，则 head 和 tail 都应为 nullptr。回想一下，nullptr 是一个 C++11 关键字，用于替换 NULL，它表示指针不指向任何节点。下面是一个创建容纳三个节点的链表的示例。每个节点存储一个字符串元素。

步骤 1：声明 head 和 tail：

Node<string>* head = **nullptr**;	列表现在为空
Node<string>* tail = **nullptr**;	head 和 tail 都是 nullptr

head 和 tail 都是 nullptr。列表为空。

步骤 2：创建第一个节点并将其插入到列表中。

在列表中插入第一个节点后，head 和 tail 指向该节点，如图 20.2 所示。

图 20.2 将第一个节点追加到列表中。head 和 tail 都指向该节点

步骤 3：创建第二个节点并将其追加到列表中。

要将第二个节点追加到列表中，需要将其与第一个节点链接，如图 20.3b 所示。新节点现在是尾部节点。所以应该移动 tail 以指向这个新节点，如图 20.3c 所示。

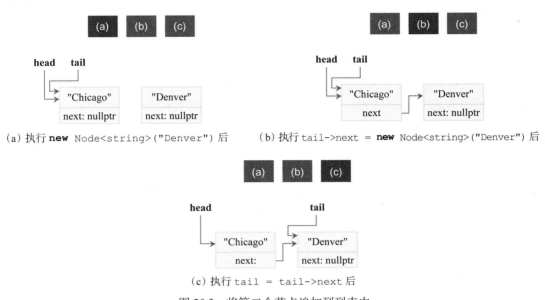

图 20.3　将第二个节点追加到列表中

步骤 4：创建第三个节点并将其追加到列表中。

要将新节点追加到列表中，需要其与最后一个节点链接，如图 20.4b 所示。新节点现在是尾部节点。所以应该移动 tail 以指向这个新节点，如图 20.4c 所示。

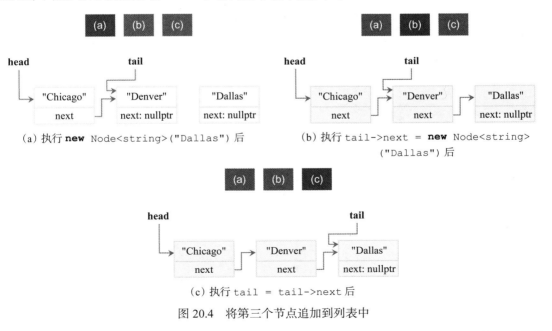

图 20.4　将第三个节点追加到列表中

每个节点都包含元素和指向下一个元素的指针。如果节点是列表中的最后一个，则其指针数据字段 next 包含值 nullptr。可以用此属性检测最后一个节点。例如，可以编写以下循环来遍历列表中的所有节点。

```
1    Node<string>* current = head;
2    while (current != nullptr)
3    {
4      cout << current->element << endl;
5      current = current->next;
6    }
```

current 指针最初指向列表中的第一个节点（第 1 行）。在循环中，检索当前节点的元素（第 4 行），然后 current 指向下一个节点（第 5 行）。循环继续直到当前节点为 nullptr 为止。

20.3 LinkedList 类

要点提示：LinkedList 类提供了在列表中存储、查找、检索和删除元素的操作。

链表（链接列表）是一种流行的数据结构，用于按顺序存储数据。例如，学生列表、可用房间列表、城市列表和书籍列表都可以使用列表存储。此处列出的操作是大多数列表的典型操作：

- 在列表中检索元素。
- 在列表中插入新元素。
- 从列表中删除元素。
- 查找列表的元素数量。
- 查找元素是否在列表中。
- 查找列表是否为空。

图 20.5 给出了 LinkedList 的类图。LinkedList 是一个类模板，其类型参数 T 表示存储在列表中的元素的类型。

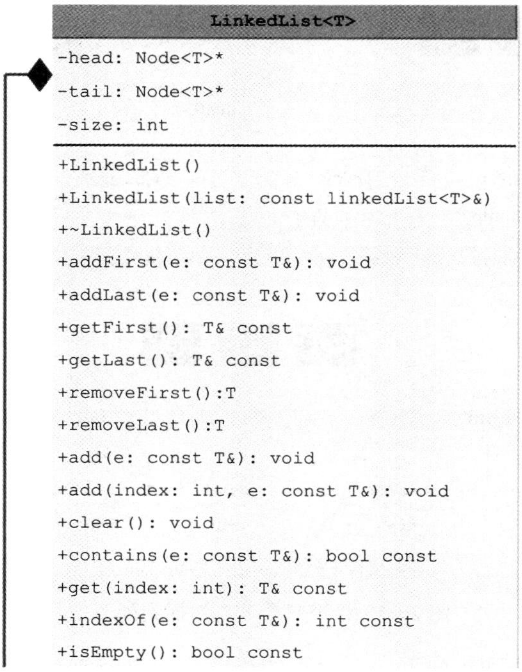

```
              LinkedList<T>
    -head: Node<T>*
    -tail: Node<T>*
    -size: int

    +LinkedList()
    +LinkedList(list: const linkedList<T>&)
    +~LinkedList()
    +addFirst(e: const T&): void
    +addLast(e: const T&): void
    +getFirst(): T& const
    +getLast(): T& const
    +removeFirst():T
    +removeLast():T
    +add(e: const T&): void
    +add(index: int, e: const T&): void
    +clear(): void
    +contains(e: const T&): bool const
    +get(index: int): T& const
    +indexOf(e: const T&): int const
    +isEmpty(): bool const
```

图 20.5 LinkedList 用节点的链接列表来实现列表

```
+lastIndexOf(e: const T&): int const
+remove(e: const T&): void
+getSize(): int const
+removeAt(index: int): T&
+set(index: int, e: const T&): T&
+begin(): Iterator<T>
+end(): Iterator<T>
```

```
Node<T>
+element: T&
+next: Node<T>*
+Node()
+Node(e: const T&)
```

图 20.5 LinkedList 用节点的链接列表来实现列表（续）

可以用 `get(int index)` 从列表中获取元素。index 从 0 开始，即列表头部的节点索引为 0。假设 LinkedList 类在头文件 LinkedList.h 中提供。我们先编写一个使用 LinkedList 类的测试程序，如 LiveExample 20.1 所示。程序用 LinkedList（第 18 行）创建一个列表。它用 add 函数向列表中添加字符串，用 remove 函数删除字符串。

LiveExample 20.1 的互动程序请访问 https://liangcpp.pearsoncmg.com/LiveRunCpp5e/faces/LiveExample.xhtml?header=off&programName=TestLinkedList&fileType=.cpp&programHeight=1060&resultHeight=310。

LiveExample 20.1　TestLinkedList.cpp

Source Code Editor:

```cpp
1  #include <iostream>
2  #include <string>
3  #include "LinkedList.h"
4  using namespace std;
5
6  void printList(const LinkedList<string>& list)
7  {
8    for (int i = 0; i < list.getSize(); i++)
9    {
10     cout << list.get(i) << " ";
11   }
12   cout << endl;
13 }
14
15 int main()
16 {
17   // Create a list for strings
18   LinkedList<string> list;
19
20   // Add elements to the list
21   list.add("America"); // Add America to the list
22   cout << "(1) ";
23   printList(list);
24
25   list.add(0, "Canada"); // Add Cananda to the beginning of the list
26   cout << "(2) ";
27   printList(list);
28
29   list.add("Russia"); // Add Russia to the end of the list
```

```cpp
30      cout << "(3) ";
31      printList(list);
32
33      list.add("France"); // Add France to the end of the list
34      cout << "(4) ";
35      printList(list);
36
37      list.add(2, "Germany"); // Add Germany to the list at index 2
38      cout << "(5) ";
39      printList(list);
40
41      list.add(5, "Norway"); // Add Norway to the list at index 5
42      cout << "(6) ";
43      printList(list);
44
45      list.add(0, "Netherlands"); // Same as list.addFirst("Netherlands")
46      cout << "(7) ";
47      printList(list);
48
49      // Remove elements from the list
50      list.removeAt(0); // Same as list.remove("Netherlands") in this case
51      cout << "(8) ";
52      printList(list);
53
54      list.removeAt(2); // Remove the element at index 2
55      cout << "(9) ";
56      printList(list);
57
58      list.removeAt(list.getSize() - 1); // Remove the last element
59      cout << "(10) ";
60      printList(list);
61
62      return 0;
63  }
```

Execution Result:

```
command>cl TestLinkedList.cpp
Microsoft C++ Compiler 2019
Compiled successful (cl is the VC++ compile/link command)

command>TestLinkedList
(1) America
(2) Canada America
(3) Canada America Russia
(4) Canada America Russia France
(5) Canada America Germany Russia France
(6) Canada America Germany Russia France Norway
(7) Netherlands Canada America Germany Russia France Norway
(8) Canada America Germany Russia France Norway
(9) Canada America Russia France Norway
(10) Canada America Russia France

command>
```

20.4　实现 `LinkedList`

要点提示：使用链接结构实现链表。

现在我们把注意力转向 `LinkedList` 类的实现。有些函数很容易实现。例如，`isEmpty()` 函数简单地返回 `head==nullptr`，`clear()` 函数则简单地销毁列表中的所有节点，并将

head 和 tail 设置为 nullptr。addLast(T element) 函数与 add(T element) 函数相同。定义两者是为了方便。

20.4.1 实现 `addFirst(T element)`

addFirst(T element) 函数可以如下实现：

```cpp
template<typename T>
void LinkedList<T>::addFirst(T element)
{
  Node<T>* newNode = new Node<T>(element);
  newNode->next = head;
  head = newNode;
  size++;

  if (tail == nullptr)
    tail = head;
}
```

addFirst(T element) 函数创建一个新的节点（第 4 行）来存储元素，并将节点插入列表的开头（第 5 行），如图 20.6b 所示。插入后，head 应指向该新元素节点（第 6 行），见图 20.6c。

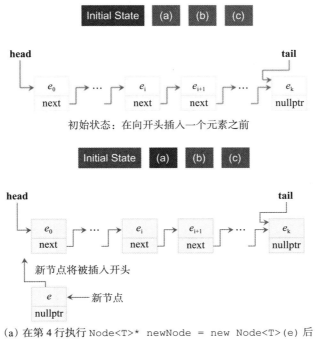

(a) 在第 4 行执行 Node<T>* newNode = new Node<T>(e) 后

图 20.6 在列表的开头插入一个新元素

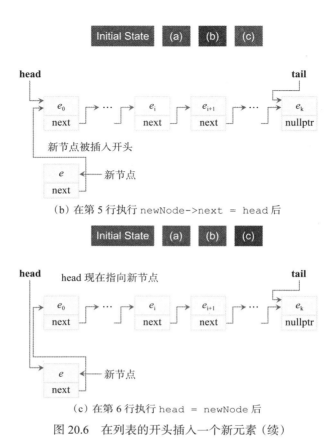

（b）在第 5 行执行 newNode->next = head 后

（c）在第 6 行执行 head = newNode 后

图 20.6 在列表的开头插入一个新元素（续）

如果列表为空（第 9 行），则 head 和 tail 都将指向此新节点（第 10 行）。创建节点后，大小应增加 1（第 7 行）。

显然，addFirst 函数需要 $O(1)$ 时间。

20.4.2 实现 `addLast(T element)`

函数 addLast(T element) 创建一个容纳一个元素的节点，并将该节点追加到列表的末尾。它可以如下实现：

```
1  template<typename T>
2  void LinkedList<T>::addLast(T element)
3  {
4    if (tail == nullptr)
5    {
6      head = tail = new Node<T>(element);
7    }
8    else
9    {
10     tail->next = new Node<T>(element);
11     tail = tail->next;
12   }
13
14   size++;
15 }
```

考虑两种情况：

1. 如果列表为空（第 4 行），则 head 和 tail 都指向这个新节点（第 6 行）；
2. 否则，在列表的末尾插入节点（第 10 行）。插入之后，tail 应该指向该新元素节点（第 11 行），如图 20.7 所示。在任何情况下，创建一个节点后，大小都应增加 1（第 14 行）。

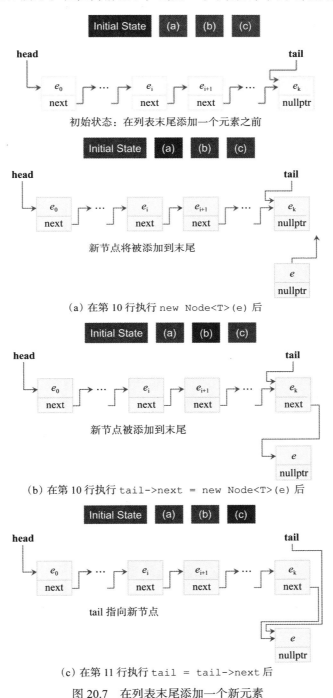

图 20.7 在列表末尾添加一个新元素

显然，addLast 函数需要 $O(1)$ 时间。

20.4.3 实现 add(int index, T element)

函数 add(int index, T element) 向列表指定索引处添加元素。它可以如下实现：

```cpp
template<typename T>
void LinkedList<T>::add(int index, T element)
{
  if (index == 0)
    addFirst(element);
  else if (index >= size)
    addLast(element);
  else
  {
    Node<T>* current = head;
    for (int i = 1; i < index; i++)
      current = current->next;
    Node<T>* temp = current->next;
    current->next = new Node<T>(element);
    (current->next)->next = temp;
    size++;
  }
}
```

考虑三种情况：

1. 如果 index 为 0，则调用 addFirst(element)（第 5 行）将元素插入列表的开头；

2. 如果 index 大于或等于列表 size，则调用 addLast(element)（第 7 行）将元素插入列表末尾；

3. 否则，创建一个新节点存储新元素，并定位插入位置。如图 20.8b 所示，新节点将插入 current 节点和 temp 节点之间。该函数将新节点赋值给 current->next，并将 temp 节点赋值给新节点的 next，如图 20.8c 所示。现在，大小增加了 1（第 16 行）。

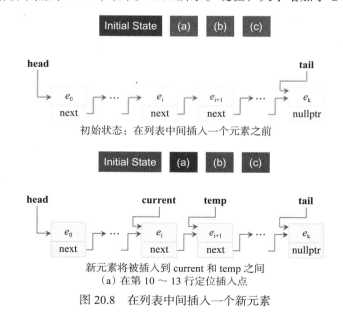

图 20.8 在列表中间插入一个新元素

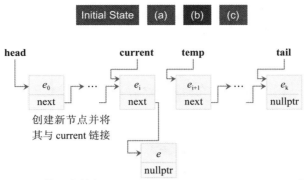

（b）在第 14 行执行 `current->next = new Node<T>(e)` 后

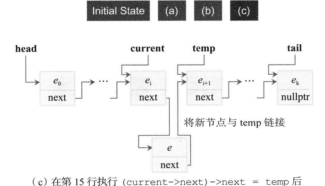

（c）在第 15 行执行 `(current->next)->next = temp` 后

图 20.8　在列表中间插入一个新元素（续）

显然，`addFirst(index, element)` 函数需要 $O(n)$ 时间。

20.4.4　实现 `removeFirst( )`

`removeFirst()` 函数可以如下实现：

```
1  template<typename T>
2  T LinkedList<T>::removeFirst() throw (runtime_error)
3  {
4    if (size == 0)
5      throw runtime_error("No elements in the list");
6    else
7    {
8      Node<T>* temp = head;
9      head = head->next;
10     if (head == nullptr) tail = nullptr;
11     size--;
12     T element = temp->element;
13     delete temp;
14     return element;
15   }
16 }
```

考虑三种情况：

1. 如果列表为空，则抛出异常（第 5 行）；
2. 否则，通过将 head 指向第二个节点（第 9 行），从列表中删除第一个节点，如图 20.9 所示；
3. 如果列表只有一个节点，那么在删除它之后，tail 应该设置为 nullptr（第 10 行）。删除后，列表大小减小 1（第 11 行）。

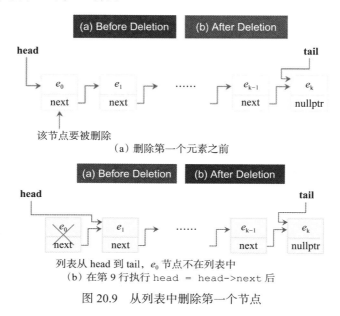

图 20.9　从列表中删除第一个节点

显然，removeFirst() 函数需要 $O(1)$ 时间。

20.4.5　实现 removeLast()

函数 removeLast() 可以如下实现：

```
1  template<typename T>
2  T LinkedList<T>::removeLast()
3  {
4    if (size == 0 || size == 1)
5      return removeFirst(); // Same as removeFirst() in this case
6    else
7    {
8      Node<T>* current = head;
9      for (int i = 0; i < size - 2; i++)
10       current = current->next;
11
12     Node<T>* temp = tail;
13     tail = current;
14     tail->next = nullptr;
15     size--;
16     T element = temp->element;
17     delete temp;
18     return element;
19   }
20 }
```

考虑两种情况：
1. 如果列表为空或只有一个元素，那么调用 `removeFirst()` 处理这种情况（第 4～5 行）；
2. 否则，定位 `current` 以指向倒数第二个节点（第 8～10 行），如图 20.10a 所示。将 `tail` 设置为 `current`（第 13 行）。`tail` 现在被重新定位，以指向倒数第二个节点，如图 20.10b 所示。删除后，列表大小减小 1（第 15 行），最后一个节点被删除，如图 20.10c 所示。返回被删除节点的元素值（第 18 行）。

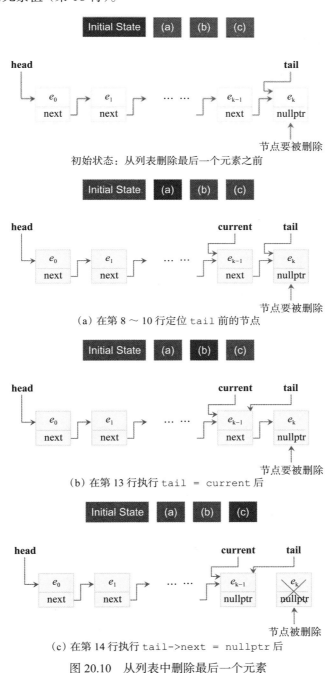

图 20.10 从列表中删除最后一个元素

由于该算法需要找到 tail 之前的指针，定位它需要 $O(n)$ 时间。所以 removeLast() 函数需要 $O(n)$ 时间。这里使用的链表称为单链表，在向前的一个方向上遍历节点。在编程练习 20.8 中，可以使用双链表实现 $O(1)$ 时间的 removeLast() 函数。

20.4.6 实现 removeAt(int index)

函数 removeAt(int index) 在指定的索引处找到节点，然后将其删除。它可以如下实现：

```
template<typename T>
T LinkedList<T>::removeAt(int index)
{
  if (index < 0 || index >= size)
    throw runtime_error("Index out of range");
  else if (index == 0)
    return removeFirst();
  else if (index == size - 1)
    return removeLast();
  else
  {
    Node<T>* previous = head;
    for (int i = 1; i < index; i++)
    {
      previous = previous->next;
    }

    Node<T>* current = previous->next;
    previous->next = current->next;
    size--;
    T element = current->element;
    delete current;
    return element;
  }
}
```

考虑四种情况：

1. 如果 index 超出列表的范围（即 index < 0 || index >= size），抛出异常（第5行）；
2. 如果 index 为 0，则调用 removeFirst() 删除第一个节点（第 7 行）；
3. 如果 index 为 size-1，则调用 removeLast() 删除最后一个节点（第 9 行）；
4. 否则，在指定 index 处定位节点。设 current 表示该节点，previous 表示其之前的节点，如图 20.11a 所示。将 current->next 赋值给 previous->next 来删除当前节点（第 19 行），如图 20.11b 所示。

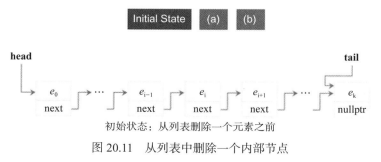

初始状态：从列表删除一个元素之前

图 20.11 从列表中删除一个内部节点

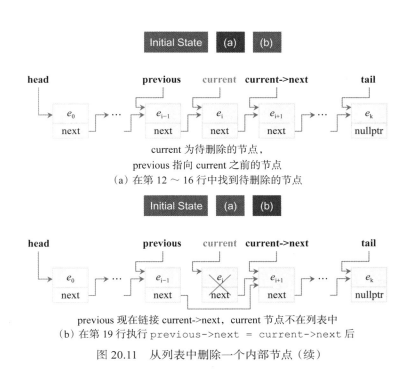

图 20.11 从列表中删除一个内部节点（续）

显然，removeAt(index) 函数需要 $O(n)$ 时间。

20.4.7 `LinkedList` 的源代码

LiveExample 20.2 给出了 `LinkedList` 的实现。

LiveExample 20.2 的互动程序请访问 https://liangcpp.pearsoncmg.com/LiveRunCpp5e/faces/LiveExample.xhtml?header=off&programName=LinkedListV2&fileType=.h&programHeight=5800&resultVisible=false。

LiveExample 20.2　LinkedList.h

Source Code Editor:

```cpp
#ifndef LINKEDLIST_H
#define LINKEDLIST_H
#include <stdexcept>
using namespace std;

template<typename T>
class Node
{
public:
  T element; // Element contained in the node
  Node<T>* next; // Pointer to the next node

  Node() // No-arg constructor
  {
    next = nullptr;
  }

  Node(const T& e) // Constructor
  {
    this->element = e;
    next = nullptr;
```

```cpp
22    }
23  };
24
25  template<typename T>
26  class Iterator : public std::iterator<std::forward_iterator_tag, T>
27  {
28  public:
29    Iterator(Node<T>* p)
30    {
31      current = p;
32    }
33
34    Iterator operator++() // Prefix ++
35    {
36      current = current->next;
37      return *this;
38    }
39
40    Iterator operator++(int dummy) // Postfix ++
41    {
42      Iterator temp(current);
43      current = current->next;
44      return temp;
45    }
46
47    T& operator*()
48    {
49      return current->element;
50    }
51
52    bool operator==(const Iterator<T>& iterator)
53    {
54      return current == iterator.current;
55    }
56
57    bool operator!=(const Iterator<T>& iterator)
58    {
59      return current != iterator.current;
60    }
61
62  private:
63    Node<T>* current;
64  };
65
66  template<typename T>
67  class LinkedList
68  {
69  public:
70    LinkedList(); // No-arg constructor
71    LinkedList(const LinkedList<T>& list); // Copy constructor
72    virtual ~LinkedList(); // Destructor
73    LinkedList<T>& operator=(const LinkedList<T>& list);
74    void addFirst(const T& e);
75    void addLast(const T& e);
76    T& getFirst() const;
77    T& getLast() const;
78    T removeFirst();
79    T removeLast();
80    void add(const T& e);
81    void add(int index, const T& e);
82    void clear();
83    bool contains(const T& e) const;
84    T& get(int index) const;
85    int indexOf(const T& e) const;
```

```cpp
 86    bool isEmpty() const;
 87    int lastIndexOf(const T& e) const;
 88    void remove(const T& e);
 89    int getSize() const;
 90    T removeAt(int index);
 91    T& set(int index, const T& e);
 92
 93    Iterator<T> begin() const
 94    {
 95      return Iterator<T>(head);
 96    }
 97
 98    Iterator<T> end() const
 99    {
100      return Iterator<T>(tail->next);
101    }
102
103  private:
104    Node<T>* head;
105    Node<T>* tail;
106    int size;
107  };
108
109  template<typename T>
110  LinkedList<T>::LinkedList()
111  {
112    head = tail = nullptr; // Initialize head and tail
113    size = 0;
114  }
115
116  template<typename T>
117  LinkedList<T>::LinkedList(const LinkedList<T>& list)
118  {
119    head = tail = nullptr;
120    size = 0;
121
122    Node<T>* current = list.head;
123    while (current != nullptr)
124    {
125      this->add(current->element);
126      current = current->next;
127    }
128  }
129
130  template<typename T>
131  LinkedList<T>::~LinkedList()
132  {
133    clear();
134  }
135
136  template<typename T>
137  LinkedList<T>& LinkedList<T>::operator=(const LinkedList<T>& list)
138  {
139    if (this != &list) // Do nothing with self-assignment
140    {
141      clear(); // Destroy this
142      head = tail = nullptr;
143      size = 0;
144
145      Node<T>* current = list.head;
146      while (current != nullptr)
147      {
148        this->add(current->element);
149        current = current->next;
```

```cpp
150        }
151      }
152
153      return *this;
154    }
155
156    template<typename T>
157    void LinkedList<T>::addFirst(const T& e)
158    {
159      Node<T>* newNode = new Node<T>(e);
160      newNode->next = head;
161      head = newNode;
162      size++;
163
164      if (tail == nullptr)
165        tail = head;
166    }
167
168    template<typename T>
169    void LinkedList<T>::addLast(const T& e)
170    {
171      if (tail == nullptr)
172      {
173        head = tail = new Node<T>(e);
174      }
175      else
176      {
177        tail->next = new Node<T>(e);
178        tail = tail->next;
179      }
180
181      size++;
182    }
183
184    template<typename T>
185    T& LinkedList<T>::getFirst() const
186    {
187      if (size == 0)
188        throw runtime_error("Index out of range");
189      else
190        return head->element;
191    }
192
193    template<typename T>
194    T& LinkedList<T>::getLast() const
195    {
196      if (size == 0)
197        throw runtime_error("Index out of range");
198      else
199        return tail->element;
200    }
201
202    template<typename T>
203    T LinkedList<T>::removeFirst()
204    {
205      if (size == 0)
206        throw runtime_error("No elements in the list");
207      else
208      {
209        Node<T>* temp = head;
210        head = head->next;
211        if (head == nullptr) tail = nullptr;
212        size--;
213        T element = temp->element;
214        delete temp;
```

```cpp
215      return element;
216    }
217  }
218
219  template<typename T>
220  T LinkedList<T>::removeLast()
221  {
222    if (size == 0 || size == 1)
223      return removeFirst();
224    else
225    {
226      Node<T>* current = head;
227      for (int i = 0; i < size - 2; i++)
228        current = current->next;
229
230      Node<T>* temp = tail;
231      tail = current;
232      tail->next = nullptr;
233      size--;
234      T element = temp->element;
235      delete temp;
236      return element;
237    }
238  }
239
240  template<typename T>
241  void LinkedList<T>::add(const T& e)
242  {
243    addLast(e);
244  }
245
246  template<typename T>
247  void LinkedList<T>::add(int index, const T& e)
248  {
249    if (index == 0)
250      addFirst(e);
251    else if (index >= size)
252      addLast(e);
253    else
254    {
255      Node<T>* current = head;
256      for (int i = 1; i < index; i++)
257        current = current->next;
258      Node<T>* temp = current->next;
259      current->next = new Node<T>(e);
260      (current->next)->next = temp;
261      size++;
262    }
263  }
264
265  template<typename T>
266  void LinkedList<T>::clear()
267  {
268    while (head != nullptr)
269    {
270      Node<T>* temp = head;
271      head = head->next;
272      delete temp;
273    }
274
275    tail = nullptr;
276    size = 0;
277  }
278
```

```cpp
template<typename T>
T& LinkedList<T>::get(int index) const
{
  if (index < 0 || index > size - 1)
    throw runtime_error("Index out of range");

  Node<T>* current = head;
  for (int i = 0; i < index; i++)
    current = current->next;

  return current->element;
}

template<typename T>
int LinkedList<T>::indexOf(const T& e) const
{
  // Implement it in this exercise
  Node<T>* current = head;
  for (int i = 0; i < size; i++)
  {
    if (current->element == e)
      return i;
    current = current->next;
  }

  return -1;
}

template<typename T>
bool LinkedList<T>::isEmpty() const
{
  return head == nullptr;
}

template<typename T>
int LinkedList<T>::getSize() const
{
  return size;
}

template<typename T>
T LinkedList<T>::removeAt(int index)
{
  if (index < 0 || index >= size)
    throw runtime_error("Index out of range");
  else if (index == 0)
    return removeFirst();
  else if (index == size - 1)
    return removeLast();
  else
  {
    Node<T>* previous = head;
    for (int i = 1; i < index; i++)
    {
      previous = previous->next;
    }

    Node<T>* current = previous->next;
    previous->next = current->next;
    size--;
    T element = current->element;
    delete current;
    return element;
  }
}
```

```
344
345   // The functions remove(const T& e), lastIndexOf(const T& e),
346   // contains(const T& e), and set(int index, const T& e) are
347   // left as an exercise
348
349   #endif
```

链表含有 `Node` 类中定义的节点（第 6～23 行）。可以获得遍历链表中元素的迭代器。`Iterator` 类（第 25～64 行）将在 20.5 节中讨论。

`LinkedList` 类的头在第 66～107 行中定义。无参数构造函数（第 109～114 行）构造了一个空链表，`head` 和 `tail` 为 `nullptr`，`size` 为 0。

复制构造函数（第 116～128 行）通过从现有列表复制内容来创建新的链表。这通过将现有链表中的元素插入到新链表中来完成（第 122～127 行）。

析构函数（第 130～134 行）通过调用 `clear` 函数（第 265～277 行）从链表中删除所有节点，`clear` 函数从列表中删除所有的节点（第 272 行）。

赋值运算符（=）（第 136～154 行）首先销毁本列表，然后创建要复制列表的副本。注意，复制构造函数、析构函数和赋值运算符需要同时实现。

函数 `addFirst(T element)`（第 156～166 行）、`addLast(T element)`（第 168～182 行）、`removeFirst()`（第 202～217 行）、`removeLast()`（第 219～238 行）、`add(T element)`（第 240～244 行）、`add(int index, T element)`（第 246～263 行）和 `removeAt(int index)`（第 319～343 行）的实现在 20.4.1～20.4.6 节中进行了讨论。

函数 `getFirst()` 和 `getLast()`（第 184～200 行）分别返回列表中的第一个和最后一个元素。

函数 `get(int index)` 返回指定索引处的元素（第 279～290 行）。

`lastIndexOf(T element)`、`remove(T element)`、`contains(T element)` 和 `set(int index, Object o)`（第 345～347 行）的实现被省略，留作练习。

20.4.8 `LinkedList` 的时间复杂度

表 20.1 总结了 `LinkedList` 函数的复杂度。

注意，实现 `LinkedList` 可以不使用 `size` 数据字段。但是，`size()` 函数将花费 $O(n)$ 时间。可以用数组、向量或链表来存储元素。如果事先不知道元素的数量，那么用向量或链表会更高效，因为它们可以动态地增长和收缩。如果应用程序需要频繁地在任意位置插入和删除元素，则用链表会更高效，因为将元素插入数组或向量需要移动插入点之后的所有元素。如果应用程序中的元素数量是固定的，并且应用程序不需要随机插入和删除元素，那么使用

表 20.1 `LinkedList` 中函数的时间复杂度

函数	时间
add(e: T)	$O(1)$
add(index: int, e: T)	$O(n)$
clear()	$O(n)$
contains(e: T)	$O(n)$
get(index: int)	$O(n)$
indexOf(e: T)	$O(n)$
isEmpty()	$O(1)$
lastIndexOf(e: T)	$O(n)$
remove(e: T)	$O(n)$
getSize()	$O(1)$
remove(index: int)	$O(n)$
set(index: int, e: T)	$O(n)$
addFirst(e: T)	$O(1)$
removeFirst()	$O(1)$
addLast(e: T)	$O(1)$
removeLast()	$O(n)$

数组简单且高效。

20.5 迭代器

要点提示：迭代器是为遍历各种类型容器中的元素提供统一方式的对象。

迭代器是 C++ 中的一个重要概念。标准模板库（STL）使用迭代器访问容器中的元素。STL 将在第 22 章和第 23 章中介绍。本节定义了一个迭代器类，并创建了一个用于遍历链表中元素的迭代器对象。这里的目标有两个：（1）考虑一个定义迭代器类的例子；（2）熟悉迭代器以及使用它们遍历容器中的元素。

迭代器可视为封装的指针。在链表中，可以用指针遍历列表。但是迭代器的函数比指针多。迭代器是对象。迭代器包含访问和操作元素的函数。迭代器通常包含重载运算符，如表 20.2 所示。

用于遍历链表中元素的迭代器 Iterator 类如图 20.12 所示。

表 20.2 迭代器中的典型重载运算符

运算符	描述
++	将迭代器前进到下一个元素
*	访问迭代器所指向的元素
==	测试两个迭代器是否指向同一个元素
!=	测试两个迭代器是否指向不同的元素

```
            Iterator<T>
-current: Node<T>*

+Iterator(p: Node<T>*)
+operator++(): Iterator<T>
+operator*(): T
+operator==(itr: Iterator<T>&): bool
+operator!=(itr: Iterator<T>&): bool
```

图 20.12 迭代器用函数封装指针

这个类在 LiveExample 20.2 中的第 25 ～ 64 行中实现。由于构造函数及其他函数都很短，所以它们被实现为内联函数。Iterator 类使用 current 数据字段指向正在遍历的节点（第 63 行）。构造函数（第 29 ～ 32 行）创建一个指向指定节点的迭代器。递增运算符的前置和后置形式（第 34 ～ 45 行）的实现移动 current 指向列表中的下一个节点（第 36、43 行）。注意，后置递增运算符使迭代器递增（第 43 行），但返回原始迭代器（第 42、44 行）。* 运算符返回迭代器指向的元素（第 49 行）。== 运算符测试当前迭代器是否与另一个迭代器相同（第 54 行）。

迭代器与实际存储元素的容器对象一起使用。容器类应该为返回迭代器提供 begin() 和 end() 函数，如表 20.3 所示。

表 20.3 返回迭代器的常用函数

运算符	描述
begin()	返回一个迭代器，该迭代器指向容器中的第一个元素
end()	返回一个迭代器，该迭代器表示容器中最后一个元素之后的位置。这个迭代器可以用来测试是否遍历了容器中的所有元素

为了从 LinkedList 中获得迭代器，在 LiveExample 20.2 的第 93 ～ 101 行中定义并实现了以下两个函数

```
Iterator<T> begin() const;
Iterator<T> end() const;
```

函数 begin() 返回列表中第一个元素的迭代器,函数 end() 返回表示列表中最后一个元素之后的位置的迭代器。

LiveExample 20.3 给出了一个使用迭代器遍历链表中的元素并以大写形式显示字符串的示例。该程序在第 17 行为字符串创建一个 LinkedList,将四个字符串添加到链表中(第 20 ~ 23 行),使用迭代器遍历链表中的所有元素并以大写字母显示(第 26 ~ 30 行)。

LiveExample 20.3 的互动程序请访问 https://liangcpp.pearsoncmg.com/LiveRunCpp5e/faces/LiveExample.xhtml?header=off&programName=TestIterator&fileType=.cpp&programHeight=580&resultHeight=160。

LiveExample 20.3　TestIterator.cpp

```cpp
#include <iostream>
#include <string>
#include "LinkedList.h"
using namespace std;

string toUpperCase(string s)
{
  for (int i = 0; i < s.length(); i++)
    s[i] = toupper(s[i]);

  return s;
}

int main()
{
  // Create a list for strings
  LinkedList<string> list;

  // Add elements to the list
  list.add("America");
  list.add("Canada");
  list.add("Russia");
  list.add("France");

  // Traverse a list using iterators
  for (Iterator<string> iterator = list.begin();
    iterator != list.end(); iterator++)
  {
    cout << toUpperCase(*iterator) << " ";
  }

  return 0;
}
```

```
command>cl TestIterator.cpp
Microsoft C++ Compiler 2019
Compiled successful (cl is the VC++ compile/link command)
```

```
command>TestIterator
AMERICA CANADA RUSSIA FRANCE

command>
```

注意：迭代器的功能类似于指针。它可以使用指针、数组索引或其他数据结构来实现。迭代器的抽象省去了实现的细节。第 22 章将介绍 STL 中的迭代器。STL 迭代器为访问容器中的元素提供了一个统一的接口，因此，就像访问链表中的元素一样，可以使用迭代器访问向量或集合中的元素。

迭代器可用来高效遍历容器中的元素。注意，printList 函数花费 $O(n^2)$ 时间，因为 get(index) 函数需要 $O(n)$ 时间获得列表中第 i 个索引处的元素。因此，应该使用迭代器重写 printList 函数以提高其效率。

```cpp
void printList(const LinkedList<string>& list)
{
  Iterator<string> current = list.begin();
  while (current != list.end())
  {
    cout << *current << " ";
    ++current;
  }
  cout << endl;
}
```

注意，Iterator 类派生自 std::iterator<std::forward_iterator_tag, T>。让 Iterator 成为 std::iterator<std::forward_iterator_tag, T> 的子类不是必需的，但这样做可以针对 LinkedList 元素调用 C++STL 库函数。这些函数使用迭代器遍历容器中的元素。LiveExample 20.4 给出了一个使用 C++STL 库函数 max_element 和 min_element 返回链表中的最大和最小元素的示例。

LiveExample 20.4 的互动程序请访问 https://liangcpp.pearsoncmg.com/LiveRunCpp5e/faces/LiveExample.xhtml?header=off&programName=TestSTLAlgorithm&fileType=.cpp&programHeight=440&resultHeight=170。

LiveExample 20.4　TestSTLAlgorithm.cpp

Source Code Editor:
```cpp
 1  #include <iostream>
 2  #include <algorithm>
 3  #include <string>
 4  #include "LinkedList.h"
 5  using namespace std;
 6
 7  int main()
 8  {
 9    // Create a list for strings
10    LinkedList<string> list;
11
12    // Add elements to the list
13    list.add("America");
14    list.add("Canada");
15    list.add("Russia");
16    list.add("France");
17
18    cout << "The max element in the list is: " <<
```

```
19        *max_element(list.begin(), list.end()) << endl;
20
21    cout << "The min element in array1: " <<
22        *min_element(list.begin(), list.end()) << endl;
23
24    return 0;
25 }
```

Execution Result:
```
command>cl TestSTLAlgorithm.cpp
Microsoft C++ Compiler 2019
Compiled successful (cl is the VC++ compile/link command)

command>TestSTLAlgorithm
The max element in the list is: Russia
The min element in array1: America

command>
```

`max_element` 和 `min_element` 函数在 11.8 节中介绍。这些函数在参数中使用指针。迭代器就像指针。方便起见，可以将迭代器作为指针来调用。`max_element(iterator1, iterator2)` 返回 `iterator1` 和 `iterator2 - 1` 之间最大元素的迭代器。

20.6 C++11 `foreach` 循环

要点提示：可以使用 `foreach` 循环遍历集合中的元素。

`foreach` 循环是一种常见的计算机语言特性，用于按顺序遍历集合中的元素。C++11 中支持此特性。我们已经用 `foreach` 循环遍历过数组或向量中的元素。例如，以下代码显示数组 `myList` 中的所有元素：

```cpp
double myList[] = {3.3, 4.5, 1};
for (double& e: myList)
{
    cout << e << endl;
}
```

可以将代码解读为"对于 `myList` 中的每个元素 e，执行以下操作"。注意，变量 e 必须声明为与 `myList` 中元素相同的类型。

事实上，可以在任何带有 `begin()` 和 `end()` 函数的集合上使用 `foreach`，这些函数返回 `std::iterator` 类型的开始迭代器和结束迭代器。`foreach` 循环的语法为

```cpp
for (elementType& element: collection)
{
    // Process the element
}
```

在 C++ 中，向量是一个带有迭代器的集合。因此，可以使用 `foreach` 循环遍历向量中的所有元素。例如，以下代码遍历字符串向量中的所有元素。

```cpp
vector<string> names;
names.push_back("Atlanta");
names.push_back("New York");
names.push_back("Kansas");
for (string& s: names)
```

```
{
  cout << s << endl;
}
```

由于列表是一个带有迭代器的集合，所以可以用 foreach 循环重写 LiveExample 20.3，如 LiveExample 20.5 所示。

LiveExample 20.5 的互动程序请访问 https://liangcpp.pearsoncmg.com/LiveRunCpp5e/faces/LiveExample.xhtml?header=off&programName=TestForeachLoop&fileType=.cpp&programHeight=560&resultHeight=160。

LiveExample 20.5　TestForeachLoop.cpp

```cpp
#include <iostream>
#include <string>
#include "LinkedList.h"
using namespace std;

string toUpperCase(string& s)
{
  for (int i = 0; i < s.length(); i++)
    s[i] = toupper(s[i]);

  return s;
}

int main()
{
  // Create a list for strings
  LinkedList<string> list;

  // Add elements to the list
  list.add("America");
  list.add("Canada");
  list.add("Russia");
  list.add("France");

  // Traverse a list using iterators
  for (string& s: list)
  {
    cout << toUpperCase(s) << " ";
  }

  return 0;
}
```

Execution Result:

```
command>cl TestForeachLoop.cpp
Microsoft C++ Compiler 2019
Compiled successful (cl is the VC++ compile/link command)

command>TestForeachLoop
AMERICA CANADA RUSSIA FRANCE

command>
```

程序创建一个列表（第17行），并将四个字符串添加到列表中（第20～23行）。foreach循环用于遍历列表中的字符串（第26～29行）。

20.7 链表的变体

要点提示：可以使用各种类型的链表组织某些应用程序的数据。

前面几节介绍的链表称为**单向链表**。它包含指向列表第一个节点的指针，按顺序每个节点包含指向下一个节点的指针。链表的几种变体在某些应用程序中很有用。

循环单向链表的不同之处在于，最后一个节点的指针指向第一个节点，如图 20.13a 所示。注意，循环链表不需要 tail。循环链表的一个很好的应用场景是在为多个用户提供服务的分时操作系统中。系统从循环列表中选择一个用户，并将少量的 CPU 时间授予该用户，然后转到列表中的下一个用户。

双向链表包含具有两个指针的节点。如图 20.13b 所示，一个指针指向下一个节点，另一个指针指向上一个节点。简单起见，这两个指针被称为前向指针和后向指针。因此，双向链表可以向前和向后遍历。

循环双向链表最后一个节点的前向指针指向第一个节点，第一个指针的后向指针指向最后一个节点，如图 20.13c 所示。

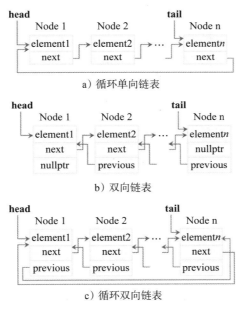

a) 循环单向链表

b) 双向链表

c) 循环双向链表

图 20.13 链表可能以各种形式出现

这些链表的实现留作练习。

注意：在单向链表中，removeLast() 花费 $O(n)$ 时间。在双向链表中，removeLast() 可以在 $O(1)$ 时间内实现。

20.8 队列

要点提示：队列是一种先进先出的数据结构。

队列表示一种等待列表。它可以被视为一种特殊类型的列表，其元素插入到末尾（尾部），并从开头（头部）访问和删除。

有两种方法可以设计队列类。

使用组合：可以将链表声明为队列类中的数据字段，如图 20.14a 所示。
使用继承：可以通过扩展链表类来定义队列类，如图 20.14b 所示。

图 20.14　队列可以使用组合或继承来实现

这两种设计都很好，但使用组合更好，因为它使你能够定义一个全新的队列类，而不会从链表中继承不必要和不合适的函数。图 20.15 显示了队列的 UML 类图。它的实现在 LiveExample 20.6 中给出。

```
            Queue<T>
-list: LinkedList<T>
+enqueue(element: const T&): void
+dequeue(): T
+getSize(): int const
```

图 20.15　Queue 用一个链表来提供先进先出的数据结构

LiveExample 20.6 的互动程序请访问 https://liangcpp.pearsoncmg.com/LiveRunCpp5e/faces/LiveExample.xhtml?header=off&programName=Queue&fileType=.h&programHeight=750&resultVisible=false。

LiveExample 20.6　Queue.h

Source Code Editor:

```cpp
#ifndef QUEUE_H
#define QUEUE_H
#include "LinkedList.h"
#include <stdexcept>
using namespace std;

template<typename T>
class Queue
{
public:
  Queue();
  void enqueue(const T& element);
  T dequeue();
  int getSize() const;

private:
  LinkedList<T> list;
};

template<typename T>
Queue<T>::Queue()
{
}

template<typename T>
void Queue<T>::enqueue(const T& element)
{
  list.addLast(element);
}
```

```cpp
31  template<typename T>
32  T Queue<T>::dequeue()
33  {
34    return list.removeFirst();
35  }
36
37  template<typename T>
38  int Queue<T>::getSize() const
39  {
40    return list.getSize();
41  }
42
43  #endif
```

第 17 行创建一个链表以存储队列中的元素。enqueue(T element) 函数（第 25～29 行）将元素添加到队列的尾部。dequeue() 函数（第 31～35 行）从队列的开头删除一个元素，并返回删除的元素。getSize() 函数（第 37～41 行）返回队列中的元素数量。

LiveExample 20.7 给出了一个用 Queue 类创建 int 值队列（第 17 行）和字符串队列（第 24 行）的示例。它用 enqueue 函数向队列添加元素（第 19、25～27 行），用 dequeue 函数从队列中删除 int 值和字符串。

LiveExample 20.7 的互动程序请访问 https://liangcpp.pearsoncmg.com/LiveRunCpp5e/faces/LiveExample.xhtml?header=off&programName=TestQueue&fileType=.cpp&programHeight=560&resultHeight=180。

LiveExample 20.7　TestQueue.cpp

```cpp
#include <iostream>
#include "Queue.h"
#include <string>
using namespace std;

template<typename T>
void printQueue(Queue<T>& queue)
{
  while (queue.getSize() > 0)
    cout << queue.dequeue() << " ";
  cout << endl;
}

int main()
{
  // Queue of int values
  Queue<int> intQueue;
  for (int i = 0; i < 10; i++)
    intQueue.enqueue(i); // Add i to the queue

  printQueue(intQueue);

  // Queue of strings
  Queue<string> stringQueue;
  stringQueue.enqueue("New York");
  stringQueue.enqueue("Boston");
  stringQueue.enqueue("Denver");

  printQueue(stringQueue);
```

```
30
31    return 0;
32 }
```

Execution Result:
```
command>cl TestQueue.cpp
Microsoft C++ Compiler 2019
Compiled successful (cl is the VC++ compile/link command)

command>TestQueue
0 1 2 3 4 5 6 7 8 9
New York Boston Denver

command>
```

20.9 优先级队列

要点提示：优先级队列中的元素被分配了优先级。具有最高优先级的元素先从优先级队列中删除。

常规队列是一种先进先出的数据结构。元素被追加到队列的末尾，并从开头处删除。在优先级队列中，为元素分配了优先级。首先访问或删除具有最高优先级的元素。例如，医院的急诊室为患者分配优先号码，首先治疗具有最高优先级的患者。

优先级队列可以使用堆来实现，其中根是队列中优先级最高的元素。堆在 19.6 节中介绍。优先级队列的类图如图 20.16 所示。它的实现在 LiveExample 20.8 中给出。

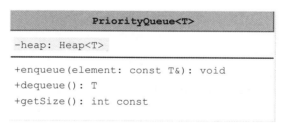

图 20.16 优先级队列使用堆来提供最高优先级元素先进出的数据结构

LiveExample 20.8 的互动程序请访问 https://liangcpp.pearsoncmg.com/LiveRunCpp5e/faces/LiveExample.xhtml?header=off&programName=PriorityQueue&fileType=.h&programHeight=710&resultVisible=false。

LiveExample 20.8　PriorityQueue.h

Source Code Editor:
```
1  #ifndef PRIORITYQUEUE_H
2  #define PRIORITYQUEUE_H
3  #include "Heap.h"
4
5  template<typename T>
6  class PriorityQueue
7  {
8  public:
```

```
 9      PriorityQueue();
10      void enqueue(const T& element);
11      T dequeue();
12      int getSize() const;
13
14    private:
15      Heap<T> heap;
16    };
17
18    template<typename T>
19    PriorityQueue<T>::PriorityQueue()
20    {
21    }
22
23    template<typename T>
24    void PriorityQueue<T>::enqueue(const T& element)
25    {
26      heap.add(element);
27    }
28
29    template<typename T>
30    T PriorityQueue<T>::dequeue()
31    {
32      return heap.remove();
33    }
34
35    template<typename T>
36    int PriorityQueue<T>::getSize() const
37    {
38      return heap.getSize();
39    }
40
41    #endif
```

Answer　Reset

LiveExample 20.9 给出了患者优先级队列的示例。第 6～38 行定义了 Patient 类。第 43 行创建一个优先级队列。第 44～47 行创建了四个具有相关优先级值的患者并将其入队。第 51 行将患者从队列中移出。

LiveExample 20.9 的互动程序请访问 https://liangcpp.pearsoncmg.com/LiveRunCpp5e/faces/LiveExample.xhtml?header=off&programName=TestPriorityQueue&fileType=.cpp&programHeight=970&resultHeight=160。

LiveExample 20.9　TestPriorityQueue.cpp

Source Code Editor:
```
 1    #include <iostream>
 2    #include "PriorityQueue.h"
 3    #include <string>
 4    using namespace std;
 5
 6    class Patient
 7    {
 8    public:
 9      Patient(const string& name, int priority)
10      {
11        this->name = name;
12        this->priority = priority;
13      }
14
```

```cpp
15      bool operator<(const Patient& secondPatient)
16      {
17          return (this->priority < secondPatient.priority);
18      }
19
20      bool operator>(const Patient& secondPatient)
21      {
22          return (this->priority > secondPatient.priority);
23      }
24
25      string getName() const
26      {
27          return name;
28      }
29
30      int getPriority() const
31      {
32          return priority;
33      }
34
35  private:
36      string name;
37      int priority;
38  };
39
40  int main()
41  {
42      // Queue of patients
43      PriorityQueue<Patient> patientQueue;
44      patientQueue.enqueue(Patient("John", 2));
45      patientQueue.enqueue(Patient("Jim", 1));
46      patientQueue.enqueue(Patient("Tim", 5));
47      patientQueue.enqueue(Patient("Cindy", 7));
48
49      while (patientQueue.getSize() > 0)
50      {
51          Patient element = patientQueue.dequeue(); // Get a patient from the queue
52          cout << element.getName() << " (priority: " <<
53              element.getPriority() << ") ";
54      }
55
56      return 0;
57  }
```

Automatic Check | Compile/Run | Reset | Answer   Choose a Compiler: VC++

Execution Result:

```
command>cl TestPriorityQueue.cpp
Microsoft C++ Compiler 2019
Compiled successful (cl is the VC++ compile/link command)

command>TestPriorityQueue
Cindy (priority: 7) Tim (priority: 5) John (priority: 2) Jim (priority: 1)

command>
```

在 Patient 类中定义了运算符 < 和 >，因此可以比较两个患者。堆中的元素可以是任意类类型，只要该类型元素可以用 < 和 > 运算符进行比较。

关键术语

circular doubly linked list（循环双向链表）
circular singly linked list（循环单向链表）
dequeue（出队）
doubly linked list（双向链表）
enqueue（入队）

linked list（链表）
priority queue（优先级队列）
queue（队列）
singly linked list（单向链表）

章节总结

1. 链表能够动态地增长和收缩。链表中的节点使用 new 运算符动态创建，并使用 delete 运算符销毁。
2. 队列表示一种等待列表。它可以被视为一种特殊类型的列表，其元素插入到末尾（尾部），并从开头（头部）访问和删除。
3. 如果事先不知道元素的数量，那么使用链表会更高效，它可以动态地增长和收缩。
4. 如果你的应用程序需要频繁地在任意位置插入和删除元素，则使用链表存储元素会更高效，因为将元素插入数组需要移动数组中插入点之后的所有元素。
5. 如果需要以先进先出的方式处理元素，则使用队列来存储元素。优先级队列可以用堆来实现，其中根是队列中优先级最高的元素。

编程练习

20.2～20.4 节

*20.1 （实现 remove(T element)）在 LiveExample 20.2 中省略了 remove(T element) 的实现。请实现它。用 https://liangcpp.pearsoncmg.com/test/Exercise20_01.txt 中的代码完成你的程序。

*20.2 （实现 lastIndexOf(T element)）在 LiveExample 20.2 中省略了 lastIndexOf(T element) 的实现。请实现它。

*20.3 （实现 contains(T element)）在 LiveExample 20.2 中省略了 contains(T element) 的实现。请实现它。

*20.4 （实现 set(int index, T element)）在 LiveExample 20.2 中省略了 set(int index, T element) 的实现。请实现它。

*20.5 （实现 reverse() 函数）在 LinkedList 类中添加一个名为 reverse() 的新成员函数，用于反转列表中的节点。

*20.6 （实现 sort() 函数）在 LinkedList 类中添加一个名为 sort() 的新成员函数，用于重新排列列表，对节点中的元素进行排序。

20.7 （在 LinkedList 中添加集合型操作）在 LinkedList 中添加并实现以下函数：

```
// Add the elements in otherList to this list.
void addAll(const LinkedList<T>& otherList)

// Remove all the elements in otherList from this list
void removeAll(const LinkedList<T>& otherList)

// Retain the elements in this list if they are also in otherList
void retainAll(const LinkedList<T>& otherList)
```

添加三个函数运算符 +、- 和 ^，用于集合的并、差和交。重载 = 运算符以实现列表的深层复制。添加用于访问/修改元素的 [] 运算符。用模板 https://liangcpp.pearsoncmg.com/test/Exercise20_07.txt 中的代码完成程序。

20.6 节

*20.8 (创建一个双向链表) 正文中的 LinkedList 类是一个单向链表,它支持对列表进行单向遍历。修改 Node 类添加新的字段名 previous 以指向列表中前面的节点,如下所示:

```
template<typename T>
class Node
{
public:
  T element;  // Element contained in the node
  Node<T>* previous; // Pointer to the previous node
  Node<T>* next; // Pointer to the next node
  Node() // No-arg constructor
  {
    previous = nullptr;
    next = nullptr;
  }
  Node(T element) // Constructor
  {
    this->element = element;
    previous = nullptr;
    next = nullptr;
  }
};
```

利用双向链表简化函数 add(int index, T element) 和 removeAt(int index) 的实现。修改 Iterator 类实现递减运算符 --,该运算符移动迭代器以指向前一个元素。

20.7 ～ 20.8 节

20.9 (使用继承实现 Stack) 在 LiveExample 12.7 中,Stack 是用组合实现的。创建一个扩展 LinkedList 的新栈类。

20.10 (使用继承实现 Queue) 在 LiveExample 20.6 中,Queue 是用组合实现的。创建一个扩展 LinkedList 的新队列类。

20.11 (实现 ArrayList) 本书介绍了链表,并给出了链表的实现方法。请你实现一个 ArrayList 来执行列表操作。ArrayList 中的元素存储在一个动态创建的数组中。如果 ArrayList 中元素的实际大小超过了数组容量,则创建一个容量是当前数组两倍的新数组,并将当前数组的内容复制到此新数组中。ArrayList 类的 UML 图如图 20.17 所示。

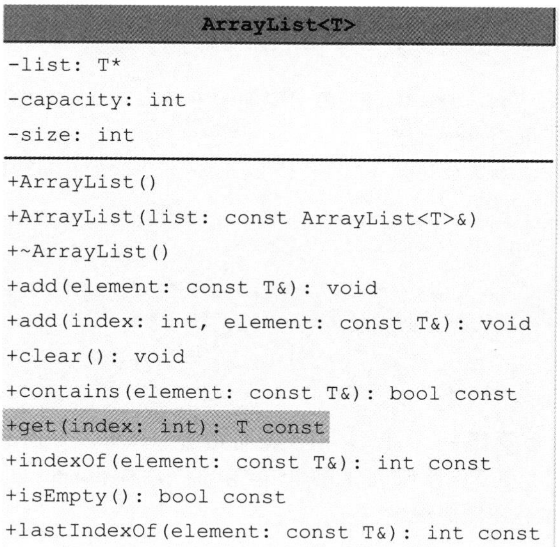

```
              ArrayList<T>
-list: T*
-capacity: int
-size: int
─────────────────────────────────────
+ArrayList()
+ArrayList(list: const ArrayList<T>&)
+~ArrayList()
+add(element: const T&): void
+add(index: int, element: const T&): void
+clear(): void
+contains(element: const T&): bool const
+get(index: int): T const
+indexOf(element: const T&): int const
+isEmpty(): bool const
+lastIndexOf(element: const T&): int const
```

图 20.17 ArrayList 使用数组实现列表

```
+remove(element: const T&): void
+getSize(): int const
+removeAt(index: int): T
+set(index: int, element: const T&): T
+begin(): Iterator<T>
+end(): Iterator<T>
Implement the [] operator for
accessing an element in the list
```

图 20.17 ArrayList 使用数组实现列表（续）

用 https://liangcpp.pearsoncmg.com/test/Exercise20_11.txt 中的代码测试 ArrayList。

第 21 章

Introduction to C++ Programming and Data Structures, Fifth Edition

二叉查找树

学习目标

1. 熟悉二叉查找树（21.2 节）。
2. 使用链接数据结构表示二叉树（21.3 节）。
3. 访问二叉查找树中的节点（21.4 节）。
4. 在二叉查找树中查找元素（21.5 节）。
5. 在二叉查找树中插入一个元素（21.6 节）。
6. 中序、后序和前序遍历二叉树（21.7 节）。
7. 定义并实现 BST 类（21.8 节）。
8. 删除二叉查找树中的节点（21.9 节）。
9. 设计和实现遍历二叉树元素的迭代器（21.10 节）。
10. 用二叉树实现霍夫曼编码以压缩数据（21.11 节）。

21.1 简介

要点提示：对于搜索、插入和删除操作，二叉查找树比列表更高效。

上一章给出了链表的实现。在这些数据结构中搜索、插入和删除操作的时间复杂度是 $O(n)$。本章提出一种新的数据结构，称为二叉查找树，它平均需要 $O(\log n)$ 的时间来搜索、插入和删除元素。

21.2 二叉查找树基础知识

要点提示：二叉查找树可以用链接结构来实现。

回想一下，列表、栈和队列是由一系列元素组成的线性结构。**二叉树**是一种层次结构。它要么是空的，要么由一个称为根的元素和两个不同的二叉树（称为左子树和右子树）组成，这两个子树中的一个或两个都可能是空的，如图 21.1a 所示。二叉树的示例如图 21.1b ～ c 所示。

路径的长度是路径中边的数量。节点的深度是从根到节点的路径长度。给定深度的所有节点的集合有时被称为树的一层。兄弟节点是共享同一父节点的节点。一个节点的左（右）子树的根称为该节点的左（右）子节点。没有子节点的节点称为叶子。一棵空树的高度是 −1。非空树的高度是从根节点到其最远叶子的路径长度。考虑图 21.1b 中的树，从节点 60 到 45 的路径的长度为 2。节点 60 的深度为 0，节点 55 的深度为 1，节点 45 的深度为 2。这棵树的高度是 2。节点 45 和 57 是同层节点。节点 45、57、67 和 107 处于同一层。

一种称为**二叉查找树**（BST）的特殊类型的二叉树是一种很有用的二叉树。BST（没有重复元素）具有这样的一个属性，即对于树中的每个节点，其左子树中任何节点的值都小于该节点的值，其右子树中任何节点的值都大于该节点的值。图 21.1 中的二叉树都是 BST。本节重点关注 BST。

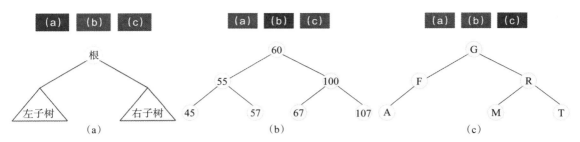

图 21.1 二叉树的每个节点都有零、一或两个子树

21.3 表示二叉查找树

要点提示：二叉树可以使用链接结构来实现。

二叉树可以使用一组链接节点来表示。每个节点包含一个值和两个名为 `left` 和 `right` 的链接，这两个链接分别指向左子节点和右子节点，如图 21.2 所示。

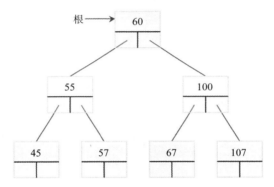

图 21.2 二叉树可以使用一组链接节点来表示

节点可以如下所示定义为一个类：

```cpp
template<typename T>
class TreeNode
{
public:
  T element; // Element contained in the node
  TreeNode<T>* left; // Pointer to the left child
  TreeNode<T>* right; // Pointer to the right child
  TreeNode(T element) // Constructor
  {
    this->element = element;
    left = nullptr;
    right = nullptr;
  }
};
```

我们用变量 `root` 指树的根节点。如果树为空，则根为 `nullptr`。以下代码创建了图 21.1b 中树的前三个节点：

```cpp
// Create the root node
TreeNode<int>* root = new TreeNode<int>(60);
// Create the left child node
root->left = new TreeNode<int>(55);
// Create the right child node
root->right = new TreeNode<int>(100);
```

21.4 访问二叉查找树中的节点

要点提示：可以从根开始访问二叉查找树。

假设创建了一个具有三个节点的树，如前一节所述。可以通过 root 指针访问树中的节点。以下是显示根及其左右节点处的元素的语句。

```cpp
// Display the root element
cout << root->element << endl;
// Display the element in the left child of the root
cout << (root->left)->element << endl;
// Display the element in the right child of the root
cout << (root->right)->element << endl;
```

21.5 查找元素

要点提示：BST 实现了类似于二分查找的高效查找。

在 BST 中查找元素，要从根开始，从根向下扫描，直到找到匹配项或到达空子树。该算法如 Listing 21.1 所示。让 current 指向根（第 4 行）。重复以下步骤，直到 current 为 nullptr（第 6 行）或元素匹配 current->element（第 16 行）。

- 如果 element 小于 current->element，则将 current->left 赋值给 current（第 9 行）。
- 如果 element 大于 current->element，则将 current->right 赋值给 current（第 13 行）。
- 如果 element 等于 current->element，则返回 true（第 16 行）。

如果 current 为 nullptr，则子树为空，元素不在树中（第 18 行）。

Listing 21.1 在 BST 中查找元素

```cpp
1   template<typename T>
2   bool search(T element)
3   {
4     TreeNode<T>* current = root; // Start from the root
5   
6     while (current != nullptr)
7       if (element < current->element)
8       {
9         current = current->left; // Go left
10      }
11      else if (element > current->element)
12      {
13        current = current->right; // Go right
14      }
15      else // Element matches current->element
16        return true; // Element is found
17  
18    return false; // Element is not in the tree
19  }
```

21.6 将元素插入二叉查找树

要点提示：新元素作为叶子节点插入。

要在 BST 中插入元素，需要找到插入位置。关键思想是找到新节点的父节点。Listing 21.2 给出了算法。

Listing 21.2 在 BST 中插入元素

```cpp
template<typename T>
bool insert(T element)
{
  if (root == nullptr)
    root = new TreeNode<T>(element);
  else
  {
    // Locate the parent node
    current = root;
    while (current != nullptr)
      if (element < current->element)
      {
        parent = current;
        current = current->left;
      }
      else if (element > current->element)
      {
        parent = current;
        current = current->right;
      }
      else
        return false; // Duplicate node not inserted

    // Create the new node and attach it to the parent node
    if (element < parent->element)
      parent->left = new TreeNode<T>(element);
    else
      parent->right = new TreeNode<T>(element);

    return true; // Element inserted
  }
}
```

如果树为空，则用新元素创建一个根节点（第 4 ~ 5 行）。否则，定位新元素节点的父节点（第 9 ~ 22 行）。为元素创建一个新节点，并将该节点链接到其父节点（第 25 ~ 28 行）。如果新元素小于父元素，则新元素的节点将是父元素的左子节点（第 26 行）。如果新元素大于父元素，则新元素的节点是父元素的右子节点（第 28 行）。

例如，要将 101 插入图 21.2 的树中，算法中的 while 循环完成后，parent 指向 107 的节点，如图 21.3a 所示。101 的新节点成为 parent 节点的左子节点。要将 59 插入到树中，算法中的 while 完成后，parent 节点指向 57 的节点，如图 21.3b 所示。59 的新节点成为 parent 节点的右子节点。

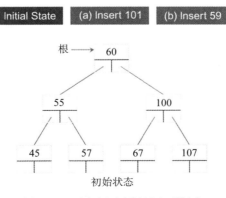

图 21.3 两个新元素被插入到树中

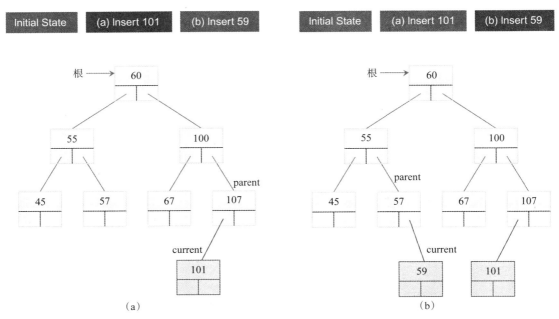

图 21.3 两个新元素被插入到树中（续）

21.7 树的遍历

要点提示：中序、后序、前序、深度优先和广度优先是遍历二叉树中元素的常用方法。

树的遍历是访问树中每个节点一次的过程。有几种方法可以遍历一棵树。本节介绍**中序遍历**、**前序遍历**、**后序遍历**、深度优先遍历和广度优先遍历。

中序遍历（中根遍历），首先递归地访问当前节点的左子树，然后递归地访问当前节点，最后递归地访问当前节点的右子树。中序遍历按递增顺序显示 BST 中的所有节点，如图 21.4a 所示。

后序遍历（后根遍历），首先递归地访问当前节点的左子树，然后递归地访问当前节点的右子树，最后访问当前节点本身，如图 21.4b 所示。

前序遍历（先根遍历），首先递归地访问当前节点，然后递归地访问当前节点的左子树，最后递归地访问当前节点的右子树。深度优先遍历与前序遍历相同，如图 21.4c 所示。

注意：可以按照前序遍历的方式插入元素来重建二叉查找树。重建的树保留了原始二叉查找树节点的父子关系。

深度优先遍历先是访问根，然后以任意顺序递归访问其左子树和右子树。前序遍历可以看作深度优先遍历的一个特例，它递归地访问它的左子树，然后访问它的右子树，如图 21.4d 所示。广度优先遍历，节点被逐层访问。首先访问根，然后从左到右访问根的所有子节点，然后从左到右访问根的孙节点，以此类推，如图 21.4e 所示。

可以用以下图 21.5 中的简单树来帮助记忆前序、后序和中序。

中序为 1 + 2，后序为 1 2 +，前序为 + 1 2（译者注：记住中根、后根和先根更容易些）。

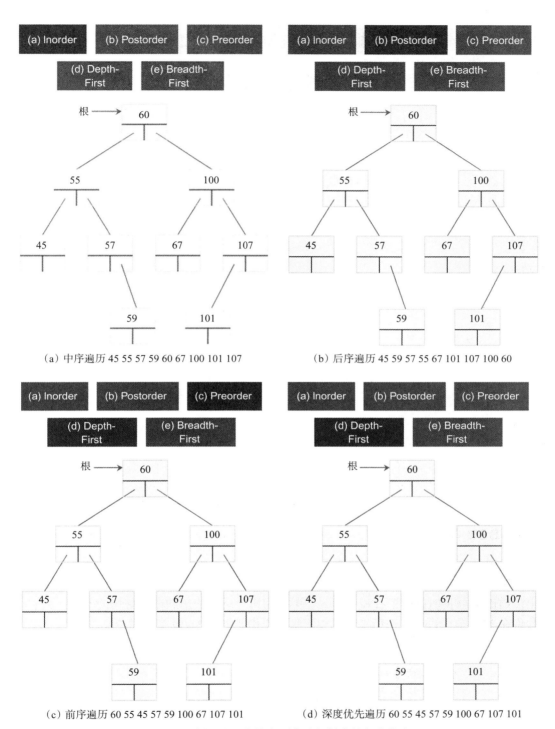

图 21.4 树的遍历按特定顺序访问树中的每个节点

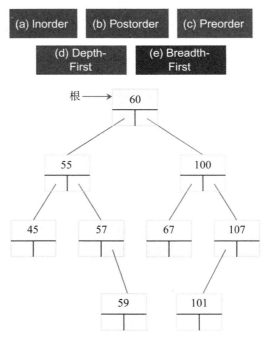

（e）广度优先遍历 60 55 100 45 57 67 107 59 101

图 21.4　树的遍历按特定顺序访问树中的每个节点（续）

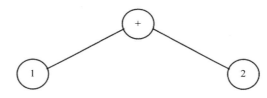

图 21.5　一个简单的树，用于记忆前序、后序和中序

21.8　BST 类

要点提示：BST 类定义了在二叉查找树中存储和操作数据的数据结构。

让我们定义一个名为 BST 的二叉查找树类，如图 21.6 所示。它的实现在 LiveExample 21.1 中给出。前面部分我们讨论了查找、插入、中序遍历、后序遍历和前序遍历的实现。remove 函数将在 21.9 节中讨论。遍历二叉树中元素的迭代器将在 21.10 节中讨论。

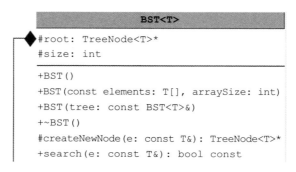

图 21.6　BST 类支持二叉查找树的许多操作

```
+insert(e: const T&): bool
+remove(e: const T&): bool
+inorder(): void const
+preorder(): void const
+postorder(): void const
+getSize(): int const
+clear(): void
+path(e: const T&): vector<TreeNode<T>*>*
+begin(): Iterator<T> const
+end(): Iterator<T> const
```

TreeNode\<T\>
element: T
left: TreeNode\<T\>*
right: TreeNode\<T\>*
TreeNode(e: const T&)

图 21.6 BST 类支持二叉查找树的许多操作（续）

LiveExample 21.1 的互动程序请访问 https://liangcpp.pearsoncmg.com/LiveRunCpp5e/faces/LiveExample.xhtml?header=off&programName=BST&fileType=.h&programHeight=6510&resultVisible=false。

LiveExample 21.1　BST.h

Source Code Editor:

```cpp
#ifndef BST_H
#define BST_H

#include <vector>
#include <stdexcept>
using namespace std;

template<typename T>
class TreeNode
{
public:
  T element; // Element contained in the node
  TreeNode<T>* left; // Pointer to the left child
  TreeNode<T>* right; // Pointer to the right child

  TreeNode(const T& e) // Constructor
  {
    this->element = e;
    left = nullptr;
    right = nullptr;
  }
};

template <typename T>
class Iterator: public std::iterator<std::forward_iterator_tag, T>
{
public:
  Iterator(TreeNode<T>* p)
  {
    if (p == nullptr)
      current = -1; // The end
    else
```

```cpp
       {
         // Get all the elements in inorder
         treeToVector(p);
         current = 0;
       }
     }

     Iterator operator++()
     {
       current++;
       if (current == v.size())
         current = -1; // The end
       return *this;
     }

     T &operator*()
     {
       return v[current];
     }

     bool operator==(const Iterator<T>& iterator) const
     {
       return current == iterator.current;
     }

     bool operator!=(const Iterator<T>& iterator) const
     {
       return current != iterator.current;
     }
   private:
     int current;
     vector<T> v;
     void treeToVector(const TreeNode<T>* p)
     {
       if (p != nullptr)
       {
         treeToVector(p->left);
         v.push_back(p->element);
         treeToVector(p->right);
       }
     }
   };

   template <typename T>
   class BST
   {
   public:
     BST(); // No-arg constructor
     BST(const T elements[], int arraySize);
     BST(const BST<T>& tree); // Copy constructor
     ~BST(); // Destructor
     bool search(const T& e) const;
     virtual bool insert(const T& e);
     virtual bool remove(const T& e);
     void inorder() const;
     void preorder() const;
     void postorder() const;
     int getSize() const;
     void clear();
     vector<TreeNode<T>*>* path(const T& e) const;

     Iterator<T> begin() const
     {
```

```cpp
 97        return Iterator<T>(root);
 98      };
 99
100      Iterator<T> end() const
101      {
102        return Iterator<T>(nullptr);
103      };
104
105    protected:
106      TreeNode<T>* root;
107      int size;
108      virtual TreeNode<T>* createNewNode(const T& e);
109
110    private:
111      void inorder(const TreeNode<T>* root) const;
112      void postorder(const TreeNode<T>* root) const;
113      void preorder(const TreeNode<T>* root) const;
114      void copy(const TreeNode<T>* root);
115      void clear(const TreeNode<T>* root);
116    };
117
118    template <typename T>
119    BST<T>::BST()
120    {
121      root = nullptr;
122      size = 0;
123    }
124
125    template <typename T>
126    BST<T>::BST(const T elements[], int arraySize)
127    {
128      root = nullptr;
129      size = 0;
130
131      for (int i = 0; i < arraySize; i++)
132      {
133        insert(elements[i]);
134      }
135    }
136
137    /* Copy constructor */
138    template <typename T>
139    BST<T>::BST(const BST<T>& tree)
140    {
141      root = nullptr;
142      size = 0;
143      copy(tree.root); // Recursively copy nodes to this tree
144    }
145
146    /* Copies the element from the specified tree to this tree */
147    template <typename T>
148    void BST<T>::copy(const TreeNode<T>* root)
149    {
150      if (root != nullptr)
151      {
152        insert(root->element);
153        copy(root->left);
154        copy(root->right);
155      }
156    }
157
158    /* Destructor */
159    template <typename T>
160    BST<T>::~BST()
```

```cpp
{
  clear();
}

/* Return true if the element is in the tree */
template <typename T>
bool BST<T>::search(const T& e) const
{
  TreeNode<T>* current = root; // Start from the root

  while (current != nullptr)
    if (e < current->element)
    {
      current = current->left; // Go Left
    }
    else if (e > current->element)
    {
      current = current->right; // Go right
    }
    else // Element e matches current.element
      return true; // Element e is found

  return false; // Element e is not in the tree
}

template <typename T>
TreeNode<T>* BST<T>::createNewNode(const T& e)
{
  return new TreeNode<T>(e);
}

// Insert element e into the binary tree
// Return true if the element is inserted successfully
// Return false if the element is already in the list
template <typename T>
bool BST<T>::insert(const T& e)
{
  if (root == nullptr)
    root = createNewNode(e); // Create a new root
  else
  {
    // Locate the parent node
    TreeNode<T>* parent = nullptr;
    TreeNode<T>* current = root;
    while (current != nullptr)
      if (e < current->element)
      {
        parent = current;
        current = current->left;
      }
      else if (e > current->element)
      {
        parent = current;
        current = current->right;
      }
      else
        return false; // Duplicate node not inserted

    // Create the new node and attach it to the parent node
    if (e < parent->element)
      parent->left = createNewNode(e);
    else
      parent->right = createNewNode(e);
  }
```

```cpp
225
226      size++;
227      return true; // Element inserted
228    }
229
230    /* Inorder traversal */
231    template <typename T>
232    void BST<T>::inorder() const
233    {
234      inorder(root);
235    }
236
237    /* Inorder traversal from a subtree */
238    template <typename T>
239    void BST<T>::inorder(const TreeNode<T>* root) const
240    {
241      if (root == nullptr) return;
242      inorder(root->left);
243      cout << root->element << " ";
244      inorder(root->right);
245    }
246
247    /* Postorder traversal */
248    template <typename T>
249    void BST<T>::postorder() const
250    {
251      postorder(root);
252    }
253
254    /** Inorder traversal from a subtree */
255    template <typename T>
256    void BST<T>::postorder(const TreeNode<T>* root) const
257    {
258      if (root == nullptr) return;
259      postorder(root->left);
260      postorder(root->right);
261      cout << root->element << " ";
262    }
263
264    /* Preorder traversal */
265    template <typename T>
266    void BST<T>::preorder() const
267    {
268      preorder(root);
269    }
270
271    /* Preorder traversal from a subtree */
272    template <typename T>
273    void BST<T>::preorder(const TreeNode<T>* root) const
274    {
275      if (root == nullptr) return;
276      cout << root->element << " ";
277      preorder(root->left);
278      preorder(root->right);
279    }
280
281    /* Get the number of nodes in the tree */
282    template <typename T>
283    int BST<T>::getSize() const
284    {
285      return size;
286    }
287
288    /* Remove all nodes from the tree */
```

```cpp
289    template <typename T>
290    void BST<T>::clear()
291    {
292      // Left as exercise
293    }
294
295    /* Return a path from the root leading to the specified element */
296    template <typename T>
297    vector<TreeNode<T>*>* BST<T>::path(const T& e) const
298    {
299      vector<TreeNode<T>*>* v = new vector<TreeNode<T>*>();
300      TreeNode<T>* current = root;
301
302      while (current != nullptr)
303      {
304        v->push_back(current);
305        if (e < current->element)
306          current = current->left;
307        else if (e > current->element)
308          current = current->right;
309        else
310          break;
311      }
312
313      return v;
314    }
315
316    /* Delete an element e from the binary tree.
317     * Return true if the element is deleted successfully
318     * Return false if the element is not in the tree */
319    template <typename T>
320    bool BST<T>::remove(const T& e)
321    {
322      // Locate the node to be deleted and also locate its parent node
323      TreeNode<T>* parent = nullptr;
324      TreeNode<T>* current = root;
325      while (current != nullptr)
326      {
327        if (e < current->element)
328        {
329          parent = current;
330          current = current->left;
331        }
332        else if (e > current->element)
333        {
334          parent = current;
335          current = current->right;
336        }
337        else
338          break; // Element e is in the tree pointed by current
339      }
340
341      if (current == nullptr)
342        return false; // Element e is not in the tree
343
344      // Case 1: current has no left children
345      if (current->left == nullptr)
346      {
347        // Connect the parent with the right child of the current node
348        if (parent == nullptr)
349        {
350          root = current->right;
351        }
```

```cpp
352      else
353      {
354        if (e < parent->element)
355          parent->left = current->right;
356        else
357          parent->right = current->right;
358      }
359
360      delete current; // Delete current
361    }
362    else
363    {
364      // Case 2: The current node has a left child
365      // Locate the rightmost node in the left subtree of
366      // the current node and also its parent
367      TreeNode<T>* parentOfRightMost = current;
368      TreeNode<T>* rightMost = current->left;
369
370      while (rightMost->right != nullptr)
371      {
372        parentOfRightMost = rightMost;
373        rightMost = rightMost->right; // Keep going to the right
374      }
375
376      // Replace the element in current by the element in rightMost
377      current->element = rightMost->element;
378
379      // Eliminate rightmost node
380      if (parentOfRightMost->right == rightMost)
381        parentOfRightMost->right = rightMost->left;
382      else
383        // Special case: parentOfRightMost->right == current
384        parentOfRightMost->left = rightMost->left;
385
386      delete rightMost; // Delete rightMost
387    }
388
389    size--;
390    return true; // Element inserted
391  }
392
393  #endif
```

Answer   Reset

　　二叉树包含 TreeNode 类中定义的节点（第 8～22 行）。可以获得遍历二叉树中元素的迭代器。Iterator 类（第 24～75 行）将在 21.10 节中讨论。

　　BST 类的头在第 77～116 行中定义。第 110～115 行定义了五个私有函数。这些是辅助函数，仅用于实现公共函数。数据字段 root 和 size（第 106～107 行）被声明为受保护的，因此可以从子类直接访问它们。后面在第 25 章中，我们将定义 AVLTree 类，它是从 BST 派生的。createNewNode 函数创建一个新节点（第 108 行）。该函数被定义为虚拟的，因为它将在 AVLTree 类中重新定义，以创建一种新类型的节点。insert 和 remove 函数（第 86～87 行）也被定义为虚拟函数，因为它们将在 AVLTree 类中被重新定义，因此可以为不同类型的树动态调用它们。

设计模式说明：工厂函数模式

　　函数 createNewNode() 的设计应用了工厂函数模式，该模式创建从函数返回的对象，而不是直接使用代码中的构造函数来创建对象。假设工厂函数返回一个类型为 A 的对象。此设计使你能够重载该函数以创建 A 的子类型的对象。BST 类中的 `createNewNode()` 函数返回 TreeNode 对象。在后面章节中，我们将重载此函数以返回 TreeNode 子类型的对象。

无参数构造函数（第 118 ～ 123 行）构造一个 `root` 为 `nullptr`、`size` 为 0 的空二叉树。构造函数（第 125 ～ 135 行）构造一个用数组元素初始化的二叉树。

复制构造函数（第 138 ～ 144 行）通过复制现有树中的内容来创建一个新的二叉树。这是通过使用复制函数将元素从现有树递归插入到新树来完成的（第 147 ～ 156 行）。

析构函数（第 159 ～ 163 行）从树中删除所有节点。

函数 `search(T element)`（第 166 ～ 184 行）搜索 BST 中的元素。如果找到元素，则返回 `true`（第 181 行），否则返回 `false`（第 183 行）。

函数 `insert(T element)`（第 195 ～ 228 行）为元素创建一个节点，并将其插入到树中。如果该树为空，则该节点成为根（第 199 行）。否则，函数会为节点找到一个合适的父节点来维护树的顺序。如果元素已经在树中，则函数返回 `false`（第 217 行）；否则返回 `true`（第 227 行）。

函数 `inorder( )`（第 231 ～ 235 行）调用 `inorder(root)` 遍历整个树。函数 `inorder(TreeNode root)` 遍历具有指定根的树。这是一个递归函数。它递归遍历左子树，然后遍历根，最后遍历右子树。当树为空时，遍历结束。

函数 `postorder()`（第 248 ～ 252 行）和函数 `preorder()`（第 265 ～ 269 行）类似，用递归实现。

函数 `path(T element)`（第 296 ～ 314 行）查找从根到包含该元素的节点或将插入元素的父节点的路径。

LiveExample 21.2 给出了一个使用 BST 中某些函数的示例。

LiveExample 21.2 的互动程序请访问 https://liangcpp.pearsoncmg.com/LiveRunCpp5e/faces/LiveExample.xhtml?header=off&programName=TestBST&fileType=.cpp&programHeight=730&resultHeight=280。

LiveExample 21.2　TestBST.cpp

Source Code Editor:

```cpp
#include <iostream>
#include <vector>
#include <string>
#include "BST.h"
using namespace std;

int main()
{
  BST<string> tree;
  tree.insert("George");
  tree.insert("Michael");
  tree.insert("Tom");
  tree.insert("Adam");
  tree.insert("Jones");
  tree.insert("Peter"); // Insert Peter
  tree.insert("Daniel");

```

```cpp
18      cout << "Inorder (sorted): ";
19      tree.inorder();
20
21      cout << "\nPostorder: ";
22      tree.postorder();
23
24      cout << "\nPreorder: ";
25      tree.preorder();
26
27      cout << "\nThe number of nodes is " << tree.getSize() << endl;
28      cout << "search(\"Jones\") " << tree.search("Jones") << endl;
29      cout << "search(\"John\") " << tree.search("John") << endl;
30
31      cout << "A path from the root to Peter is: ";
32      vector<TreeNode<string>*>* v = tree.path("Peter");
33      for (unsigned i = 0; i < (*v).size(); i++)
34        cout << (*v)[i]->element << " ";
35
36      int numbers[] = {2, 4, 3, 1, 8, 5, 6, 7};
37      BST<int> intTree(numbers, 8);
38      cout << "\nInorder (sorted): ";
39      intTree.inorder();
40
41      return 0;
42    }
```

Execution Result:

```
command>cl TestBST.cpp
Microsoft C++ Compiler 2019
Compiled successful (cl is the VC++ compile/link command)

command>TestBST
Inorder (sorted): Adam Daniel George Jones Michael Peter Tom
Postorder: Daniel Adam Jones Peter Tom Michael George
Preorder: George Adam Daniel Michael Jones Tom Peter
The number of nodes is 7
search("Jones") 1
search("John") 0
A path from the root to Peter is: George Michael Tom Peter
Inorder (sorted): 1 2 3 4 5 6 7 8

command>
```

程序使用 BST<string>（第 9 行）创建一个字符串的二叉树。该程序将字符串添加到二叉树中（第 10～16 行），并按中序（第 19 行）、后序（第 22 行）和前序（第 25 行）显示元素。getSize 和 search 函数在第 27～29 行中调用。第 31～34 行显示了从根到 Peter 的路径。

该程序为整数数组中的整数创建一个二叉树（第 36 行），并按递增顺序显示数字（第 39 行）。

插入所有字符串元素后，tree 应该如图 21.7a 所示。树 intTree 的创建如图 21.7b 所示。

如果元素以不同的顺序插入，则树看起来会有所不同。然而，只要元素集相同，中序遍历就会以相同的顺序打印元素。中序遍历显示有序列表。

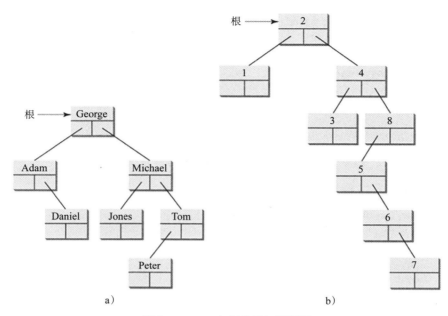

图 21.7 BST 在创建后如图所示

21.9 删除二叉查找树中的元素

要点提示：要从 BST 中删除元素，首先要在树中找到它，然后在删除元素和重新连接树前考虑两种情况——节点有或没有左子树。

函数 `insert(element)` 在 21.6 节中介绍，用于向二叉树添加元素。我们经常需要从二叉树中删除一个元素。从二叉树中删除一个元素比将一个元素添加到二叉树要复杂得多。

要从二叉树中删除元素，首先需要找到包含该元素的节点及其父节点。让 current 指向包含元素的二叉树节点，parent 指向 current 节点的父节点。current 节点可以是 parent 节点的左子节点或右子节点。考虑以下两种情况。

情况 1：current 节点没有左子节点，如图 21.8a 所示。只需将 parent 节点与 current 节点的右子节点连接，如图 21.8b 所示。

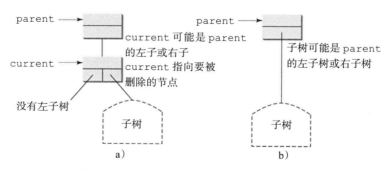

图 21.8 情况 1：current 节点没有左子节点

例如，删除图 21.9a 中的节点 10。将节点 10 的父节点与节点 10 的右子节点连接，如图 21.9b 所示。

注意：如果当前节点是一个叶子，则属于情况 1。例如，删除图 21.9a 中的元素 16，要将其右子节点连接到节点 16 的父节点。在这种情况下，节点 16 的右子节点是 nullptr。

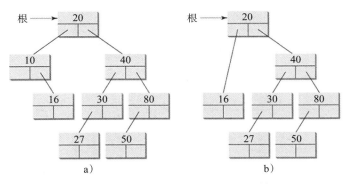

图 21.9　情况 1：从 a) 中删除节点 10 的结果是 b)

情况 2：current 节点有一个左子节点。如图 21.10a 所示，让 rightMost 指向 current 节点的左子树中包含最大元素的节点，parentOfRightMost 则指向 rightMost 节点的父节点。注意，rightMost 节点不能有右子节点，但可以有左子节点。将 current 节点的元素值替换为 rightMost 节点的值，将 parentOfRightMost 节点与 rightMost 节点的左子节点连接，并删除 rightMost 节点，如图 21.10b 所示。

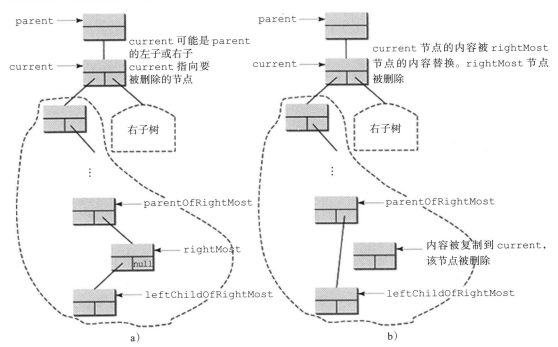

图 21.10　情况 2：current 节点有一个左子节点

例如，考虑删除图 21.11a 中的节点 20。rightMost 节点的元素值为 16。将 current 节点的元素值 20 替换为 16，并使节点 10 成为节点 14 的父节点，如图 21.11b 所示。

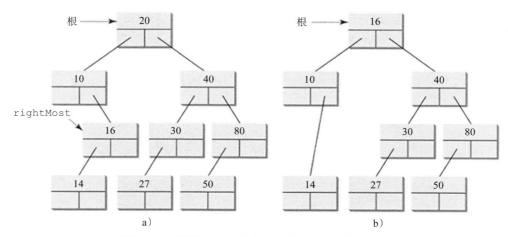

图 21.11 情况 2：从 a) 中删除节点 20 的结果是 b)

注意：如果 current 的左子没有右子，current->left 指向 current 的左子树中的大元素。在这种情况下，rightMost 是 current->left，parentOfRightMost 为 current。需要特别处理这种情况，以便将 rightMost 的右子节点与 parentOfRightMost 重新连接。

Listing 21.3 描述了从二叉树中删除元素的算法。

Listing 21.3　从二叉树中删除元素

```
bool remove(T e)
{
    在树中定位元素 e
    if 元素 e 没有被找到
        return false;
    令 current 为包含 e 的节点，parent 为 current 的父亲;
    if (current 没有左子) // 情况 1
        把 current 的右子与 parent 连接;
        现在 current 未被引用，因此它被删除了;
    else // 情况 2
        在 current 的左子树中定位最右节点 rightMost
        将 rightMost 节点的元素值复制到 current。
        将 rightMost 的父亲与 rightMost 的左子连接; 删除 rightMost。
    return true; // 元素删除成功
}
```

不能把函数命名为 delete，因为 delete 是 C++ 关键字。

LiveExample 21.1 中的第 319 ～ 391 行给出了 remove 函数的完整实现。该函数定位要删除的节点（即 current 节点），并在第 323 ～ 339 行定位其父节点（即 parent 节点）。如果 current 为 nullptr（第 341 行），则该元素不在树中。因此，该函数返回 false（第 342 行）。注意，如果 current 是 root，则 parent 是 nullptr。如果树为空，则 current 和 parent 都是 nullptr。

第 345 ～ 361 行介绍了算法的情况 1。在这种情况下，current 节点没有左子节点（即 current->left 为 nullptr）。如果 parent 为 nullptr，则将 current->right 赋值给 root（第 348 ～ 351 行）。否则，将 current->right 赋值给 parent->left 或 parent->right，具体取决于 current 是 parent（第 353 ～ 358 行）的左子还是右子。

第 362 ～ 387 行介绍了该算法的情况 2。在这种情况下，current 有一个左子节点。该算法定位 current 节点的左子树中最右的节点（命名为 rightMost）及其父节点（命名为 parentOfRightMost）（第 370 ～ 374 行）。用 rightMost 元素替换 current 元素（第 377 行）；将 rightMost->left 赋值给 parentOfRightMost->left 或 parentOfRightMost->right（第 380 ～ 384 行），具体取决于 rightMost 是 parentOfRightMost 的右子还是左子。

LiveExample 21.3 是一个从二叉树中删除元素的测试程序。

LiveExample 21.3 的互动程序请访问 https://liangcpp.pearsoncmg.com/LiveRunCpp5e/faces/LiveExample.xhtml?header=off&programName=TestBSTDelete&fileType=.cpp&programHeight=770&resultHeight=470。

LiveExample 21.3　TestBSTDelete.cpp

Source Code Editor:

```cpp
#include <iostream>
#include <vector>
#include <string>
#include "BST.h"
using namespace std;

template <typename T>
void printTree(const BST<T>& tree)
{
  // Traverse tree
  cout << "Inorder (sorted): ";
  tree.inorder();
  cout << "\nPostorder: ";
  tree.postorder();
  cout << "\nPreorder: ";
  tree.preorder();
  cout << "\nThe number of nodes is " << tree.getSize() << endl;
}

int main()
{
  BST<string> tree;
  tree.insert("George");
  tree.insert("Michael");
  tree.insert("Tom");
  tree.insert("Adam");
  tree.insert("Jones");
  tree.insert("Peter");
  tree.insert("Daniel");
  printTree(tree);

  cout << "\nAfter delete George:";
  tree.remove("George"); // Delete George
  printTree(tree);

  cout << "\nAfter delete Adam:";
  tree.remove("Adam");
  printTree(tree);

  cout << "\nAfter delete Michael:";
  tree.remove("Michael");
```

```
42        printTree(tree);
43
44        return 0;
45    }
```

Execution Result:
```
command>cl TestBSTDelete.cpp
Microsoft C++ Compiler 2019
Compiled successful (cl is the VC++ compile/link command)

command>TestBSTDelete
Inorder (sorted): Adam Daniel George Jones Michael Peter Tom
Postorder: Daniel Adam Jones Peter Tom Michael George
Preorder: George Adam Daniel Michael Jones Tom Peter
The number of nodes is 7

After delete George:Inorder (sorted): Adam Daniel Jones Michael Peter Tom
Postorder: Adam Jones Peter Tom Michael Daniel
Preorder: Daniel Adam Michael Jones Tom Peter
The number of nodes is 6

After delete Adam:Inorder (sorted): Daniel Jones Michael Peter Tom
Postorder: Jones Peter Tom Michael Daniel
Preorder: Daniel Michael Jones Tom Peter
The number of nodes is 5

After delete Michael:Inorder (sorted): Daniel Jones Peter Tom
Postorder: Peter Tom Jones Daniel
Preorder: Daniel Jones Tom Peter
The number of nodes is 4

command>
```

图 21.12～图 21.14 显示了从树中删除元素时树是如何演变的。

注意：显然，中序、前序和后序的时间复杂度是 $O(n)$，因为每个节点只遍历一次。查找、插入和删除的时间复杂度是树的高度。在最坏的情况下，树的高度是 $O(n)$。

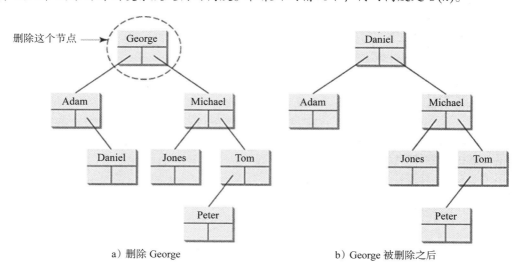

a）删除 George   b）George 被删除之后

图 21.12 删除 George 属于情况 2

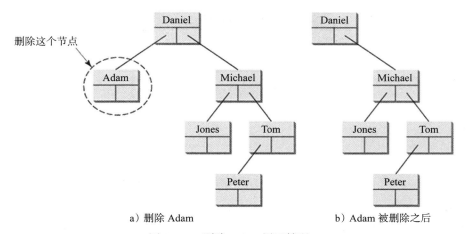

图 21.13 删除 Adam 属于情况 1

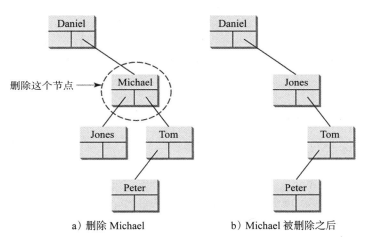

图 21.14 删除 Michael 属于情况 2

考虑以下场景。
1. 如果忘记删除已丢弃的节点，会发生什么情况？程序将继续运行，但会发生内存泄漏。
2. 如果 `rightMost` 节点是一个叶子，程序能工作吗？能。在这种情况下，`rightMost->left` 是 `nullptr`，它被赋值给 `parentOfRightMost->right`。

21.10　BST 的迭代器

要点提示：可以使用迭代器遍历二叉树。

20.5 节介绍了迭代器，并为 `LinkedList` 定义了一个 `Iterator` 类。迭代器为遍历容器元素提供统一方式非常有用。每个迭代器类都有相同的模式，所以它们非常相似。本节定义一个 `Iterator` 类，用于遍历 BST 中的元素，如图 21.15 所示。

此类在 LiveExample 21.1 中的第 24~75 行中实现。方便起见，构造函数和函数被实现为内联函数。

`Iterator` 类有两个数据字段，`v` 和 `current`。数据字段 `v`（第 65 行）被用作二叉树元素的辅助存储器。数据字段 `current` 指向 `v` 中的当前元素（第 64 行）。

```
┌─────────────────────────────────────┐
│            Iterator<T>              │
├─────────────────────────────────────┤
│ -current: int                       │
│ -v: vector<T>                       │
├─────────────────────────────────────┤
│ +Iterator(p: TreeNode<T>*)          │
│ +operator++(): Iterator<T>          │
│ +operator*(): T                     │
│ +operator==(itr: Iterator<T>&): bool│
│ +operator!=(itr: Iterator<T>&): bool│
└─────────────────────────────────────┘
```

图 21.15 该类定义了一个迭代器，用于访问 BST 中的元素

构造函数（第 28～38 行）创建一个迭代器。调用 treeToVector(p) 函数（第 35 行）将树中以 p 为根的所有元素存储到 v 中（第 66～74 行）。current 数据字段指向 v 中的当前元素。最初，它被设置为 0（第 36 行）。

函数 operator++() 将 current 移动以指向 v 中的下一个元素（第 42 行）。如果指针移过 v 的最后一个元素，则 current 设置为 -1（第 44 行）。

为了从 BST 获得迭代器，在 LiveExample 21.1 的第 95～103 行中定义并实现了以下两个函数。

```
Iterator<T> begin() const;
Iterator<T> end() const;
```

函数 begin() 返回要遍历的第一个元素的迭代器，函数 end() 返回列表最后一个元素后一个位置的迭代器。

LiveExample 21.4 给出了一个使用迭代器遍历树中的字符串元素并以大写形式显示字符串的示例。该程序在第 18 行创建一个字符串的 BST，将四个字符串添加到列表中（第 21～24 行），并使用迭代器遍历列表中的所有元素并以大写字母显示（第 27～31 行）。

LiveExample 21.4 的互动程序请访问 https://liangcpp.pearsoncmg.com/LiveRunCpp5e/faces/LiveExample.xhtml?header=off&programName=TestBSTIterator&fileType=.cpp&programHeight=680&resultHeight=190。

LiveExample 21.4　TestBSTIterator.cpp

Source Code Editor:
```cpp
 1  #include <iostream>
 2  #include <string>
 3  #include <algorithm>
 4  #include "BST.h"
 5  using namespace std;
 6
 7  string toUpperCase(string& s)
 8  {
 9    for (int i = 0; i < s.length(); i++)
10      s[i] = toupper(s[i]);
11
12    return s;
13  }
14
15  int main()
16  {
17    // Create a binary search tree for strings
```

```
18      BST<string> tree;
19
20      // Add elements to the tree
21      tree.insert("America");
22      tree.insert("Canada");
23      tree.insert("Russia");
24      tree.insert("France");
25
26      // Traverse a binary tree using iterators
27      for (Iterator<string> iterator = tree.begin();
28         iterator != tree.end(); ++iterator)
29      {
30        cout << toUpperCase(*iterator) << " ";
31      }
32
33      cout << endl << "Min element is " <<
34        *min_element(tree.begin(), tree.end()) << endl;
35      cout << "Max element is "
36        << *max_element(tree.begin(), tree.end()) << endl;
37
38      return 0;
39    }
```

Automatic Check | Compile/Run | Reset | Answer   Choose a Compiler: VC++

Execution Result:

```
command>cl TestBSTIterator.cpp
Microsoft C++ Compiler 2019
Compiled successful (cl is the VC++ compile/link command)

command>TestBSTIterator
AMERICA CANADA FRANCE RUSSIA
Min element is America
Max element is Russia

command>
```

21.11 案例研究：数据压缩

要点提示：霍夫曼编码通过用较少的位来编码出现较频繁的字符以进行数据压缩。字符的编码是根据文本中字符的出现情况，使用称为霍夫曼编码树的二叉树构建的。

压缩数据是一项常见的任务。有许多实用程序可用于压缩文件。本节介绍由大卫·霍夫曼于 1952 年发明的霍夫曼编码。

在 ASCII 码中，每个字符都用 8 位编码。如果一个文本由 100 个字符组成，则需要 800 位来表示该文本。霍夫曼编码的思想是，用较少的位来编码出现较频繁的字符，而用较多的位编码不太常用的字符，以此减小文件的总体大小。在霍夫曼编码中，字符的编码是根据字符在文本中的出现情况用二叉树构建的，这种二叉树称为**霍夫曼编码树**。假设文本是 Mississippi。其霍夫曼树如图 21.16a 所示。节点的左边和右边分别赋值为 0 和 1。每个字符都是树上的一片叶子。字符的编码由从根到叶子的路径中的边值组成，如图 21.16b 所示。由于 i 和 s 在文本中的出现次数多于 M 和 p，因此它们被赋值了较短的编码。

编码树还用于将位序列解码为字符。解码从序列的第一位开始，根据位值确定是转到树根的左分支还是右分支。然后看下一位，根据位值继续向下到左分支或右分支。到达叶子时，就找到一个字符。此时流中的下一位是下一字符的第一位。例如，流 011001 被解码为 sip，其中 01 匹配 s，1 匹配 i，以及 001 匹配 p。

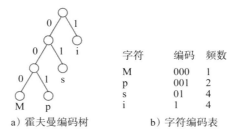

图 21.16 使用编码树根据文本中字符的出现情况构建字符的编码

基于图 21.16 中的编码方案，

Mississippi $\xrightarrow{\text{被编码为}}$ 000101011010110010011 $\xrightarrow{\text{被解码为}}$ Mississippi

要构建霍夫曼编码树，使用以下算法：

1. 从多棵树的森林开始。每棵树包含一个字符的节点。节点的权重是字符在文本中的出现频数。

2. 重复以下操作来组合树，直到只剩一棵树：选择两棵权重最小的树，创建一个新节点作为它们的父节点。新树的权重是子树的权重之和。

3. 对于每个内部节点，为其左边指定值 0，为其右边指定值 1。所有叶节点表示文本中的字符。

以下是为文本 Mississippi 构建编码树的示例。字符的频数表如图 21.16b 所示。最初，森林包含单节点树，如图 21.17a 所示。这些树被反复组合形成大树，直到只剩下一棵树，如图 21.17b～d 所示。

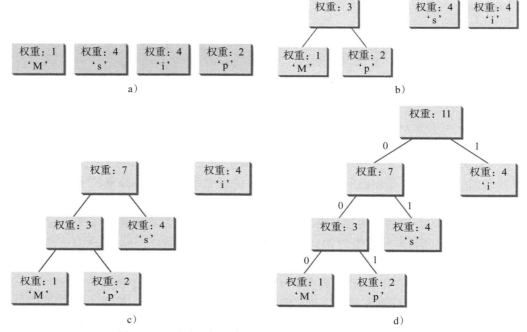

图 21.17 通过反复组合两个最小的加权树来构建编码树

值得注意的是，没有一个编码是另一个编码的前缀。此属性确保流可以被确定地解码。

算法设计说明：这里使用的算法是**贪婪算法**的一个例子。贪婪算法经常用于求解优化问题。该算法做出局部最优的选择，希望这种选择将得到全局最优解。在这种情况下，算法总是选择两个权重最小的树，并创建一个新节点作为它们的父节点。这种直观的最优局部解确实得到了构建霍夫曼树的最终最优解。

作为贪婪算法的另一个例子，可以考虑将钱换成尽可能少的硬币。贪婪算法会先取尽可能大的硬币。例如，对于 98 美分，你会用 25 美分的硬币来换 75 美分，再加上两个 10 美分硬币一共换 95 美分，再加上三个 1 美分换完 98 美分。贪婪算法找到了这个问题的最优解。但贪婪算法并不总是能找到最优结果；参见编程练习 18.16 中的最佳箱子包装问题。

LiveExample 21.5 给出了一个程序，该程序提示用户输入字符串，显示字符串中字符的频数表，并显示每个字符的霍夫曼编码。

LiveExample 21.5 的互动程序请访问 https://liangcpp.pearsoncmg.com/LiveRunCpp5e/faces/LiveExample.xhtml?header=off&programName=HuffmanCode&programHeight=2420&resultHeight=290。

LiveExample 21.5　HuffmanCode.cpp

Source Code Editor:

```cpp
#include <iostream>
#include <string>
#include <iomanip>
#include "Heap.h"
using namespace std;

/** Get the frequency of the characters in counts */
void getCharacterFrequency(const string& text, vector<int>& counts)
{
  for (int i = 0; i < text.length(); i++)
    counts[static_cast<int>(text[i])]++; // Count the character in text
}

class Node
{
public:
  char element; // Stores the character for a leaf node
  int weight; // weight of the subtree rooted at this node
  Node* left; // Reference to the left subtree
  Node* right; // Reference to the right subtree
  string code = ""; // The code of this node from the root

  /** Create an empty node */
  Node()
  {
  }

  /** Create a node with the specified weight and character */
  Node(int weight, char element)
  {
    this->weight = weight;
    this->element = element;
  }
};

/** Define a Huffman coding tree */
class Tree
{
```

```cpp
39    public:
40      Node* root; // The root of the tree
41
42      /** Create a tree with two subtrees */
43      Tree(const Tree& t1, const Tree& t2)
44      {
45        root = new Node();
46        root->left = t1.root;
47        root->right = t2.root;
48        root->weight = t1.root->weight + t2.root->weight;
49      }
50
51      /** Create a tree containing a leaf node */
52      Tree(int weight, char element)
53      {
54        root = new Node(weight, element);
55      }
56
57      int compareTo(const Tree& t)
58      {
59        if (root->weight < t.root->weight) // Purposely reverse the order
60          return 1;
61        else if (root->weight == t.root->weight)
62          return 0;
63        else
64          return -1;
65      }
66    };
67
68    bool operator<(Tree& t1, Tree& t2) // Used in the Heap class
69    {
70      return t1.compareTo(t2) < 0;
71    }
72
73    bool operator>(Tree& t1, Tree& t2) // Used in the Heap class
74    {
75      return t1.compareTo(t2) > 0;
76    }
77
78    /** Get a Huffman tree from the codes */
79    Tree getHuffmanTree(const vector<int>& counts)
80    {
81      // Create a heap to hold trees
82      Heap<Tree>* heap = new Heap<Tree>(); // Defined in LiveExample 19.5
83      for (int i = 0; i < counts.size(); i++)
84        if (counts[i] > 0)
85          heap->add(Tree(counts[i], static_cast<char>(i))); // A leaf node tree
86
87      while (heap->getSize() > 1)
88      {
89        Tree t1 = heap->remove(); // Remove the smallest weight tree
90        Tree t2 = heap->remove(); // Remove the next smallest weight
91        heap->add(Tree(t1, t2)); // Combine two trees
92      }
93
94      return heap->remove(); // The final tree
95    }
96
97    /* Recursively get codes to the leaf node */
98    void assignCode(const Node& root, vector<string>& codes)
99    {
100     if (root.left != nullptr)
101     {
102       root.left->code = root.code + "0";
103       assignCode(*(root.left), codes);
```

```cpp
104            root.right->code = root.code + "1";
105            assignCode(*(root.right), codes);
106          }
107          else
108            codes[static_cast<int>(root.element)] = root.code;
109        }
110
111      /** Get Huffman codes for the characters
112       * This method is called once after a Huffman tree is built
113       */
114      vector<string> getCode(const Node* root)
115      {
116        vector<string> codes(128);
117        assignCode(*root, codes);
118        return codes;
119      }
120
121      int main()
122      {
123        cout << "Enter a text: ";
124        string text;
125        getline(cin, text, '\n');
126
127        vector<int> counts(128); // 128 ASCII characters
128        getCharacterFrequency(text, counts); // Count frequency
129
130        cout << setw(15) << "ASCII Code" << setw(15) << "Character"
131          << setw(15) << "Frequency" << setw(15) << "Code" << endl;
132
133        Tree tree = getHuffmanTree(counts); // Create a Huffman tree
134
135        vector<string> codes = getCode(tree.root); // Get codes
136
137        for (int i = 0; i < codes.size(); i++)
138          if (counts[i] != 0) // static_cast<char>(i )is not in text if counts[i] is 0
139            cout << setw(15) << i << setw(15) << static_cast<char>(i) <<
140              setw(15) << counts[i] << setw(15) << codes[i] << endl;
141
142        return 0;
143      }
```

Enter input data for the program (Sample data provided below. You may modify it.)

Welcome

Choose a Compiler: VC++

Execution Result:

```
command>cl HuffmanCode.cpp
Microsoft C++ Compiler 2019
Compiled successful (cl is the VC++ compile/link command)

command>HuffmanCode
Enter a text: Welcome
     ASCII Code      Character      Frequency           Code
             87              W              1            110
             99              c              1            111
            101              e              2             10
            108              l              1            011
            109              m              1            010
            111              o              1             00

command>
```

程序提示用户输入文本字符串（第 124 ~ 126 行），并计算文本中字符的出现频数（第 129 行）。getCharacterFrequency 函数（第 8 ~ 12 行）创建一个数组 counts，用于统计文本中 128 个 ASCII 字符中每个字符的出现次数。如果文本中出现一次字符，则其相应计数增加 1（第 11 行）。

该程序基于 counts 获得霍夫曼编码树（第 134 行）。该树由链接的节点组成。Node 类在第 14 ~ 34 行中定义。每个节点由属性 element（存储字符）、weight（存储该节点下子树的权重）、left（链接到左子树）、right（链接到右子树）和 code（存储字符的霍夫曼编码）组成。Tree 类（第 37 ~ 66 行）包含 root 属性。从 root 开始可以访问树中的所有节点。operator< 和 operator> 函数（第 68 ~ 76 行）被定义为用 compareTo 函数比较堆中的树。compareTo 函数将此树与另一个树进行比较（第 57 ~ 65 行）。比较的顺序被特意颠倒过来（第 59 ~ 64 行），以便先从树堆中删除最小权重的树。

getHuffmanTree 函数返回一个霍夫曼编码树。最初，创建单节点树并将其添加到堆中（第 82 ~ 85 行）。在 while 循环的每次迭代中（第 87 ~ 92 行），从堆中删除两个最小权重的树，并将其组合形成一棵大树，然后将新树添加到堆中。这个过程一直持续到堆中只包含一棵树，这是文本的最终霍夫曼树。

assignCode 函数为树中的每个节点赋值编码（第 98 ~ 110 行）。getCode 函数获取叶节点中每个字符的编码（第 115 ~ 120 行）。元素 codes[i] 包含字符 (char)i 的编码，其中 i 从 0 到 127。

关键术语

binary search tree（二叉查找树）　　　　postorder traversal（后序遍历）
binary tree（二叉树）　　　　　　　　　preorder traversal（前序遍历）
inorder traversal（中序遍历）　　　　　tree traversal（树遍历）

章节总结

1. 二叉树可以用链接节点来实现。每个节点包含元素值和指向左右两个子节点的两个指针。
2. 树遍历是按照一定的顺序访问树中的每个节点一次的过程。遍历一棵树有几种方法。
3. 中序遍历首先递归地访问当前节点的左子树，然后递归地访问当前节点，最后递归地访问该节点的右子树。中序遍历按递增顺序显示二叉查找树中的所有节点。
4. 后序遍历首先递归地访问当前节点的左子树，然后递归地访问该节点的右子树，最后递归地访问当前节点。
5. 前序遍历首先递归地访问当前节点，然后递归地访问当前节点的左子树，最后递归地访问当前节点的右子树。深度优先遍历与前序遍历相同。
6. 广度优先遍历，节点被逐层访问。首先访问根，然后从左到右访问根的所有子节点，然后从左到右访问根的孙节点，以此类推。
7. 在 BST 中查找元素，从根开始，将该元素与根进行比较。如果元素小于根，则查找左子树。如果元素大于根，则查找右子树。如果元素等于根，则找到匹配项。该过程持续，直到找到匹配项或查找完毕。
8. 在 BST 中插入元素，先进行查找。如果元素已经在树中，则不能插入。否则执行查找，直到找到一个叶节点并为该元素创建一个新节点。新节点是叶节点的左子节点或右子节点。
9. 从 BST 中删除元素，先进行查找以定位该元素，并考虑两种情况来删除和重新组织树。
10. 可以定义一个迭代器遍历二叉树中的元素。

编程练习

21.2 节

*21.1 （在 BST 中查找）在 BST 类中添加一个实例函数来查找树中的元素。

```
// Search element in this binary tree
bool search(const T& element)
```

*21.2 （BST 中的广度优先遍历）在 BST 类中添加一个实例函数，以广度优先的顺序遍历树。

```
// Display the nodes in a breadth-first traversal
void breadthFirstTraversal()
```

*21.3 （BST 的高度）在 BST 类中添加一个实例函数以返回树的高度。

```
// Return the height of this binary tree.
// See Section 21.2 for the definition of the binary tree height.
int height()
```

用 https://liangcpp.pearsoncmg.com/test/Exercise21_03.txt 的模板完成代码。

**21.4 （使用栈实现中序遍历）使用栈而不是递归实现 BST 中的 inorder 函数。

**21.5 （使用栈实现前序遍历）使用栈而不是递归实现 BST 中的 preorder 函数。

**21.6 （使用栈实现后序遍历）使用栈而不是递归实现 BST 中的 postorder 函数。

***21.7 （BST 的父节点引用）如下所示通过添加对节点的父节点引用来修改 TreeNode：

TreeNode<T>
#element: T
#left: TreeNode<T>*
#right: TreeNode<T>*
#parent: TreeNode<T>*

修改 BST 类确保每个节点都具有正确的父链接。在 BST 中添加以下函数：

```
// Returns the parent for the specified node
TreeNode<T>* getParent(TreeNode<T>& node)
```

编写一个测试程序，提示用户输入 10 个整数，将它们添加到树中，从树中删除第一个整数，并显示所有叶节点的路径。

*21.8 （查找叶子）在 BST 类中添加一个实例函数，返回树中的叶子数量。

```
// Returns the number of leaf nodes
int getNumberofLeaves()
```

*21.9 （查找非叶节点）在 BST 类中添加一个实例函数，返回非叶节点的数量。

```
// Returns the number of non-leaf nodes
int getNumberofNonLeaves()
```

用 https://liangcpp.pearsoncmg.com/test/Exercise21_09.txt 的模板完成代码。

*21.10 （测试完美二叉树）完美二叉树是所有层都满的完全二叉树。在 BST 类中添加一个实例函数，如果树是完美的则返回 true。（提示：完美二叉树的节点数是 $2^{height}-1$。）

```
// Returns true if the tree is a perfect binary tree
bool isPerfectBST()
```

用 https://liangcpp.pearsoncmg.com/test/Exercise21_10.txt 的模板完成代码。

第 22 章

STL 容器

学习目标

1. 了解容器、迭代器和算法之间的关系（22.2 节）。
2. 区分序列容器、关联容器和容器适配器（22.2 节）。
3. 区分容器 `vector`、`deque`、`list`、`set`、`multiset`、`map`、`multimap`、`stack`、`queue` 和 `priority_queue`（22.2 节）。
4. 使用容器的共同特性（22.2 节）。
5. 使用迭代器访问容器中的元素（22.3 节）。
6. 区分迭代器类型：输入、输出、正向、双向和随机访问（22.3.1 节）。
7. 使用运算符操作迭代器（22.3.2 节）。
8. 从容器中获取迭代器，并了解容器支持的迭代器类型（22.3.3 节）。
9. 使用 `istream_iterator` 和 `ostream_iterator` 实现输入和输出（22.3.4 节）。
10. 声明自动类型推断的自动变量（22.4 节）。
11. 在序列容器 `vector`、`deque` 和 `list` 中存储、检索和处理元素（22.5 节）。
12. 在关联容器 `set`、`multiset`、`map` 和 `multimap` 中存储、检索和处理元素（22.6 节）。
13. 在容器适配器 `stack`、`queue` 和 `priority_queue` 中存储、检索和处理元素（22.7 节）。

22.1 简介

要点提示：STL 为经典数据结构提供了一个标准库。

第 20 章和第 21 章介绍了经典的数据结构，如链表、栈、队列、堆、优先级队列和二叉查找树。这些流行的数据结构在应用程序中被广泛使用。C++ 为这些和许多其他有用的数据结构提供了一个称为标准模板库（STL）的库。因此，我们可以使用它们以避免重复工作。我们已经学习的一个例子是 `vector` 类（12.6 节），本章介绍 STL，我们将学习如何使用其中的类简化应用程序开发。

22.2 STL 基础

要点提示：STL 包含三个主要部分：容器、迭代器和算法。

STL 是由惠普公司的 Alexander Stepanov 和 Meng Lee 在与 David Musser 合作进行的泛型程序设计研究的基础上开发的。它是用 C++ 编写的库集。STL 中的类和函数都是模板类和模板函数。

STL 包含以下三个主要部分。

- **容器**：STL 中的类是容器类。容器对象（如向量）用于存储通常称为元素的数据集。
- **迭代器**：STL 容器类广泛使用迭代器，迭代器是便于遍历容器中元素的对象。迭代器就像内置的指针，为访问和操作容器元素提供了便利的方式。

- **算法**：算法在函数中用来处理数据，如排序、查找和比较元素。STL 中大约实现了 80 个算法。它们中的大多数使用迭代器访问容器中的元素。**STL 算法**将在第 23 章中介绍。

STL 容器可分为以下三类。

- **序列容器**：序列容器（也称为顺序容器）表示线性数据结构。三个序列容器分别是 `vector`、`list` 和 `deque`（发音为 deck）。
- **关联容器**：关联容器是非线性容器，可以快速定位存储在其中的元素。这样的容器可以存储值集或键/值对。四个关联容器是 `set`、`multiset`、`map` 和 `multimap`。
- **容器适配器**：容器适配器是序列容器的受限版本。它们改编自序列容器以处理特殊情况。三个容器适配器分别是 `stack`、`queue` 和 `priority_queue`。

表 22.1 汇总了容器类及其头文件。

表 22.1 容器类

类别	容器	头文件	描述	
序列容器	vector	\<vector\>	用于直接访问任意元素，在向量末尾快速插入和删除	一级容器
	deque	\<deque\>	用于直接访问任意元素，在双端队列的前端和末尾快速插入和删除	
	list	\<list\>	用于在任意地方快速插入和删除	
关联容器	set	\<set\>	用于直接查找，没有重复的元素	
	multiset	\<set\>	与 set 相同，除了允许元素重复	
	map	\<map\>	键/值对映射，不允许重复，并使用键快速查找	
	multimap	\<map\>	与 map 相同，除了允许键重复	
容器适配器	stack	\<stack\>	后进先出的容器	
	queue	\<queue\>	先进先出的容器	
	priority_queue	\<queue\>	最高优先级的元素先删除	

C++ 没有为容器类定义基类，但所有 STL 容器都有一些共同的特性和函数。例如，每个容器都有一个无参数构造函数、一个复制构造函数、一个析构函数等。表 22.2 列出了所有容器的公共函数，表 22.3 列出了序列容器和关联容器的公共函数。这两种类型的容器也被称为**一级容器**。

表 22.2 所有容器的公共函数

函数	描述
无参数构造函数	构造空容器
有参数构造函数	除了无参数构造函数，每个容器都有几个带参数的构造函数
复制构造函数	通过从相同类型的现有容器中复制元素来创建容器
析构函数	容器销毁后执行清理
empty()	如果容器中没有元素，则返回 true
size()	返回容器中元素的个数
运算符 =	将一个容器复制到另一个容器
关系运算符(<, <=, >, >=, ==, !=)	将两个容器中的元素依次进行比较，以确定关系

表 22.3 一级容器的公共函数

函数	描述
c1.swap(c2)	交换 c1 和 c2 两个容器的元素
c.max_size()	返回容器可容纳的最大元素数

(续)

函数	描述
c.clear()	清除容器中所有元素
c.begin()	返回指向容器第一个元素的迭代器
c.end()	返回一个迭代器，它指向容器末尾的后一个位置
c.rbegin()	返回指向容器最后一个元素的迭代器，用于按相反顺序处理元素
c.rend()	返回一个迭代器，它指向容器第一个元素的前一个位置
c.erase(begin,end)	清除容器中从 begin 到 end-1 的元素。begin 和 end 都是迭代器

LiveExample 22.1 给出了一个简单的示例，演示如何创建 vector、list、deque、set、multiset、stack 和 queue。

LiveExample 22.1 的互动程序请访问 https://liangcpp.pearsoncmg.com/LiveRunCpp5e/faces/LiveExample.xhtml?header=off&programName=SimpleSTLDemo&programHeight=1840&resultHeight=1100。

LiveExample 22.1　SimpleSTLDemo.cpp

Source Code Editor:

```cpp
#include <iostream>
#include <vector>
#include <list>
#include <deque>
#include <set>
#include <stack>
#include <queue>
using namespace std;

int main()
{
    vector<int> vector1, vector2;
    list<int> list1, list2;
    deque<int> deque1, deque2;
    set<int> set1, set2;
    multiset<int> multiset1, multiset2;
    stack<int> stack1, stack2;
    queue<int> queue1, queue2;

    cout << "Vector: " << endl;
    vector1.push_back(1);
    vector1.push_back(2);
    vector2.push_back(30);
    cout << "size of vector1: " << vector1.size() << endl;
    cout << "size of vector2: " << vector2.size() << endl;
    cout << "maximum size of vector1: " << vector1.max_size() << endl;
    cout << "maximum size of vector2: " << vector2.max_size() << endl;
    vector1.swap(vector2);
    cout << "size of vector1: " << vector1.size() << endl;
    cout << "size of vector2: " << vector2.size() << endl;
    cout << "vector1 < vector2? " << (vector1 < vector2)
        << endl << endl;

    cout << "List: " << endl;
    list1.push_back(1);
    list1.push_back(2);
    list2.push_back(30);
    cout << "size of list1: " << list1.size() << endl;
    cout << "size of list2: " << list2.size() << endl;
    cout << "maximum size of list1: " << list1.max_size() << endl;
```

```cpp
41      cout << "maximum size of list2: " << list2.max_size() << endl;
42      list1.swap(list2);
43      cout << "size of list1: " << list1.size() << endl;
44      cout << "size of list2: " << list2.size() << endl;
45      cout << "list1 < list2? " << (list1 < list2) << endl << endl;
46
47      cout << "Deque: " << endl;
48      deque1.push_back(1);
49      deque1.push_back(2);
50      deque2.push_back(30);
51      cout << "size of deque1: " << deque1.size() << endl;
52      cout << "size of deque2: " << deque2.size() << endl;
53      cout << "maximum size of deque1: " << deque1.max_size() << endl;
54      cout << "maximum size of deque2: " << deque2.max_size() << endl;
55      list1.swap(list2);
56      cout << "size of deque1: " << deque1.size() << endl;
57      cout << "size of deque2: " << deque2.size() << endl;
58      cout << "deque1 < deque2? " << (deque1 < deque2) << endl << endl;
59
60      cout << "Set: " << endl;
61      set1.insert(1);
62      set1.insert(1);
63      set1.insert(2);
64      set2.insert(30);
65      cout << "size of set1: " << set1.size() << endl;
66      cout << "size of set2: " << set2.size() << endl;
67      cout << "maximum size of set1: " << set1.max_size() << endl;
68      cout << "maximum size of set2: " << set2.max_size() << endl;
69      set1.swap(set2);
70      cout << "size of set1: " << set1.size() << endl;
71      cout << "size of set2: " << set2.size() << endl;
72      cout << "set1 < set2? " << (set1 < set2) << endl << endl;
73
74      cout << "Multiset: " << endl;
75      multiset1.insert(1);
76      multiset1.insert(1);
77      multiset1.insert(2);
78      multiset2.insert(30);
79      cout << "size of multiset1: " << multiset1.size() << endl;
80      cout << "size of multiset2: " << multiset2.size() << endl;
81      cout << "maximum size of multiset1: " <<
82             multiset1.max_size() << endl;
83      cout << "maximum size of multiset2: " <<
84             multiset2.max_size() << endl;
85      multiset1.swap(multiset2);
86      cout << "size of multiset1: " << multiset1.size() << endl;
87      cout << "size of multiset2: " << multiset2.size() << endl;
88      cout << "multiset1 < multiset2? " <<
89             (multiset1 < multiset2) << endl << endl;
90
91      cout << "Stack: " << endl;
92      stack1.push(1);
93      stack1.push(1);
94      stack1.push(2);
95      stack2.push(30);
96      cout << "size of stack1: " << stack1.size() << endl;
97      cout << "size of stack2: " << stack2.size() << endl;
98      cout << "stack1 < stack2? " << (stack1 < stack2) << endl << endl;
99
100     cout << "Queue: " << endl;
101     queue1.push(1);
102     queue1.push(1);
103     queue1.push(2);
104     queue2.push(30);
```

```
105        cout << "size of queue1: " << queue1.size() << endl;
106        cout << "size of queue2: " << queue2.size() << endl;
107        cout << "queue1 < queue2? " << (queue1 < queue2) << endl << endl;
108
109        return 0;
110   }
```

Execution Result:

```
command>cl SimpleSTLDemo.cpp
Microsoft C++ Compiler 2019
Compiled successful (cl is the VC++ compile/link command)

command>SimpleSTLDemo
Vector:
size of vector1: 2
size of vector2: 1
maximum size of vector1: 1073741823
maximum size of vector2: 1073741823
size of vector1: 1
size of vector2: 2
vector1 < vector2? 0

List:
size of list1: 2
size of list2: 1
maximum size of list1: 357913941
maximum size of list2: 357913941
size of list1: 1
size of list2: 2
list1 < list2? 0

Deque:
size of deque1: 2
size of deque2: 1
maximum size of deque1: 1073741823
maximum size of deque2: 1073741823
size of deque1: 2
size of deque2: 1
deque1 < deque2? 1

Set:
size of set1: 2
size of set2: 1
maximum size of set1: 214748364
maximum size of set2: 214748364
size of set1: 1
size of set2: 2
set1 < set2? 0

Multiset:
size of multiset1: 3
size of multiset2: 1
maximum size of multiset1: 214748364
maximum size of multiset2: 214748364
size of multiset1: 1
size of multiset2: 3
multiset1 < multiset2? 0

Stack:
size of stack1: 3
size of stack2: 1
stack1 < stack2? 1
```

```
Queue:
size of queue1: 3
size of queue2: 1
queue1 < queue2? 1

command>
```

每个容器都有一个无参数构造函数。该程序在第 12 ~ 18 行用这些容器的无参数构造函数创建 vector、list、deque、set、multiset、stack 和 queue。

该程序在第 21 ~ 23、35 ~ 37 和 48 ~ 50 行用 push_back(element) 函数将元素追加到 vector、list、deque 中；在第 61 ~ 64 行和第 75 ~ 78 行用 insert(element) 函数将元素插入 set 和 multiset 中；在第 92 ~ 95 行和第 101 ~ 104 行用 push(element) 函数将元素压入 stack 和 queue。

在第 61 ~ 62 行中，整数 1 被插入 set1 中两次。由于 Set 不允许重复元素，因此在第 63 行将 2 插入 set1 后，set1 包含 {1,2}。multiset 允许重复，因此在第 75 ~ 77 行将 1、1 和 2 插入 multiset1 之后，multiset1 包含 {1,1,2}。

所有容器都支持关系运算符。程序在第 31、45、58 和 72 行中比较两个相同类型的容器。

22.3 STL 迭代器

要点提示：STL 迭代器为遍历容器元素提供了一种统一的方法。

迭代器在一级容器中被广泛用于访问和操作元素。正如在表 22.3 中看到的，一级容器中的几个函数（例如，begin() 和 end()）与迭代器有关。20.5 节介绍了在容器中实现迭代器的示例。在阅读本节之前，你会发现复习 20.5 节很有帮助。

函数 begin() 返回指向容器第一个元素的迭代器，函数 end() 返回表示容器最后一个元素后一个位置的迭代器，如图 22.1 所示。

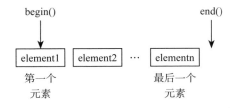

图 22.1 end() 表示最后一个元素的后一个位置

通常用以下循环遍历容器中的所有元素：

```
for (iterator p = c.begin(); p != c.end(); p++)
{
  processing *p; // *p is the current element
}
```

每个容器都有自己的迭代器类型。抽象隐藏了详细的实现，并为在所有容器上使用迭代器提供了一种统一的方式。迭代器在所有容器中的使用方式相同，因此，如果你知道如何将迭代器与某个容器类一起使用，则可以将其应用于任何其他容器。

LiveExample 22.2 演示了如何在向量和集合中使用迭代器。

LiveExample 22.2 的互动程序请访问 https://liangcpp.pearsoncmg.com/LiveRunCpp5e/faces/LiveExample.xhtml?header=off&programName=IteratorDemo&programHeight=670&resultHeight=180。

LiveExample 22.2　IteratorDemo.cpp

```cpp
#include <iostream>
#include <vector>
#include <set>
using namespace std;

int main()
{
    vector<int> intVector;
    intVector.push_back(10);
    intVector.push_back(40);
    intVector.push_back(50);
    intVector.push_back(20);
    intVector.push_back(30);

    vector<int>::iterator p1;
    cout << "Traverse the vector: ";
    for (p1 = intVector.begin(); p1 != intVector.end(); p1++)
    {
        cout << *p1 << " ";
    }

    set<int> intSet;
    intSet.insert(10);
    intSet.insert(40);
    intSet.insert(50);
    intSet.insert(20);
    intSet.insert(30);

    set<int>::iterator p2;
    cout << "\nTraverse the set: ";
    for (p2 = intSet.begin(); p2 ! = intSet.end( );p2++)
    {
        cout << *p2 << " ";
    }
    cout << endl;

    return 0;
}
```

Execution Result:

```
command>cl IteratorDemo.cpp
Microsoft C++ Compiler 2019
Compiled successful (cl is the VC++ compile/link command)

command>IteratorDemo
Traverse the vector: 10 40 50 20 30
Traverse the set: 10 20 30 40 50

command>
```

该程序创建了一个 int 值的向量（第 8 行），向其追加五个数字（第 9 ~ 13 行），并用迭代器遍历向量（第 15 ~ 20 行）。

迭代器 p1 在第 15 行中声明：

```
vector<int>::iterator p1;
```

每个容器都有自己的迭代器类型。这里 vector<int>::iterator 表示 vector<int> 类中的迭代器类型。

表达式（第 17 行）

```
p1 = intVector.begin();
```

获得指向向量 intVector 中第一个元素的迭代器，并将迭代器赋值给 p1。

表达式（第 17 行）

```
p1 != intVector.end();
```

检查 p1 是否已经经过了容器中最后一个元素。

表达式（第 17 行）

```
p1++
```

将迭代器移动到下一个元素。

表达式（第 19 行）

```
*p1
```

返回 p1 所指向的元素。

类似地，该程序创建一个 int 值的集合（第 22 行），插入五个数字（第 23 ~ 27 行），并使用迭代器遍历该集合（第 29 ~ 34 行）。注意，集合中的元素是经过排序的，因此程序在样例输出中显示 10、20、30、40 和 50。

从这个例子可以看到，迭代器的功能就像一个指针。迭代器变量指向容器中的一个元素。使用递增运算符 (p++) 可以将迭代器移动到下一个元素，可以用解引用运算符 (*p) 访问该元素。

22.3.1 迭代器的类型

每个容器都有自己的迭代器类型。迭代器可分为以下五类。

- **输入迭代器**：输入迭代器用于从容器中读取元素。它一次只能向前移动一个元素。
- **输出迭代器**：输出迭代器用于将元素写入容器。它一次只能向前移动一个元素。
- **正向迭代器**：正向迭代器结合了输入迭代器和输出迭代器的所有功能，支持读取和写入操作。
- **双向迭代器**：双向迭代器是一种正向迭代器。而且，它还可以向后移动。
- **随机访问迭代器**：随机访问迭代器是双向迭代器。而且，它可以按任意顺序访问任何元素，即它可以向前或向后跳跃多个元素。

vector 和 deque 容器支持随机访问迭代器，list、set、multiset、map 和 multimap 容器支持双向迭代器。注意，stack、queue 和 priority_queue 不支持迭代器，如表 22.4 所示。

表 22.4 容器支持的迭代器类型

STL 容器	支持的迭代器类型
vector	随机访问迭代器
deque	随机访问迭代器
list	双向迭代器
set	双向迭代器
multiset	双向迭代器
map	双向迭代器
multimap	双向迭代器
stack	不支持迭代器
queue	不支持迭代器
priority_queue	不支持迭代器

22.3.2 迭代器运算符

我们可以通过重载运算符来操作迭代器，以移动其位置、访问元素以及进行比较。表 22.5 显示了迭代器支持的运算符。

表 22.5 迭代器支持的运算符

运算符	描述
所有迭代器	
++p	前置递增迭代器
p++	后置递增迭代器
输入迭代器	
*p	解引用迭代器（用作右值）
p1 == p2	如果 p1 和 p2 指向同一个元素，则结果为 true
p1 != p2	如果 p1 和 p2 指向不同的元素，则结果为 true
输出迭代器	
*p	解引用迭代器（用作左值）
双向迭代器	
--p	前置递减迭代器
p--	后置递减迭代器
随机访问迭代器	
p += i	迭代器 p 增加 i 个位置
p -= i	迭代器 p 减少 i 个位置
p + i	返回 p 后面的第 i 个位置的迭代器
p - i	返回 p 前面的第 i 个位置的迭代器
p1 < p2	如果 p1 在 p2 之前，则返回 true
p1 <= p2	如果 p1 在 p2 之前或等于 p2，则返回 true
p1 > p2	如果 p1 在 p2 之后，则返回 true
p1 >= p2	如果 p1 在 p2 之后或等于 p2，则返回 true
p[i]	返回距离位置 p 偏移 i 处的元素

所有迭代器都支持前置递增运算符和后置递增运算符（++p 和 p++）。输入迭代器还支持作为右值的解引用运算符（*）、相等测试运算符（==）和不相等测试运算符（!=）。输出迭代器还支持作为左值的解引用运算符（*）。正向迭代器支持输入和输出迭代器中提供的所

有函数。除了正向迭代器中的所有函数外,双向迭代器还支持前置递减和后置递减运算符。随机访问迭代器支持此表中列出的所有运算符。

LiveExample 22.3 演示了如何在迭代器上使用这些运算符。

LiveExample 22.3 的互动程序请访问 https://liangcpp.pearsoncmg.com/LiveRunCpp5e/faces/LiveExample.xhtml?header=off&programName=IteratorOperatorDemo&fileType=.cpp&programHeight=510&resultHeight=230。

LiveExample 22.3　IteratorOperatorDemo.cpp

Source Code Editor:

```cpp
#include <iostream>
#include <vector>
using namespace std;

int main()
{
  vector<int> intVector;
  intVector.push_back(10);
  intVector.push_back(20);
  intVector.push_back(30);
  intVector.push_back(40);
  intVector.push_back(50);
  intVector.push_back(60);

  vector<int>::iterator p1 = intVector.begin();
  for (; p1 != intVector.end(); p1++)
  {
    cout << *p1 << " ";
  }

  cout << endl << *(--p1) << endl; // The element before p1
  cout << *(p1 - 3) << endl; // The 3rd element before p1
  cout << p1[-3] << endl; // The 3rd element from the end
  *p1 = 1234;
  cout << *p1 << endl;

  return 0;
}
```

Execution Result:

```
command>cl IteratorOperatorDemo.cpp
Microsoft C++ Compiler 2019
Compiled successful (cl is the VC++ compile/link command)

command>IteratorOperatorDemo
10 20 30 40 50 60
60
30
30
1234

command>
```

vector 类包含随机访问迭代器。该程序创建一个向量(第 7 行),在其中追加六个元素(第 8～13 行),并在第 15 行中获得迭代器 p1。由于 vector 类包含随机访问迭代器,表 22.5 中的所有运算符都可以应用于 p1。随机访问迭代器是双向的。第 21 行中的迭代器向后移动。随机访问迭代器可以移动到任何位置来访问元素。第 22 ～ 23 行中的语句将迭代器移动到当前位置前面三个位置。

22.3.3 预定义迭代器

STL 容器使用 `typedef` 关键字来预定义迭代器的同义词。预定义的迭代器有 `iterator`、`const_iterator`、`reverse_iterator` 和 `const_reverse_iterator`。为了一致，在每个一级容器中都定义了这些迭代器。因此，可以在应用程序中统一使用它们。

例如

```
vector<int>::iterator p1;
```

将 `p1` 定义为 `vector<int>` 容器的迭代器。

```
list<int>::iterator p2;
```

将 `p2` 定义为 `list<int>` 容器的迭代器。

注意：由于 `iterator` 是在类（如 `vector`）中定义的 `typedef`，因此需要作用域解析运算符来引用它。

`const_iterator` 与 `iterator` 相同，只是不能通过 `const_iterator` 修改元素。`const_iterator` 是只读的。LiveExample 22.4 显示了 `iterator` 和 `const_iterator` 之间的区别。

LiveExample 22.4 的互动程序请访问 https://liangcpp.pearsoncmg.com/LiveRunCpp5e/faces/LiveExample.xhtml?header=off&programName=ConstIteratorDemo&programHeight=370&resultHeight=140。

LiveExample 22.4　ConstIteratorDemo.cpp

```cpp
#include <iostream>
#include <vector>
using namespace std;

int main()
{
    vector<int> intVector;
    intVector.push_back(10);

    vector<int>::iterator p1 = intVector.begin();
    vector<int>::const_iterator p2 = intVector.begin();

    *p1 = 123; // OK
    *p2 = 123; // Not allowed

    cout << *p1 << endl;
    cout << *p2 << endl;

    return 0;
}
```

Execution Result:
```
c:\example>cl ConstIteratorDemo.cpp
Microsoft C++ Compiler 2017
ConstIteratorDemo.cpp
```

```
ConstIteratorDemo.cpp(14): error 'p2': you cannot assign to a const variable
c:\example>
```

由于p2是const_iterator(第11行),因此无法通过p2修改元素。

可以使用反向迭代器反向遍历容器。LiveExample 22.5演示了如何使用reverse_iterator。LiveExample 22.5的互动程序请访问https://liangcpp.pearsoncmg.com/LiveRunCpp5e/faces/LiveExample.xhtml?header=off&programName=ReverseIteratorDemo&programHeight=350&resultHeight=160。

LiveExample 22.5 ReverseIteratorDemo.cpp

Source Code Editor:
```cpp
#include <iostream>
#include <vector>
using namespace std;

int main()
{
  vector<int> intVector;
  intVector.push_back(10);
  intVector.push_back(30);
  intVector.push_back(20);

  vector<int>::reverse_iterator p1 = intVector.rbegin();
  for (; p1 != intVector.rend(); p1++)
  {
    cout << *p1 << " ";
  }

  return 0;
}
```

Execution Result:
```
command>cl ReverseIteratorDemo.cpp
Microsoft C++ Compiler 2019
Compiled successful (cl is the VC++ compile/link command)

command>ReverseIteratorDemo
20 30 10

command>
```

该程序在第12行声明了一个reverse_iterator p1。函数rbegin()返回一个reverse_iterator,它指向容器的最后一个元素(第12行)。函数rend()返回一个reverse_iterator,它指向反向容器第一个元素之后的下一个元素(第13行)。

const_reverse_iterator与reverse_iterator相同,只是不能通过const_reverse_iterator修改元素。const_reverse_iterator是只读的。

22.3.4 istream_iterator 和 ostream_iterator

迭代器可以用来对元素进行排序。可以用迭代器对容器中的元素以及输入/输出流中的元素进行排序。LiveExample 22.6演示了如何使用istream_iterator从输入流输入数据,以及如何使用ostream_iterator将数据输出到输出流。程序提示用户输入三个整数，并显示最大的整数。

LiveExample 22.6 的互动程序请访问 https://liangcpp.pearsoncmg.com/LiveRunCpp5e/faces/LiveExample.xhtml?header=off&programName=InputOutputStreamIteratorDemo&fileType=.cpp&programHeight=390&resultHeight=180。

LiveExample 22.6　InputOutputStreamIteratorDemo.cpp

Source Code Editor:

```cpp
#include <iostream>
#include <iterator>
#include <algorithm>
using namespace std;

int main()
{
  cout << "Enter three numbers: ";
  istream_iterator<int> inputIterator(cin);
  ostream_iterator<int> outputIterator(cout);

  int number1 = *inputIterator;
  inputIterator++;
  int number2 = *inputIterator;
  inputIterator++;
  int number3 = *inputIterator;

  cout << "The largest number is ";
  *outputIterator = max(max(number1, number2), number3);

  return 0;
}
```

Enter input data for the program (Sample data provided below. You may modify it.)

```
34 12 23
```

[Automatic Check] [Compile/Run] [Reset] [Answer]　　　Choose a Compiler: VC++

Execution Result:

```
command>cl InputOutputStreamIteratorDemo.cpp
Microsoft C++ Compiler 2019
Compiled successful (cl is the VC++ compile/link command)

command>InputOutputStreamIteratorDemo
Enter three numbers: 34 12 23
The largest number is 34

command>
```

istream_iterator 和 ostream_iterator 位于 <iterator> 头文件中，因此它在第 2 行中被包含进来。第 9 行创建 istream_iterator inputIterator 用于从 cin 对象中读取整数。第 10 行创建 ostream_iterator outputIterator 用于将整数写入 cout 对象。

解引用运算符应用于 inputIterator（第 12 行）以从 cin 中读取一个整数，迭代器移动到输入流中的下一个数字（第 13 行）。解引用运算符应用于 outputIterator（第 19 行），将一个整数写入 cout。这里 *outputIterator 是一个左值。

此示例演示如何使用 istream_iterator 和 ostream_iterator。使用 istream_iterator 和 ostream_iterator 看起来好像是一种不自然的控制台输入和输出方式。在下一章介绍 STL 算法时，你将看到使用这些迭代器的真正好处。

22.4 C++11 自动类型推断

要点提示：新的 C++11 auto 关键字可以用来声明变量。变量的类型由编译器根据赋给变量的值的类型自动确定，这被称为自动类型推断。

在 C++11 中，如果编译器能够从初始化语句中确定变量的类型，则可以简单地使用 auto 关键字声明它，如下例所示：

```
auto x = 3;
auto y = x;
```

编译器根据赋给 x 的整数值 3，自动确定变量 x 的类型为 int，并由于 int 变量 x 被赋值给 y，而确定变量 y 的类型为 int，这被称为**自动类型推断**。

前面的代码不是使用自动类型的典型例子。自动类型的真正用途是替换长类型。例如

```
vector<int>::reverse_iterator p1 = intVector.rbegin();
```

最好替换为

```
auto p1 = intVector.rbegin();
```

新代码更短、更简单，而且程序员不用编写冗长而笨拙的类型声明。

警告：我们故意将 auto 的引入推迟到现在，是为了避免这个有用的特性被不当使用。不要将其用于简单类型，通常它仅用于与 STL 相关联的复杂类型。

22.5 序列容器

要点提示：序列容器维护容器中元素的顺序。

STL 提供了三个序列容器：vector、list 和 deque。vector 容器和 deque 容器用数组实现，list 容器用链表实现。

- 如果将元素添加到 vector 末尾，则 vector 是高效的，但在除 vector 末尾以外的任何位置插入或删除元素开销都很大。
- deque 类似于 vector。它是一个支持对元素进行随机访问的序列。像 vector 一样，在 deque 的末尾插入和删除元素效率很高。deque 和 vector 之间的主要区别在于，deque 对于在 deque 开头插入和删除元素也是高效的。但在 deque 中间插入或删除元素仍然开销很大。
- list 是用双向链表实现的。这适用于需要使用迭代器在 list 任何位置频繁插入和删除的应用程序。

vector 的成本最小，deque 的成本略高于 vector，list 的成本最大。表 22.2 和表 22.3 列出了所有容器和一级容器的公共函数。除了这些常见的函数外，每个序列容器还具有表 22.6 中所示的函数。

表 22.6 序列容器的公共函数

函数	描述
assign(n, element)	为容器中指定元素的 *n* 个副本赋值
assign(begin, end)	为迭代器 begin 到 end-1 指定范围内的元素赋值
push_back(element)	向容器追加元素

（续）

函数	描述
pop_back()	删除容器最后一个元素
front()	返回容器第一个元素
back()	返回容器最后一个元素
insert(position,element)	在迭代器指定位置处插入元素

22.5.1 序列容器：vector

如表 22.2 所示，每个容器都有一个无参数构造函数、一个复制构造函数和一个析构函数，并支持函数 `empty()`、`size()` 和关系运算符。每个一级容器都包含函数 `swap`、`max_size`、`clear`、`begin`、`end`、`rbegin`、`rend` 和 `erase`，如表 22.3 所示，以及迭代器支持的运算符，如表 22.5 所示。每个序列容器都包含 `assign`、`push_back`、`pop_back`、`front`、`back` 和 `insert` 函数，如表 22.6 所示。除了这些常见函数外，`vector` 类还包含表 22.7 中所示的函数。

表 22.7 vector 类特有的函数

函数	描述
vector(n, element)	构造一个向量，用同一元素的 n 个副本填充
vector(begin, end)	构造一个向量，初始化从迭代器 begin 到 end-1 的元素
vector(size)	构造具有指定大小的向量
at(index)	返回指定索引处的元素

LiveExample 22.7 演示了如何使用 vector 中的函数。

LiveExample 22.7 的互动程序请访问 https://liangcpp.pearsoncmg.com/LiveRunCpp5e/faces/LiveExample.xhtml?header=off&programName=VectorDemo&fileType=.cpp&programHeight=740&resultHeight=240。

LiveExample 22.7 VectorDemo.cpp

Source Code Editor:

```cpp
#include <iostream>
#include <vector>
using namespace std;

int main()
{
  double values[] = {1, 2, 3, 4, 5, 6, 7};
  vector<double> doubleVector(values, values + 7);

  cout << "Initial contents in doubleVector: ";
  for (int i = 0; i < doubleVector.size(); i++)
    cout << doubleVector[i] << " ";

  doubleVector.assign(4, 11.5);

  cout << "\nAfter the assign function, doubleVector: ";
  for (int i = 0; i < doubleVector.size(); i++)
    cout << doubleVector[i] << " ";

  doubleVector.at(0) = 22.4; // Assign 22.4 to doubleVector at index 0
  cout << "\nAfter the at function, doubleVector: ";
  for (int i = 0; i < doubleVector.size(); i++)
    cout << doubleVector[i] << " ";
```

```
25      auto itr = doubleVector.begin();
26      doubleVector.insert(itr + 1, 555);
27      doubleVector.insert(itr + 1, 666);
28      cout << "\nAfter the insert function, doubleVector: ";
29      for (int i = 0; i < doubleVector.size(); i++)
30          cout << doubleVector[i] << " ";
31
32      doubleVector.erase(itr + 2, itr + 4);
33      cout << "\nAfter the erase function, doubleVector: ";
34      for (int i = 0; i < doubleVector.size(); i++)
35          cout << doubleVector[i] << " ";
36
37      doubleVector.clear(); // Delete all in doubleVector
38      cout << "\Size is " << doubleVector.size() << endl;
39      cout << "Is empty? " <<
40          (doubleVector.empty() ? "true" : "false") << endl;
41
42      return 0;
43  }
```

Execution Result:

```
command>cl VectorDemo.cpp
Microsoft C++ Compiler 2019
Compiled successful (cl is the VC++ compile/link command)

command>VectorDemo
Initial contents in doubleVector: 1 2 3 4 5 6 7
After the assign function, doubleVector: 11.5 11.5 11.5 11.5
After the at function, doubleVector: 22.4 11.5 11.5 11.5
After the insert function, doubleVector: 22.4 666 555 11.5 11.5 11.5
After the erase function, doubleVector: 22.4 666 11.5 11.5 Size is 0
Is empty? true

command>
```

该程序在第 7 行创建一个由七个元素组成的数组，并用数组中的元素创建一个向量。可以用指针访问数组。指针类似于迭代器，因此 values 指向数组中的第一个元素，values+7 指向数组最后一个元素的后一个位置。

程序用 for 循环显示向量中的所有元素（第 17～18 行）。下标运算符 [] (第 18 行) 可以访问容器 vector 或 deque 中的元素。

程序将 22.4 赋值给向量的第一个元素（第 20 行）。

```
doubleVector.at(0) = 22.4;
```

此语句与下面的语句相同：

```
doubleVector[0] = 22.4;
```

迭代器可指定容器中的位置。在第 25 行获得一个迭代器，一个新元素被插入位置 itr+1（第 26 行），而另一个新元素则被插入相同的位置 itr+1（第 27 行）。程序删除从 itr+2 到 itr+4-1 的元素（第 32 行）。

22.5.2 序列容器：deque

deque 一词代表双端队列。deque 提供了高效的操作来支持在两端进行插入和删除。除了所有序列容器的公共函数外，deque 类还包含表 22.8 中所示的函数。

表 22.8　deque 类特有的函数

函数	描述
deque(n, element)	构造一个双端队列,用同一元素的 n 个副本填充
deque(begin, end)	构造一个双端队列,初始化从迭代器 begin 到 end-1 的元素
deque(size)	构造具有指定大小的双端队列
at(index)	返回指定索引处的元素
push_front(element)	将元素插入队首
pop_front()	删除队首的元素

LiveExample 22.8 演示了如何使用 deque 中的函数。

LiveExample 22.8 的互动程序请访问 https://liangcpp.pearsoncmg.com/LiveRunCpp5e/faces/LiveExample.xhtml?header=off&programName=DequeDemo&fileType=.cpp&programHeight=950&resultHeight=290。

LiveExample 22.8　DequeDemo.cpp

Source Code Editor:

```cpp
#include <iostream>
#include <deque>
using namespace std;

int main()
{
  double values[] = {1, 2, 3, 4, 5, 6, 7};
  deque<double> doubleDeque(values, values + 7);

  cout << "Initial contents in doubleDeque: ";
  for (int i = 0; i < doubleDeque.size(); i++)
    cout << doubleDeque[i] << " ";

  doubleDeque.assign(4, 11.5);
  cout << "\nAfter the assign function, doubleDeque: ";
  for (int i = 0; i < doubleDeque.size(); i++)
    cout << doubleDeque[i] << " ";

  doubleDeque.at(0) = 22.4; // Assign 22.4 to doubleVector at index 0
  cout << "\nAfter the at function, doubleDeque: ";
  for (int i = 0; i < doubleDeque.size(); i++)
    cout << doubleDeque[i] << " ";

  deque<double>::iterator itr = doubleDeque.begin();
  doubleDeque.insert(itr + 1, 555);
  cout << "\nAfter the insert function, doubleDeque: ";
  for (int i = 0; i < doubleDeque.size(); i++)
    cout << doubleDeque[i] << " ";

  itr = doubleDeque.begin();
  doubleDeque.erase(itr + 2, itr + 4);
  cout << "\nAfter the erase function, doubleDeque: ";
  for (int i = 0; i < doubleDeque.size(); i++)
    cout << doubleDeque[i] << " ";

  doubleDeque.clear(); // Delete all in doubleVector
  cout << "\nAfter the clear function, doubleDeque: ";
  cout << "Size is " << doubleDeque.size() << endl;
  cout << "Is empty? " <<
      (doubleDeque.empty() ? "true" : "false") << endl;

  doubleDeque.push_front(10.10);
  doubleDeque.push_front(11.15);
```

```
44      doubleDeque.push_front(12.34);
45      cout << "After the insertion, doubleDeque: ";
46      for (int i = 0; i < doubleDeque.size(); i++)
47        cout << doubleDeque[i] << " ";
48
49      doubleDeque.pop_front();
50      doubleDeque.pop_back();
51      cout << "\nAfter the pop functions, doubleDeque: ";
52      for (int i = 0; i < doubleDeque.size(); i++)
53        cout << doubleDeque[i] << " ";
54
55      return 0;
56    }
```

Execution Result:

```
command>cl DequeDemo.cpp
Microsoft C++ Compiler 2019
Compiled successful (cl is the VC++ compile/link command)

command>DequeDemo
Initial contents in doubleDeque: 1 2 3 4 5 6 7
After the assign function, doubleDeque: 11.5 11.5 11.5 11.5
After the at function, doubleDeque: 22.4 11.5 11.5 11.5
After the insert function, doubleDeque: 22.4 555 11.5 11.5 11.5
After the erase function, doubleDeque: 22.4 555 11.5
After the clear function, doubleDeque: Size is 0
Is empty? true
After the insertion, doubleDeque: 12.34 11.15 10.1
After the pop functions, doubleDeque: 11.15

command>
```

deque 类包含 vector 类中的所有函数。因此，可以在使用 vector 的任何位置使用 deque，除了在双端队列内容发生更改时才需要获得新的迭代器。在第 30 行中获得一个新的迭代器，然后用它删除双端队列中的元素（第 30 行）。

添加和删除开头的元素，双端队列比向量更高效。第 42 ～ 44 行中，push_front() 函数将元素添加到双端队列的开头，pop_front() 函数将元素从双端队列的开头删除（第 49 行），pop_back() 函数则将元素从双端队列的末尾删除（第 50 行）。

22.5.3 序列容器：`list`

list 类被实现为双向链表。它支持用迭代器在列表的任何位置高效插入和删除。除了所有序列容器的公共函数外，list 类还包含表 22.9 中所示的函数。

表 22.9 `list` 类特有的函数

函数	描述
list(n, element)	构造一个列表，用同一元素的 n 个副本填充
list(begin, end)	构造一个列表，初始化从迭代器 begin 到 end-1 的元素
list(size)	构造以指定大小初始化的列表
push_front(element)	将元素插入队首
pop_front()	删除队首的元素
remove(element)	删除与指定元素相等的所有元素
remove_if(oper)	删除所有使 oper(element) 为 true 的元素
splice(pos, list2)	list2 中的所有元素都移动到此列表指定位置之前。调用此函数后，list2 为空

(续)

函数	描述
splice(pos1, list2, pos2)	list2 中从 pos2 开始的所有元素都移到这个列表 pos1 之前
splice(pos1, list2, begin, end)	list2 中从迭代器 begin 到 end-1 的所有元素都移动到这个列表 pos1 之前
sort()	将列表中的元素按递增顺序排序
sort(oper)	对列表中的元素排序。排序标准由 oper 函数指定
merge(list2)	假设这个列表和 list2 中的元素是已经排序好的。将 list2 归并到这个列表中。归并后,list2 为空
merge(list2,oper)	假设这个列表和 list2 中的元素是根据排序标准 oper 排序的,将 list2 归并到这个列表中
reverse()	使这个列表中的元素以相反顺序排列

vector 和 deque 的迭代器是随机访问的,但 list 的是双向的。不能用下标运算符 [] 访问 list 中的元素,因为 list 不支持随机迭代器。LiveExample 22.9 演示了如何使用 list 中的函数。

LiveExample 22.9 的互动程序请访问 https://liangcpp.pearsoncmg.com/LiveRunCpp5e/faces/LiveExample.xhtml?header=off&programName=ListDemo&fileType=.cpp&programHeight=1640&resultHeight=420。

LiveExample 22.9 ListDemo.cpp

Source Code Editor:

```cpp
#include <iostream>
#include <list>
using namespace std;

int main()
{
  int values[] = {1, 2, 3, 4};
  list<int> intList(values, values + 4);

  cout << "Initial contents in intList: ";
  for (int& e: intList)
    cout << e << " ";

  intList.assign(4, 11);
  cout << "\nAfter the assign function, intList: ";
  for (int& e: intList)
    cout << e << " ";

  auto itr = intList.begin();
  itr++;
  intList.insert(itr, 555);
  intList.insert(itr, 666);
  cout << "\nAfter the insert function, intList: ";
  for (int& e: intList)
    cout << e << " ";

  auto beg = intList.begin();
  itr++;
  intList.erase(beg, itr);
  cout << "\nAfter the erase function, intList: ";
  for (int& e: intList)
    cout << e << " ";

  intList.clear();
```

```cpp
35      cout << "\nAfter the clear function, intList: ";
36      cout << "Size is " << intList.size() << endl;
37      cout << "Is empty? " <<
38         (intList.empty() ? "true" : "false");
39
40      intList.push_front(10);
41      intList.push_front(11);
42      intList.push_front(12);
43      cout << "\nAfter the push functions, intList: ";
44      for (int& e: intList)
45         cout << e << " ";
46
47      intList.pop_front();
48      intList.pop_back();
49      cout << "\nAfter the pop functions, intList: ";
50      for (int& e: intList)
51         cout << e << " ";
52
53      int values1[] = {7, 3, 1, 2};
54      list<int> list1(values1, values1 + 4);
55      list1.sort();
56      cout << "\nAfter the sort function, list1: ";
57      for (int& e: list1)
58         cout << e << " ";
59
60      list<int> list2(list1);
61      list1.merge(list2);
62      cout << "\nAfter the merge function, list1: ";
63      for (int& e: list1)
64         cout << e << " ";
65      cout << "\nSize of list2 is " << list2.size();
66
67      list1.reverse(); // Reverse list1
68      cout << "\nAfter the reverse function, list1: ";
69      for (int& e: list1)
70         cout << e << " ";
71
72      list1.push_back(7);
73      list1.push_back(1);
74      cout << "\nAfter the push functions, list1: ";
75      for (int& e: list1)
76         cout << e << " ";
77
78      list1.remove(7); // Remove number 7 from list1
79      cout << "\nAfter the remove function, list1: ";
80      for (int& e: list1)
81         cout << e << " ";
82
83      list2.assign(7, 2);
84      cout << "\nAfter the assign function, list2: ";
85      for (int& e: list2)
86         cout << e << " ";
87
88      auto p = list2.begin();
89      p++;
90      list2.splice(p, list1);
91      cout << "\nAfter the splice function, list2: ";
92      for (int& e: list2)
93         cout << e << " ";
94      cout << "\nAfter the splice function, list1's size is "
95         << list1.size();
96
97      return 0;
98   }
```

```
Automatic Check   Compile/Run   Reset   Answer          Choose a Compiler: VC++

Execution Result:
command>cl ListDemo.cpp
Microsoft C++ Compiler 2019
Compiled successful (cl is the VC++ compile/link command)

command>ListDemo
Initial contents in intList: 1 2 3 4
After the assign function, intList: 11 11 11 11
After the insert function, intList: 11 555 666 11 11 11
After the erase function, intList: 11 11
After the clear function, intList: Size is 0
Is empty? true
After the push functions, intList: 12 11 10
After the pop functions, intList: 11
After the sort function, list1: 1 2 3 7
After the merge function, list1: 1 1 2 2 3 3 7 7
Size of list2 is 0
After the reverse function, list1: 7 7 3 3 2 2 1 1
After the push functions, list1: 7 7 3 3 2 2 1 1 7 1
After the remove function, list1: 3 3 2 2 1 1 1
After the assign function, list2: 2 2 2 2 2 2
After the splice function, list2: 2 3 3 2 2 1 1 1 2 2 2 2 2 2
After the splice function, list1's size is 0

command>
```

该程序在第 7～12 行创建一个列表 intList，并显示其内容。

该程序将值为 11 的四个元素赋值给 intList（第 14 行），将 555 和 666 插入迭代器 itr 指定的位置（第 21～22 行），删除从迭代器 beg 到迭代器 itr 的元素（第 29 行），清空列表（第 34 行），并压入和弹出元素（第 40～51 行）。该程序创建了一个列表 list1（第 54 行），对其进行排序（第 55 行），并将其与 list2 归并（第 61 行）。归并后，list2 为空。

程序将 list1 中的元素顺序颠倒（第 67 行），并从 list1 中删除所有值为 7 的元素（第 78 行）。

该程序应用 splice 函数将 list1 中所有元素移动到 list2 中迭代器 p 之前的位置（第 90 行）。之后，list1 为空。

22.6 关联容器

要点提示：存储在关联容器中的元素可以通过键访问。

STL 提供了四个关联容器：set、multiset、map 和 multimap。这些容器效率极高。项的查找、插入和删除为 $O(1)$ 时间。

关联容器中的元素是根据某种排序标准进行排序的。默认情况下，使用 <= 运算符对元素进行排序。set 存储键的集合。map 存储一组键/值对的集合。键/值对中的第一个元素是键，第二个元素是值。

set 和 multiset 容器是相同的，只是 multiset 允许重复的键，而 set 不允许。map 和 multimap 是相同的，只是 multimap 允许重复的键，而 map 不允许。

表 22.2 和表 22.3 列出了所有容器和一级容器的公共函数。除了这些常见函数外，每个关联容器还支持表 22.10 中所示的函数。

表 22.10 关联容器的公共函数

函数	描述
`find(key)`	返回一个迭代器，它指向容器中具有指定键的元素
`lower_bound(key)`	返回一个迭代器，它指向容器中具有指定键的第一个元素
`upper_bound(key)`	返回一个迭代器，它指向容器中具有指定键的最后一个元素之后的下一个元素
`count(key)`	返回具有指定键的元素在容器中出现的次数

22.6.1 关联容器：`set` 和 `multiset`

元素是存储在 `set/multiset` 容器中的键。`multiset` 允许重复的键，但 `set` 不允许。LiveExample 22.10 演示了如何使用 `set` 和 `multiset` 容器。

LiveExample 22.10 的互动程序请访问 https://liangcpp.pearsoncmg.com/LiveRunCpp5e/faces/LiveExample.xhtml?header=off&programName=SetDemo&fileType=.cpp&programHeight=650&resultHeight=230。

LiveExample 22.10　SetDemo.cpp

Source Code Editor:

```cpp
#include <iostream>
#include <set>
using namespace std;

int main()
{
  int values[] = {3, 5, 1, 7, 2, 2};
  multiset<int> set1(values, values + 6);

  cout << "Initial contents in set1: ";
  for (int e: set1)
    cout << e << " ";

  set1.insert(555);
  set1.insert(1); // Add 1 to set1
  cout << "\nAfter the insert function, set1: ";
  for (int e: set1)
    cout << e << " ";

  auto p = set1.lower_bound(2);
  cout << "\nLower bound of 2 in set1: " << *p;
  p = set1.upper_bound(2);
  cout << "\nUpper bound of 2 in set1: " << *p;

  p = set1.find(2); // Find 2 in set1
  if (p == set1.end())
    cout << "\n2 is not in set1";
  else
    cout << "\nThe number of 2's in set1: " << set1.count(2);

  set1.erase(2); // Delete 2 from set1
  cout << "\nAfter the erase function, set1: ";
  for (int e: set1)
    cout << e << " ";

  return 0;
}
```

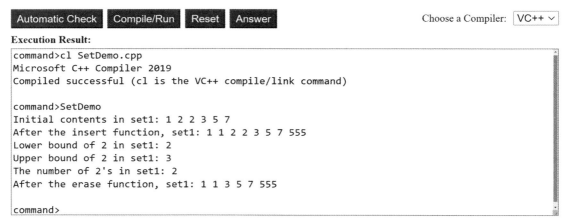

该程序在第 7 ～ 12 行创建一个集合 set1 并显示其内容。默认情况下，集合中的元素按递增顺序排序。要指定递减顺序，可以将第 8 行替换为

```
multiset<int, greater<int>> set1(values, values + 6);
```

程序插入键 555 和 1（第 14 ～ 15 行），并显示集合中的新元素（第 16 ～ 18 行）。调用 lower_bound(2)（第 20 行）返回指向容器中第一次出现的 2 的迭代器，调用 upper_bound(2)（第 22 行）返回指向容器中最后一次出现的 2 之后的下一个元素的迭代器。因此，第 23 行中的 *p 显示元素 3。

调用 find(2)（第 25 行）返回指向容器中第一次出现的 2 的迭代器。如果容器中没有这样的元素，则返回的迭代器为 end()（第 26 行）。

调用 erase(2)（第 31 行）从集合中删除所有键值为 2 的元素。

此示例创建了一个 multiset。可以用 set 替换 multiset：

```
set<int> set1(values, values + 6);
```

请用此新语句跟踪程序。

22.6.2 关联容器：map 和 multimap

映射 map/multimap 中的每个元素都是一对（也称为一项）。该对中的第一个值是键，第二个值与键相关联，如图 22.2 所示。

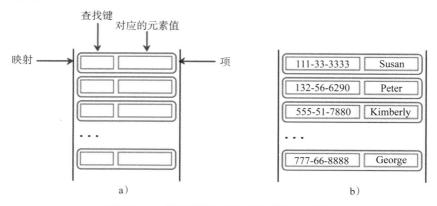

图 22.2 由键/值对组成的项存储在映射中

map/multimap 可使用键快速访问值。multimap 允许重复的键，但 map 不允许。LiveExample 22.11 演示了如何使用 map 和 multimap 容器。

LiveExample 22.11 的互动程序请访问 https://liangcpp.pearsoncmg.com/LiveRunCpp5e/faces/LiveExample.xhtml?header=off&programName=MapDemo&fileType=.cpp&programHeight=610&resultHeight=340。

LiveExample 22.11 MapDemo.cpp

Source Code Editor:

```cpp
#include <iostream>
#include <map>
#include <string>
using namespace std;

int main()
{
  map<int, string> map1;
  map1.insert(map<int, string>::value_type(100, "John Smith"));
  map1.insert(map<int, string>::value_type(101, "Peter King"));
  map1[102] = "Jane Smith";
  map1[103] = "Jeff Reed"; // Add (103, "Jeff Reed") to map1

  cout << "Initial contents in map1:\n";
  map<int, string>::iterator p;
  for (p = map1.begin(); p != map1.end(); p++)
    cout << p->first << " " << p->second << endl;

  cout << "Enter a key to serach for the name: ";
  int key;
  cin >> key;
  p = map1.find(key);

  if (p == map1.end())
    cout << "  Key " << key << " not found in map1";
  else
    cout << "  " << p->first << " " << p->second << endl;

  map1.erase(103); // Delete key 103 from map1
  cout << "\nAfter the erase function, map1:\n";
  for (p = map1.begin(); p != map1.end(); p++)
    cout << p->first << " " << p->second << endl;

  return 0;
}
```

Enter input data for the program (Sample data provided below. You may modify it.)

105

Automatic Check | Compile/Run | Reset | Answer   Choose a Compiler: VC++

Execution Result:

```
command>cl MapDemo.cpp
Microsoft C++ Compiler 2019
Compiled successful (cl is the VC++ compile/link command)

command>MapDemo
Initial contents in map1:
100 John Smith
101 Peter King
```

```
102 Jane Smith
103 Jeff Reed
Enter a key to serach for the name: 105
  Key 105 not found in map1
After the erase function, map1:
100 John Smith
101 Peter King
102 Jane Smith

command>
```

该程序用其无参数构造函数创建映射 map1（第 8 行），并将键 / 值对插入 map1 中（第 9 ~ 12 行）。可以使用 insert 函数（第 9 ~ 10 行）或语法 map1[key]=value（第 11 ~ 12 行）将一个对插入映射中。

要用 insert 函数插入一个对，必须用 value_type(key, value) 函数创建一个对。

map1[key]=value 语法比 insert 函数更短，可读性更强。如果键不在映射中，则该语法会向映射插入一个新的对。如果键已经在映射中，则该语法会用新的键 / 值对替换现有的键 / 值对。给定一个键，也可以用 map1[key] 来检索该键的值。

程序提示用户输入一个键（第 19 ~ 21 行）。调用 find(key) 返回迭代器，该迭代器指向具有指定键的元素（第 22 行）。对由键和值组成，可以使用 p->first 和 p->second 访问（第 27 行）。

调用 erase(103) 删除具有键 103 的元素（第 29 行）。

此示例创建了一个 map。可以按如下方式将 map 替换为 multimap：

```
multimap<int, string> map1;
```

该程序的运行方式与使用 map 的方式完全相同。

22.7 容器适配器

要点提示：stack、queue 和 priority_queue 在 STL 中被称为容器适配器。

STL 提供了三个容器适配器：stack、queue 和 priority_queue。它们之所以被称为适配器，是因为它们是基于序列容器改编的，用于处理特殊情况。STL 使程序员能够为容器适配器选择合适的序列容器。例如，可以创建一个具有底层数据结构 vector、deque 或 list 的栈。

容器适配器没有迭代器。表 22.2 列出了所有容器的公共函数。除了这些常见的函数外，每个容器适配器还支持插入和删除元素的 push 和 pop 函数。

22.7.1 容器适配器：stack

stack 是一个后进先出的容器。可以选择用 vector、deque 或 list 来构造 stack。默认情况下，栈是用 deque 实现的。表 22.11 列出了 stack 上常用的函数。

LiveExample 22.12 给出了使用 stack 的示例。

表 22.11 stack 的函数

函数	描述
push(element)	在栈顶插入元素
pop()	删除栈顶元素
top()	返回栈顶元素并且不删除它
size()	返回栈的大小
empty()	如果栈为空则返回 true

LiveExample 22.12 的互动程序请访问 https://liangcpp.pearsoncmg.com/LiveRunCpp5e/faces/LiveExample.xhtml?header=off&programName=StackDemo&fileType=.cpp&programHeight=600&resultHeight=180。

LiveExample 22.12　StackDemo.cpp

Source Code Editor:

```cpp
#include <iostream>
#include <stack>
#include <vector>
using namespace std;

template<typename T>
void printStack(T& stack)
{
  while (!stack.empty())
  {
    cout << stack.top() << " ";
    stack.pop(); // Remove an element from stack
  }
}

int main()
{
  stack<int> stack1;
  stack<int, vector<int>> stack2;

  for (int i = 0; i < 8; i++)
  {
    stack1.push(i);
    stack2.push(i);
  }

  cout << "Contents in stack1: ";
  printStack(stack1);

  cout << "\nContents in stack2: ";
  printStack(stack2);

  return 0;
}
```

Execution Result:

```
command>cl StackDemo.cpp
Microsoft C++ Compiler 2019
Compiled successful (cl is the VC++ compile/link command)

command>StackDemo
Contents in stack1: 7 6 5 4 3 2 1 0
Contents in stack2: 7 6 5 4 3 2 1 0

command>
```

该程序在第 18 行使用默认实现创建 `stack`，在第 19 行使用 `vector` 实现创建 `stack`。该程序将 0 到 7 压入 `stack1` 和 `stack2`（第 21 ~ 25 行），并调用 `printStack(stack1)`

和 `printStack(stack2)` 来显示和删除 `stack1` 和 `stack2` 中的所有元素。

22.7.2 容器适配器：`queue`

`queue` 是一个先进先出的容器。可以选择用 `deque` 或 `list` 来构造 `queue`。默认情况下，`queue` 是用 `deque` 实现的。表 22.12 列出了 `queue` 中常用的函数。

表 22.12 `queue` 的函数

函数	描述
`push(element)`	将元素追加到队列
`pop()`	从队列中删除一个元素
`front()`	返回队首元素并且不删除它
`back()`	返回队尾元素并且不删除它
`size()`	返回队列的大小
`empty()`	如果队列为空则返回 `true`

LiveExample 22.13 给出了使用 `queue` 的示例。

LiveExample 22.13 的互动程序请访问 https://liangcpp.pearsoncmg.com/LiveRunCpp5e/faces/LiveExample.xhtml?header=off&programName=QueueDemo&fileType=.cpp&programHeight=590&resultHeight=180。

LiveExample 22.13 QueueDemo.cpp

Source Code Editor:

```cpp
#include <iostream>
#include <queue>
#include <list>
using namespace std;

template<typename T>
void printQueue(T& queue)
{
  while (!queue.empty())
  {
    cout << queue.front() << " ";
    queue.pop(); // Remove an element from the queue
  }
}

int main()
{
  queue<int> queue1;
  queue<int, list<int>> queue2;

  for (int i = 0; i < 8; i++)
  {
    queue1.push(i);
    queue2.push(i);
  }

  cout << "Contents in queue1: ";
  printQueue(queue1);

  cout << "\nContents in queue2: ";
```

```
31      printQueue(queue2);
32
33      return 0;
34  }
```

Execution Result:
```
command>cl QueueDemo.cpp
Microsoft C++ Compiler 2019
Compiled successful (cl is the VC++ compile/link command)

command>QueueDemo
Contents in queue1: 0 1 2 3 4 5 6 7
Contents in queue2: 0 1 2 3 4 5 6 7

command>
```

该程序在第 18 行使用默认实现创建 queue，在第 19 行使用 list 实现创建 queue。

该程序将 0 到 7 的数字追加到 queue1 和 queue2（第 21 ~ 25 行），并调用 printQueue (queue1) 和 printQueue(queue2) 来显示和删除 queue1 和 queue2 中的所有元素。

22.7.3 容器适配器：`priority_queue`

在**优先级队列**中，会给元素分配优先级。具有最高优先级的元素会被优先访问或删除。

可以选择用 vector 或 deque 来构建 priority_queue。默认情况下，priority_queue 用 vector 实现。例如，可以用以下语句创建 int 值的优先级队列：

```
priority_queue<int> priority_queue1;
priority_queue<int, deque<int>> priority_queue2;
```

默认情况下，用 <= 运算符比较元素。最大的值被赋予最高的优先级。可以指定 >= 运算符来构造优先级队列，这样可以为最小值分配最高优先级。

```
priority_queue<int, deque<int>, greater<int>> priority_queue3;
```

priority_queue 类使用与 stack 类中相同的 push、pop、top、size 和 empty 函数。

LiveExample 22.14 给出了使用 priority_queue 的示例。

LiveExample 22.14 的互动程序请访问 https://liangcpp.pearsoncmg.com/LiveRunCpp5e/faces/LiveExample.xhtml?header=off&programName=PriorityQueueDemo&fileType=.cpp&programHeight=610&resultHeight=180。

LiveExample 22.14　PriorityQueueDemo.cpp

Source Code Editor:
```
1  #include <iostream>
2  #include <functional>
3  #include <queue>
4  #include <deque>
5  using namespace std;
6
```

```cpp
 7  template<typename T>
 8  void printQueue(T& pQueue)
 9  {
10    while (!pQueue.empty())
11    {
12      cout << pQueue.top() << " ";
13      pQueue.pop();// Remove an element from the queue
14    }
15  }
16
17  int main()
18  {
19    priority_queue<int> queue1;
20    priority_queue<int, deque<int>, greater<int>> queue2;
21
22    queue1.push(7); queue2.push(7);
23    queue1.push(4); queue2.push(4);
24    queue1.push(9); queue2.push(9);
25    queue1.push(2); queue2.push(2);
26    queue1.push(1); queue2.push(1);
27
28    cout << "Contents in queue1: ";
29    printQueue(queue1);
30
31    cout << "\nContents in queue2: ";
32    printQueue(queue2);
33
34    return 0;
35  }
```

Execution Result:

```
command>cl PriorityQueueDemo.cpp
Microsoft C++ Compiler 2019
Compiled successful (cl is the VC++ compile/link command)

command>PriorityQueueDemo
Contents in queue1: 9 7 4 2 1
Contents in queue2: 1 2 4 7 9

command>
```

该程序在第 19 行使用默认实现创建 `priority_queue`，在第 20 行用带有 > 运算符的 `deque` 实现创建 `priority_queue`。

程序将数字插入 `queue1` 和 `queue2`（第 22 ～ 26 行），并调用 `printQueue(queue1)` 和 `printQueue(queue2)` 来显示和删除 `queue1` 和 `queue2` 中的所有元素。在 `queue1` 中，最大值具有最高优先级，但在 `queue2` 中，最小值具有最高优先级。

关键术语

associative container（关联容器）
auto type inference（自动类型推断）
bidirectional iterator（双向迭代器）
container（容器）
container adapter（容器适配器）

deque（双端队列）
first-class container（一级容器）
forward iterator（正向迭代器）
input iterator（输入迭代器）
istream_iterator（输入流迭代器）

iterator（迭代器）
list（列表）
map（映射）
multimap（多重映射）
multiset（多重集合）
ostream_iterator（输出流迭代器）
output iterator（输出迭代器）
priority_queue（优先级队列）
queue（队列）
random-access iterator（随机访问迭代器）
sequence container（顺序容器）
set（集合）
STL algorithm STL（算法）
vector（向量）

章节总结

1. 标准模板库（STL）包含有用的数据结构。不用重复开发就可以使用它们。
2. 容器对象（如向量）用于存储通常称为元素的数据集。
3. STL 容器类广泛使用迭代器，迭代器是便于遍历容器元素的对象。迭代器就像内置的指针，提供了一种便于访问和操作容器中的元素的方式。
4. 序列容器（也称为顺序容器）表示线性数据结构。三个序列容器分别是 `vector`、`list` 和 `deque`。
5. 关联容器是非线性容器，可以快速定位存储在容器中的元素。关联容器可以存储值集或键/值对。四个关联容器分别是 `set`、`multiset`、`map` 和 `multimap`。
6. 容器适配器是序列容器的受限版本。它们改编自序列容器以处理特殊情况。三个容器适配器分别是 `stack`、`queue` 和 `priority_queue`。
7. 迭代器是指针的抽象形式，事实上，它通常是用指针实现的。每个容器都有自己的迭代器类型。该抽象隐藏了详细的实现，并为在所有容器上使用迭代器提供了一种统一的方式。
8. 迭代器可以分为五类：输入迭代器、输出迭代器、正向迭代器、双向迭代器和随机访问迭代器。
9. 输入迭代器用于从容器中读取元素。
10. 输出迭代器用于将元素写入容器。
11. 正向迭代器结合了输入迭代器和输出迭代器的所有功能，以支持读取和写入操作。
12. 双向迭代器是一种具有向后移动能力的正向迭代器。
13. 随机访问迭代器是一种双向迭代器，能够以任意顺序访问任意元素。
14. 迭代器类型决定了可以使用哪些运算符。`vector` 和 `deque` 容器支持随机访问迭代器，`list`、`set`、`multiset`、`map` 和 `multimap` 容器支持双向迭代器。`stack`、`queue` 和 `priority_queue` 不支持迭代器。
15. 如果在 `vector` 上追加元素，那么 `vector` 是高效的。在 `vector` 中插入或删除元素的成本很高。
16. `deque` 类似于 `vector`，但在 `deque` 的前端和尾部插入和删除元素的效率都很高。在 `deque` 中间插入或删除元素的成本仍然很高。
17. 链表适用于需要在列表任意位置频繁插入和删除的应用程序。
18. `set` 和 `multiset` 容器相同，只是 `multiset` 允许重复的键，而 `set` 不允许。
19. `map` 和 `multimap` 相同，只是 `multimap` 允许重复的键，而 `map` 不允许。

编程练习

互动程序请访问 https://liangcpp.pearsoncmg.com/CheckExerciseCpp/faces/CheckExercise5e.xhtml?chapter=22&programName=Exercise22_01。

*22.1 （最大值和最小值）实现以下函数，查找一级容器中的最大值和最小值元素。

```
template<typename ElementType, typename ContainerType>
ElementType maxElement(const ContainerType& container)
template<typename ElementType, typename ContainerType>
ElementType minElement(const ContainerType& container)
```

用 https://liangcpp.pearsoncmg.com/test/Exercise22_01.txt 的模板完成代码。

*22.2 （值的位置）实现以下函数，查找一级容器中指定值的位置。如果没有匹配项，则返回 -1。

```
template<typename ElementType, typename ContainerType>
int find(const ContainerType& container,
    const ElementType& value)
```

*22.3 （值的出现次数）实现以下函数，查找一级容器中指定值的出现次数：

```
template<typename ElementType, typename ContainerType>
int countElement(ContainerType& container, const ElementType& value)
```

*22.4 （反序容器）实现以下函数，将容器中的元素以相反顺序排列：

```
template<typename ContainerType>
void reverse(const ContainerType& container)
```

*22.5 （删除元素）实现以下函数，从一级容器中删除指定值。仅删除容器中第一次出现的匹配值。

```
template<typename ElementType, typename ContainerType>
void remove(ContainerType& container, const ElementType& value)
```

*22.6 （替换元素）实现以下函数，用新值替换给定元素：

```
template<typename ElementType, typename ContainerType>
void replace(ContainerType& container,
    const ElementType& oldValue, const ElementType& newValue)
```

**22.7 （两个集合的并集）实现以下数学集合并集函数，将两个集合 s1 和 s2 合并为一个新的集合 s3：

```
template<typename ElementType>
void setUnion(const set<ElementType>& s1,
    const set<ElementType>& s2, set<ElementType>& s3)
```

**22.8 （两个集合的差）实现以下数学集合差函数，用 s1 和 s2 的差生成新的集合 s3：

```
template<typename ElementType>
void difference(set<ElementType>& s1, set<ElementType>& s2, set<ElementType>& s3)
```

**22.9 （按升序显示非重复单词）编写一个程序，从文本文件中读取单词，并按升序显示所有非重复单词。（提示：使用一个集合来存储所有单词。）

**22.10 （按升序显示重复单词）编写一个程序，从文本文件中读取单词，并按升序显示所有单词（允许重复）。（提示：使用一个多重集合来存储所有单词。）

**22.11 （统计 C++ 源代码中的关键字）编写一个程序，读取 C++ 源代码文件并报告文件中的关键字数量。（提示：创建一个集合来存储所有 C++ 关键字。）

**22.12 （统计输入数字的出现次数）编写一个程序，读取未指定数量的整数，并求每个数字的出现次数。当输入为 0 时，输入结束。（提示：使用一个映射来存储对。对中的第一个元素是输入的数字，第二个元素跟踪该数字的出现次数。）

```
Sample Run for Exercise22_12.cpp
Enter input data for the program (Sample data provided below. You may modify it.)
2 3 4 1 2 34 4 3 0

Show the Sample Output Using the Preceeding Input    Reset

Execution Result:
command>Exercise22_12
```

```
Enter numbers (ending with 0): 2 3 4 1 2 34 4 3 0
number of occurrences for 1 is 1
number of occurrences for 2 is 2
number of occurrences for 3 is 2
number of occurrences for 4 is 2
number of occurrences for 34 is 1

command>
```

**22.13 （统计单词的出现次数）编写一个程序，提示用户输入文件名，统计文件中单词的出现次数，并按单词的升序显示单词及其出现次数。单词用空格隔开。该程序使用映射存储单词及其出现次数。对于每个单词，检查它是否已经是映射中的一个键。如果不是，将键和值 1 添加到映射中。否则，将映射中单词（键）的值增加 1。

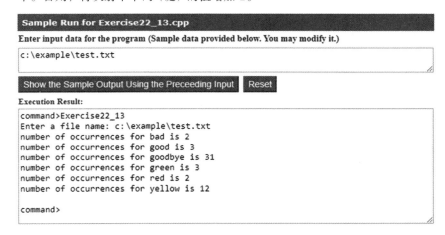

**22.14 （使用映射猜首府）重写编程练习 8.15，将州及其首府对存储在映射中。程序提示用户输入州名并显示该州的首府。

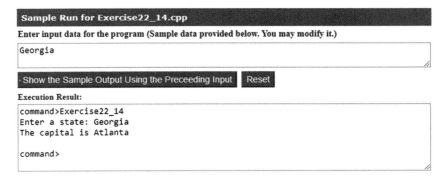

**22.15 （统计 C++ 源文件中每个关键字的出现次数）编写一个程序，读取 C++ 源代码文件并报告文件中每个关键字的出现次数。

```
Enter a file name: c:\example\Welcome.cpp
number of occurrences for int is 1
number of occurrences for namespace is 1
number of occurrences for return is 1
number of occurrences for using is 1

command>
```

*22.16 （统计辅音和元音）编写一个程序，提示用户输入文本文件名，并显示文件中元音和辅音的数量。使用集合来存储元音 A、E、I、O 和 U。

*22.17 （减法测验）重写编程练习 12.34，将答案存储在集合中，而不是向量中。

第 23 章

Introduction to C++ Programming and Data Structures, Fifth Edition

STL 算法

学习目标

1. 将各种类型的迭代器与 STL 算法一起使用（23.1～23.20 节）。
2. 探索 STL 算法的四种类型：非修改性算法、修改性算法、数值算法和堆算法（23.2 节）。
3. 使用算法 `copy` 复制内容（23.3 节）。
4. 使用算法 `fill`，`fill_n` 填充值（23.4 节）。
5. 将函数作为参数传递（23.5 节）。
6. 使用算法 `generate` 和 `generate_n`（23.6 节）。
7. 使用算法 `remove`、`remove_if`、`remove_copy`，以及 `remove_copy_if`（23.7 节）。
8. 使用布尔函数来指定 STL 算法的标准（23.8 节）。
9. 使用算法 `replace`、`replace_if`、`replace_copy`、`replace_copy_if`、`find`、`find_if`、`find_end`，以及 `find_first_of`（23.8～23.9 节）。
10. 使用算法 `search`、`search_n`、`sort`、`binary_search`、`adjacent_find`、`merge`，以及 `inplace_merge`（23.10～23.12 节）。
11. 在 STL 算法中使用函数对象（23.11 节）。
12. 使用算法 `reverse`、`reverse_copy`、`rotate`、`rotate_copy`、`swap`、`iter_swap`，以及 `swap_ranges`（23.13～23.15 节）。
13. 使用算法 `count`、`count_if`、`max_element`、`min_element`、`random_shuffle`、`for_each`，以及 `transform`（23.16～23.19 节）。
14. 使用集合算法 `includes`、`set_union`、`set_difference`、`set_intersection`，以及 `set_symmetric_difference`（23.20 节）。
15. 使用数值算法 `accumulate`、`adjacent_difference`、`inner_product`，以及 `partial_sum`（23.21 节）。
16. 使用 lambda 表达式来简化编码（23.22 节）。
17. 使用新的 C++11 STL 数值算法 `all_of`、`any_of`、`copy_if`、`copy_n`、`find_if_not` 和 `none_of`（23.23 节）。

23.1 简介

要点提示：STL 提供了通过迭代器操作容器元素的算法。

我们常常需要在容器中查找元素，用新元素替换容器中的元素，删除容器中的元素，用一些元素填充容器，反转容器中的元素，在容器中找到最大或最小元素，对容器中的元素进行排序。这些函数对所有容器都是通用的。STL 不是在每个容器中实现它们，而是将它们作为泛型算法来使用，这些算法可以应用于各种容器和数组。算法通过迭代器对元素进行操作。例如，如下所示用迭代器在向量和集合中应用 STL 算法 `max_element`:

```cpp
vector<int> v; v.push_back(4); v.push_back(14); v.push_back(1);
set<int> s; s.insert(4); s.insert(14); s.insert(1);
cout << "The max element in the vector is " <<
    *max_element(v.begin(), v.end()) << endl;
cout << "The max element in the set is " <<
    *max_element(s.begin(), s.end()) << endl;
```

在 STL 之前，算法是在具有继承性和多态性的类中实现的。STL 将算法与容器分离。这使得算法能够通过迭代器通用地应用于所有容器。STL 使算法和容器易于维护。

注意：操作、算法和函数这三个术语是可互换的。函数是操作，函数是使用算法实现的。

注意：迭代器是指针的一种泛化。指针本身就是迭代器。因此，数组指针可以被视为迭代器。迭代器通常与容器一起使用，但有些迭代器（如 istream_iterator 和 ostream_iterator）与容器无关。

23.2 算法类型

要点提示：STL 算法可分为非修改性算法、修改性算法、数值算法和堆算法。

STL 提供了大约 80 种算法。它们可分为四组：

- **非修改性算法**：非修改性算法不会更改容器中的内容。它们从元素中获取信息。表 23.1 列出了非修改性算法。
- **修改性算法**：修改性算法通过插入、删除、重新排列和更改元素的值来修改容器中的元素。表 23.2 列出了这些算法。
- **数值算法**：数值算法提供四种数值运算，用于计算累加、相邻差、内积以及部分和。表 23.3 列出了这些算法。
- **堆算法**：堆算法提供了四种操作，用于创建堆、从堆中删除元素和向堆中插入元素以及对堆进行排序。见表 23.4。

表 23.1 非修改性算法

adjacent_find	find	lower_bound	search
binary_search	find_end	mismatch	search_n
count	find_first_of	max	upper_bound
count	find_if	max_element	
equal	for_each	min	
equal_range	includes	min_element	

表 23.2 修改性算法

copy	prev_permutation	rotate_copy
copy_backward	random_shuffle	set_difference
fill	remove	set_intersection
fill_n	remove_copy	set_symmetric_difference
generate	remove_copy_if	set_union
generate_n	remove_if	sort
inplace_merge	replace	stable_partition
iter_swap	replace_copy	stable_sort
merge	replace_copy_if	swap
next_permutation	replace_if	swap_ranges

(续)

nth_element	reverse	transform
aprtial_sort	reverse_copy	unique
partial_sort_copy	rotate	unique_copy
partition		

表 23.3 数值算法

accumulate	adjacent_difference	inner_product	partial_sum

表 23.4 堆算法

make_heap	pop_heap	push_heap	sort_heap

数值算法包含在 `<numeric>` 头文件中，所有其他算法都包含在 `<algorithm>` 头文件中。所有的算法都是通过迭代器运行的。回想一下，STL 定义了五种类型的迭代器：输入、输出、正向、双向和随机访问。容器 `vector` 和 `deque` 以及数组支持随机访问迭代器，`list`、`set`、`multiset`、`map` 和 `multimap` 支持双向迭代器。大多数算法都需要一个正向迭代器。如果一个算法使用弱迭代器，它可以自动使用更强的迭代器。

许多算法在由两个迭代器指向的元素序列上进行操作。第一个迭代器指向序列的第一个元素，第二个迭代器指向序列最后一个元素之后的元素。

本章余下部分给出了演示一些常用算法的示例。

23.3 `copy` 函数

要点提示：`copy` 函数可将序列中的元素从一个容器复制到另一个容器。

`copy` 函数的语法为

```cpp
template<typename InputIterator, typename OutputIterator>
OutputIterator copy(InputIterator begin, InputIterator end,
  OutputIterator targetPosition)
```

函数将源容器 `begin` 到 `end-1` 范围内的元素复制到从 `targetPosition` 开始的目标容器，其中 `begin` 和 `end` 是源容器中的迭代器，`targetPosition` 是目标容器中的迭代器。该函数返回一个迭代器，该迭代器指向最后复制的元素的下一个位置。

LiveExample 23.1 演示了如何使用 `copy` 函数。

LiveExample 23.1 的互动程序请访问 https://liangcpp.pearsoncmg.com/LiveRunCpp5e/faces/LiveExample.xhtml?header=off&programName=CopyDemo&programHeight=610&resultHeight=210。

LiveExample 23.1 CopyDemo.cpp

Source Code Editor:

```cpp
1  #include <iostream>
2  #include <algorithm>
3  #include <vector>
4  #include <list>
5  #include <iterator>
6  using namespace std;
7
8  int main()
9  {
```

```
10      int values[] = {1, 2, 3, 4, 5, 6};
11      vector<int> intVector(5);
12      list<int> intList(5);
13
14      copy(values + 2, values + 4, intVector.begin());
15      copy(values, values + 5, intList.begin()); // Copy values to intList
16
17      cout << "After initial copy intVector: ";
18      for (int e: intVector)
19          cout << e << " ";
20
21      cout << "\nAfter initial copy intList: ";
22      for (int e: intList)
23          cout << e << " ";
24
25      intVector.insert(intVector.begin(), 747);
26      ostream_iterator<int> output(cout, " ");
27      cout << "\nAfter the insertion function, intVector: ";
28      copy(intVector.begin(),intVector.end(),output); // Copy intVector to output
29
30      copy(intVector.begin(), intVector.begin() + 4, intList.begin());
31      cout << "\nAfter the copy function, intList: ";
32      copy(intList.begin(), intList.end(), output);
33
34      return 0;
35  }
```

Execution Result:

```
command>cl CopyDemo.cpp
Microsoft C++ Compiler 2019
Compiled successful (cl is the VC++ compile/link command)

command>CopyDemo
After initial copy intVector: 3 4 0 0 0
After initial copy intList: 1 2 3 4 5
After the insertion function, intVector: 747 3 4 0 0 0
After the copy function, intList: 747 3 4 0 5

command>
```

程序创建一个数组（第 10 行）、一个向量（第 11 行）和一个列表（第 12 行）。copy 函数将元素 values[2] 和 values[3] 复制到向量的开头（第 14 行），将元素 values[0]、values[1]、values[2]、values[3] 和 values[4] 复制到列表的开头（第 15 行）。

你可以从数组复制元素到容器中，还可以将元素从容器复制到数组或输出流。该程序创建一个名为 output 的输出迭代器，用于将数据写入 cout（第 26 行）。构造函数中的 " " 用于分离发送到 cout 的数据。如果将 " " 替换为 "B"，你将看到屏幕上显示的数据由字符 B 分隔。

程序向向量中插入新元素（第 25 行），创建输出流迭代器（第 26 行），并将向量复制到输出流迭代器（第 28 行）：

```
copy(intVector.begin(), intVector.end(), output);
```

列表中的元素以类似方式显示（第 32 行）。

提示：在 22.3.4 节中引入了 ostream_iterator。使用 copy 函数将元素从容器写入输出流是很方便的。

警告：在将 n 个元素从源复制到目标之前，目标中的元素必须已经存在。否则，可能会发生运行时错误。例如，以下代码将导致运行时错误，因为向量为空：

```cpp
int values[] = {1, 2, 3, 4, 5, 6};
vector<int> intVector;
copy(values + 2, values + 4, intVector.begin()); // Error
```

23.4 `fill` 和 `fill_n`

要点提示：`fill` 和 `fill_n` 函数可以用指定值填充容器。

`fill` 函数的语法如下所示：

```cpp
template<typename ForwardIterator, typename T>
void fill(ForwardIterator begin, ForwardIterator end, const T& value)
```

`fill_n` 函数用以下语法将指定值填充到容器中从迭代器 `begin` 到 `begin+n-1` 的位置：

```cpp
template<typename ForwardIterator, typename size, typename T>
void fill_n(ForwardIterator begin, size n, const T& value)
```

LiveExample 23.2 演示了如何使用这两个函数。

LiveExample 23.2 的互动程序请访问 https://liangcpp.pearsoncmg.com/LiveRunCpp5e/faces/LiveExample.xhtml?header=off&programName=FillDemo&programHeight=490&resultHeight=210。

LiveExample 23.2　FillDemo.cpp

Source Code Editor:

```cpp
#include <iostream>
#include <algorithm>
#include <list>
#include <iterator>
using namespace std;

int main()
{
  int values[] = {1, 2, 3, 4, 5, 6};
  list<int> intList(values, values + 6);

  ostream_iterator<int> output(cout, " ");
  cout << "Initial contents, values: ";
  copy(values, values + 6, output);
  cout << "\nInitial contents, intList: ";
  copy(intList.begin(), intList.end(), output);

  fill(values + 2, values + 4, 88); // Fill 88 from values + 2 to values + 3
  fill_n(intList.begin(), 2, 99);   // Fill two 99 from intList.begin()

  cout << "\nAfter the fill function, values: ";
  copy(values, values + 6, output);
  cout << "\nAfter the fill_n function, intList: ";
  copy(intList.begin(), intList.end(), output);

  return 0;
}
```

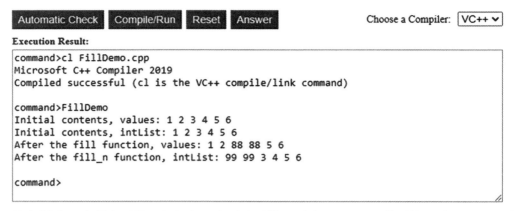

程序创建一个数组（第9行）和一个列表（第10行）。fill 函数（第18行）用88填充数组的 values[2] 和 values[3]。fill_n 函数（第19行）用值99填充从列表头开始的2个元素。

23.5 将函数作为参数传递

要点提示：在 C++ 中，可以将函数作为函数的参数进行传递。

许多 STL 函数都包含作为函数的参数。传递函数参数就是传递函数的地址。我们用三个示例来展示这个功能。

LiveExample 23.3 中给出了第一个示例，它定义了一个带有函数参数的函数，该函数参数有自己的参数。

LiveExample 23.3 的互动程序请访问 https://liangcpp.pearsoncmg.com/LiveRunCpp5e/faces/LiveExample.xhtml?header=off&programName=FunctionWithFunctionParameter1&fileType=.cpp&programHeight=620&resultHeight=180。

LiveExample 23.3　FunctionWithFunctionParameter1.cpp

Source Code Editor:

```cpp
#include <iostream>
using namespace std;

int f1(int value)
{
  return 2 * value;
}

int f2(int value)
{
  return 3 * value;
}

void m(int t[], int size, int f(int)) // Function parameter
{
  for (int i = 0; i < size; i++)
    t[i] = f(t[i]);
}

int main()
{
  int list1[] = {1, 2, 3, 4};
  m(list1, 4, f1);
```

```
24      for (int i = 0; i < 4; i++)
25        cout << list1[i] << " ";
26      cout << endl;
27
28      int list2[] = {1, 2, 3, 4};
29      m(list2, 4, f2); // Apply list2 with function f2
30      for (int i = 0; i < 4; i++)
31        cout << list2[i] << " ";
32      cout << endl;
33
34      return 0;
35    }
```

Execution Result:

```
command>cl FunctionWithFunctionParameter1.cpp
Microsoft C++ Compiler 2019
Compiled successful (cl is the VC++ compile/link command)

command>FunctionWithFunctionParameter1
2 4 6 8
3 6 9 12

command>
```

函数 m 有三个参数。最后一个参数是一个函数，它接受一个 int 参数并返回一个 int 值。语句

```
m(list1, 4, f1);
```

用三个参数调用函数 m。最后一个参数是函数 f1。语句

```
m(list2, 4, f2);
```

用三个参数调用函数 m。最后一个参数是函数 f2。

注意第 14 行

```
void m(int t[], int size, int f(int))
```

与下面语句相同：

```
void m(int t[], int size, int (*f)(int))
```

第 17 行可以写成 t[i]=(*f)(t[i])，第 23 行可以写成 m(list1, 4, *f1)，因为函数参数总是由函数的地址传递。

LiveExample 23.4 中给出了第二个示例，它定义了一个带有函数参数的函数，该函数参数没有参数。

LiveExample 23.4 的互动程序请访问 https://liangcpp.pearsoncmg.com/LiveRunCpp5e/faces/LiveExample.xhtml?header=off&programName=FunctionWithFunctionParameter2&programHeight=440&resultHeight=160。

LiveExample 23.4　FunctionWithFunctionParameter2.cpp

```cpp
#include <iostream>
using namespace std;

int nextNum()
{
   static int n = 20;
   return n++;
}

void m(int t[],int size, int f()) // No-arg function
{
   for (int i = 0; i < size; i++)
      t[i] = f();
}

int main()
{
   int list[4];
   m(list, 4, nextNum);
   for (int i = 0; i < 4; i++)
      cout << list[i] << " ";
   cout << endl;

   return 0;
}
```

Execution Result:

```
command>cl FunctionWithFunctionParameter2.cpp
Microsoft C++ Compiler 2019
Compiled successful (cl is the VC++ compile/link command)

command>FunctionWithFunctionParameter2
20 21 22 23

command>
```

函数 m 有三个参数。最后一个参数是一个不接受参数并返回 int 值的函数。

LiveExample 23.5 中给出了第三个示例，它定义了一个带有函数参数的函数，该函数参数没有返回值。

LiveExample 23.5 的互动程序请访问 https://liangcpp.pearsoncmg.com/LiveRunCpp5e/faces/LiveExample.xhtml?header=off&programName=FunctionWithFunctionParameter3&programHeight=380&resultHeight=160。

LiveExample 23.5　FunctionWithFunctionParameter3.cpp

```cpp
#include <iostream>
using namespace std;

void print(int number)
{
   cout << 2 * number << " ";
}

void m(int t[],int size, void f(int)) // void function
```

```
10  {
11      for (int i = 0; i < size; i++)
12          f(t[i]);
13  }
14
15  int main()
16  {
17      int list[4] = {1, 2, 3, 4};
18      m(list, 4, print);
19
20      return 0;
21  }
```

Execution Result:
```
command>cl FunctionWithFunctionParameter3.cpp
Microsoft C++ Compiler 2019
Compiled successful (cl is the VC++ compile/link command)

command>FunctionWithFunctionParameter3
2 4 6 8

command>
```

函数 m 有三个参数。最后一个参数是一个没有返回值的函数。

注意：也可以将函数地址传递给指针。例如，以下语句将 LiveExample 23.3 中的 f1 函数传递给指针 p：

```
int (*p)(int) = f1;
```

下面的语句将 LiveExample 23.4 中的 nextNum 函数传递给指针 p：

```
int (*p)(int) = nextNum;
```

下面的语句将 LiveExample 23.5 中的 print 函数传递给指针 p：

```
void (*p)(int) = print;
```

23.6 generate 和 generate_n

要点提示：函数 generate 和 generate_n 用函数返回的值填充序列。

generate 函数填充序列中的值，generate_n 函数填充序列中的 *n* 个值。这些函数的语法如下：

```
template<typename ForwardIterator, typename function>
void generate(ForwardIterator begin, ForwardIterator end, function gen)

template<typename ForwardIterator, typename size, typename function>
void generate_n(ForwardIterator begin, size n, function gen)
```

LiveExample 23.6 演示了如何使用这两个函数。

LiveExample 23.6 的互动程序请访问 https://liangcpp.pearsoncmg.com/LiveRunCpp5e/faces/LiveExample.xhtml?header=off&programName=GenerateDemo&fileType=.cpp&programHeight=580&resultHeight=210。

LiveExample 23.6　GenerateDemo.cpp

Source Code Editor:

```cpp
#include <iostream>
#include <algorithm>
#include <list>
#include <iterator>
using namespace std;

int nextNum()
{
  static int n = 20;
  return n++;
}

int main()
{
  int values[] = {1, 2, 3, 4, 5, 6};
  list<int> intList(values, values + 6);

  ostream_iterator<int> output(cout, " ");
  cout << "Initial contents, values: ";
  copy(values, values + 6, output);
  cout << "\nInitial contents, intList: ";
  copy(intList.begin(), intList.end(), output);

  generate(values + 2, values + 4, nextNum); // Use generate
  generate_n(intList.begin(), 2, nextNum); // Use generate_n

  cout << "\nAfter the generate function, values: ";
  copy(values, values + 6, output);
  cout << "\nAfter the generate_n function, intList: ";
  copy(intList.begin(), intList.end(), output);

  return 0;
}
```

Execution Result:

```
command>cl GenerateDemo.cpp
Microsoft C++ Compiler 2019
Compiled successful (cl is the VC++ compile/link command)

command>GenerateDemo
Initial contents, values: 1 2 3 4 5 6
Initial contents, intList: 1 2 3 4 5 6
After the generate function, values: 1 2 20 21 5 6
After the generate_n function, intList: 22 23 3 4 5 6

command>
```

程序创建一个数组（第 15 行）和一个列表（第 16 行）。generate 函数（第 24 行）用 nextNum 函数生成的值填充数组元素 values[2] 和 values[3]。注意，n 是一个静态局部变量（第 9 行），因此它的值在程序的生存期内是持久的。调用 nextNum() 第一次返回 20，第二次返回 21，下一次调用的返回值总是多 1。

23.7　remove、remove_if、remove_copy 和 remove_copy_if

要点提示：STL 提供了四个删除容器元素的函数：remove、remove_if、remove_copy，以及 remove_copy_if。

函数 remove 用以下语法从序列范围 [begin，end) 中删除与指定值匹配的元素：

```
template<typename ForwardIterator, typename T>
ForwardIterator remove(ForwardIterator begin,
  ForwardIterator end, const T& value)
```

函数 remove_if 用以下语法从序列范围 [begin，end) 中删除所有使 boolFunction(element) 为 true 的元素：

```
template<typename ForwardIterator, typename boolFunction>
ForwardIterator remove_if(ForwardIterator beg,
  ForwardIterator end, boolFunction f)
```

remove 和 remove_if 都返回一个迭代器，该迭代器指向新元素范围内最后一个元素之后的位置。

函数 remove_copy 用以下语法将序列范围 [begin，end) 内的所有其他元素复制到目标容器，除了值与指定值匹配的元素：

```
template<typename InputIterator, typename OutputIterator, typename T>
OutputIterator remove_copy(InputIterator beg, InputIterator end,
  OutputIterator targetPosition, const T& value)
```

函数 remove_copy_if 用以下语法将序列范围 [beg，end) 内的所有元素复制到目标容器，但使 boolFunction(element) 为 true 的元素除外：

```
template<typename InputIterator, typename OutputIterator,
  typename boolFunction>
OutputIterator remove_copy_if(InputIterator begin, InputIterator end,
  OutputIterator targetPosition, boolFunction f)
```

remove_copy 和 remove_copy_if 都返回一个迭代器，该迭代器指向复制的最后一个元素之后的位置。

注意：某些 STL 算法支持传递布尔函数的指针。布尔函数用于检查元素是否满足条件。例如，可以定义一个名为 greaterThan3(int element) 的函数来检查元素是否大于 3。

注意：这四个函数不会改变容器的大小。元素会被移动到容器的开头。例如，假设一个列表包含元素 {1, 2, 3, 4, 5,6}。删除 2 后，列表包含 {1, 3, 4, 5, 6, 6}。注意，最后一个元素是 6。

注意：remove_copy 和 remove_copy_if 函数将新内容复制到目标容器，但不更改源容器。

LiveExample 23.7 演示了如何使用函数 remove 和 remove_if。

LiveExample 23.7 的互动程序请访问 https://liangcpp.pearsoncmg.com/LiveRunCpp5e/faces/LiveExample.xhtml?header=off&programName=RemoveDemo&fileType=.cpp&programHeight=560&resultHeight=210。

LiveExample 23.7 RemoveDemo.cpp

Source Code Editor:
```
1  #include <iostream>
2  #include <algorithm>
3  #include <list>
4  #include <iterator>
5  using namespace std;
```

```cpp
 6
 7  bool greaterThan3(int value)
 8  {
 9      return value > 3;
10  }
11
12  int main()
13  {
14      int values[] = {1, 7, 3, 4, 3, 6, 1, 2};
15      list<int> intList(values, values + 8);
16
17      ostream_iterator<int> output(cout, " ");
18      cout << "Initial contents, values: ";
19      copy(values, values + 8, output);
20      cout << "\nInitial contents, intList: ";
21      copy(intList.begin(), intList.end(), output);
22
23      remove(values, values + 8, 3);
24      remove_if(intList.begin(), intList.end(), greaterThan3);
25
26      cout << "\nAfter the remove function, values: ";
27      copy(values, values + 8, output);
28      cout << "\nAfter the remove_if function, intList: ";
29      copy(intList.begin(), intList.end(), output);
30
31      return 0;
32  }
```

Execution Result:

```
command>cl RemoveDemo.cpp
Microsoft C++ Compiler 2019
Compiled successful (cl is the VC++ compile/link command)

command>RemoveDemo
Initial contents, values: 1 7 3 4 3 6 1 2
Initial contents, intList: 1 7 3 4 3 6 1 2
After the remove function, values: 1 7 4 6 1 2 1 2
After the remove_if function, intList: 1 3 3 1 2 6 1 2

command>
```

程序创建一个数组（第 14 行）、一个列表（第 15 行），并显示其初始内容（第 17 ～ 21 行）。数组是 {1, 7, 3, 4, 3, 6, 1, 2}。remove 函数（第 23 行）从数组中删除 3。新数组是 {1, 7, 4, 6, 1, 2, 1, 2}。

remove_if 函数（第 24 行）删除列表中所有使 greaterThan3(element) 为 true 的元素。在调用函数之前，列表为 {1, 7, 3, 4, 3, 6, 1, 2}。调用函数后，列表变为 {1, 3, 3, 1, 2, 6, 1, 2}。

LiveExample 23.8 演示了如何使用函数 remove_copy 和 remove_copy_if。

LiveExample 23.8 的互动程序请访问 https://liangcpp.pearsoncmg.com/LiveRunCpp5e/faces/LiveExample.xhtml?header=off&programName=RemoveCopyDemo&fileType=.cpp&programHeight=680&resultHeight=250。

LiveExample 23.8　RemoveCopyDemo.cpp

Source Code Editor:

```cpp
#include <iostream>
#include <algorithm>
#include <list>
#include <iterator>
using namespace std;

bool greaterThan3(int value)
{
  return value > 3;
}

int main()
{
  int values[] = {1, 7, 3, 4, 3, 6, 1, 2};
  list<int> intList(values, values + 8);

  ostream_iterator<int> output(cout, " ");
  cout << "Initial contents, values: ";
  copy(values, values + 8, output);
  cout << "\nInitial contents, intList: ";
  copy(intList.begin(), intList.end(), output);

  int newValues[] = {9, 9, 9, 9, 9, 9, 9, 9};
  list<int> newIntList(values, values + 8);
  remove_copy(values, values + 8, newValues, 3); // Use remove_copy
  remove_copy_if(intList.begin(), intList.end(), newIntList.begin(),
    greaterThan3); // Use remove_copy_if

  cout << "\nAfter the remove_copy function, values: ";
  copy(values, values + 8, output);
  cout << "\nAfter the remove_copy_if function, intList: ";
  copy(intList.begin(), intList.end(), output);
  cout << "\nAfter the remove_copy function, newValues: ";
  copy(newValues, newValues + 8, output);
  cout << "\nAfter the remove_copy_if function, newIntList: ";
  copy(newIntList.begin(), newIntList.end(), output);

  return 0;
}
```

Execution Result:

```
command>cl RemoveCopyDemo.cpp
Microsoft C++ Compiler 2019
Compiled successful (cl is the VC++ compile/link command)

command>RemoveCopyDemo
Initial contents, values: 1 7 3 4 3 6 1 2
Initial contents, intList: 1 7 3 4 3 6 1 2
After the remove_copy function, values: 1 7 3 4 3 6 1 2
After the remove_copy_if function, intList: 1 7 3 4 3 6 1 2
After the remove_copy function, newValues: 1 7 4 6 1 2 9 9
After the remove_copy_if function, newIntList: 1 3 3 1 2 6 1 2

command>
```

remove_copy 函数（第25行）从数组 values 中删除 3，并将其余值复制到数组 newValues 中。原始数组的内容不会更改。在复制之前，数组 values 为 {1, 7, 3, 4, 3, 6, 1,

2}，newValues 为 {9, 9, 9, 9, 9, 9, 9, 9}。复制后，数组 newValues 变为 {1, 7, 4, 6, 1, 2, 9, 9}。

remove_copy_if 函数（第 26 行）删除列表中所有使 greaterThan3(element) 为 true 的元素，并将其余元素复制到列表 newIntList 中。原始列表的内容没有更改。在复制之前，列表 intList 是 {1, 7, 3, 4, 3, 6, 1, 2}，newIntList 是 {1, 7, 3, 4, 3, 6, 1, 2}。复制后，列表 newIntList 变为 {1, 3, 3, 1, 2, 6, 1, 2}。

23.8 replace、replace_if、replace_copy 和 replace_copy_if

要点提示：STL 提供了四种替换容器元素的函数：replace、replace_if、replace_copy 和 replace_copy_if。

replace 函数用以下语法将序列范围 [begin, end) 内所有出现的给定值替换为新值：

```cpp
template<typename ForwardIterator, typename T>
void replace(ForwardIterator begin, ForwardIterator end,
  const T& oldValue, const T& newValue)
```

replace_if 函数用以下语法将使得 boolFunction(element) 为 true 的所有出现的元素替换为新值：

```cpp
template<typename ForwardIterator, typename boolFunction, typename T>
void replace_if(ForwardIterator begin, ForwardIterator end,
  boolFunction f, const T& newValue)
```

函数 replace_copy 用以下语法将出现的给定值替换为新值，并将结果复制到目标容器：

```cpp
template<typename ForwardIterator, typename OutputIterator,
  typename T>
ForwardIterator replace_copy(ForwardIterator begin,
  ForwardIterator end, OutputIterator targetPosition,
  T& oldValue, T& newValue)
```

函数 replace_copy_if 用以下语法将使得 boolFunction(element) 为 true 的出现的元素替换为新值，并将序列范围 [begin, end) 内的所有元素复制到目标容器：

```cpp
template<typename ForwardIterator, typename OutputIterator,
  typename boolFunction, typename T>
ForwardIterator replace_copy_if(ForwardIterator begin,
  ForwardIterator end, OutputIterator targetPosition,
  boolFunction f, const T& newValue)
```

replace_copy 和 replace_copy_if 都返回一个迭代器，该迭代器指向复制的最后一个元素之后的位置。

注意：replace_copy 和 replace_copy_if 函数将新内容复制到目标容器，但不更改源容器。

LiveExample 23.9 演示了如何使用函数 replace 和 replace_if。

LiveExample 23.9 的互动程序请访问 https://liangcpp.pearsoncmg.com/LiveRunCpp5e/faces/LiveExample.xhtml?header=off&programName=ReplaceDemo&fileType=.cpp&programHeight=570&resultHeight=210。

LiveExample 23.9 ReplaceDemo.cpp

Source Code Editor:

```cpp
#include <iostream>
#include <algorithm>
#include <list>
#include <iterator>
using namespace std;

bool greaterThan3(int value)
{
    return value > 3;
}

int main()
{
    int values[] = {1, 7, 3, 4, 3, 6, 1, 2};
    list<int> intList(values, values + 8);

    ostream_iterator<int> output(cout, " ");
    cout << "Initial contents, values: ";
    copy(values, values + 8, output);
    cout << "\nInitial contents, intList: ";
    copy(intList.begin(), intList.end(), output);

    replace(values, values + 8, 3, 747); // Use replace
    replace_if(intList.begin(), intList.end(), greaterThan3, 747); // Use replace_if

    cout << "\nAfter the replace function, values: ";
    copy(values, values + 8, output);
    cout << "\nAfter the replace_if function, intList: ";
    copy(intList.begin(), intList.end(), output);

    return 0;
}
```

Execution Result:

```
command>cl ReplaceDemo.cpp
Microsoft C++ Compiler 2019
Compiled successful (cl is the VC++ compile/link command)

command>ReplaceDemo
Initial contents, values: 1 7 3 4 3 6 1 2
Initial contents, intList: 1 7 3 4 3 6 1 2
After the replace function, values: 1 7 747 4 747 6 1 2
After the replace_if function, intList: 1 747 3 747 3 747 1 2

command>
```

程序创建一个数组（第14行）、一个列表（第15行），并显示其初始内容（第17～21行）。数组是{1, 7, 3, 4, 3, 6, 1, 2}。replace函数（第23行）将数组中的3替换为747。新的数组变为{1, 7, 747, 4, 747, 6, 1, 2}。

replace_if函数（第24行）替换列表中所有使得greaterThan3(element)为true的元素。在调用函数之前，列表为{1, 7, 3, 4, 3, 6, 1, 2}。调用该函数后，列表变为{1, 747, 3, 747, 3, 747, 1, 2}。

LiveExample 23.10演示了如何使用函数replace_copy和replace_copy_if。

LiveExample 23.10 的互动程序请访问 https://liangcpp.pearsoncmg.com/LiveRunCpp5e/faces/LiveExample.xhtml?header=off&programName=ReplaceCopyDemo&fileType=.cpp&programHeight=680&resultHeight=250。

LiveExample 23.10　ReplaceCopyDemo.cpp

Source Code Editor:

```cpp
#include <iostream>
#include <algorithm>
#include <list>
#include <iterator>
using namespace std;

bool greaterThan3(int value)
{
    return value > 3;
}

int main()
{
    int values[] = {1, 7, 3, 4, 3, 6, 1, 2};
    list<int> intList(values, values + 8);

    ostream_iterator<int> output(cout, " ");
    cout << "Initial contents, values: ";
    copy(values, values + 8, output);
    cout << "\nInitial contents, intList: ";
    copy(intList.begin(), intList.end(), output);

    int newValues[] = {9, 9, 9, 9, 9, 9, 9, 9};
    list<int> newIntList(values, values + 8);
    replace_copy(values + 2, values + 5, newValues, 3, 88); // Use replace_copy
    replace_copy_if(intList.begin(), intList.end(),
        newIntList.begin(), greaterThan3, 88); // Use replace_copy_if

    cout << "\nAfter the replace function, values: ";
    copy(values, values + 8, output);
    cout << "\nAfter the replace_if function, intList: ";
    copy(intList.begin(), intList.end(), output);
    cout << "\nAfter the replace_copy function, newValues: ";
    copy(newValues, newValues + 8, output);
    cout << "\nAfter the replace_copy_if function, newIntList: ";
    copy(newIntList.begin(), newIntList.end(), output);

    return 0;
}
```

Execution Result:

```
command>cl ReplaceCopyDemo.cpp
Microsoft C++ Compiler 2019
Compiled successful (cl is the VC++ compile/link command)

command>ReplaceCopyDemo
Initial contents, values: 1 7 3 4 3 6 1 2
Initial contents, intList: 1 7 3 4 3 6 1 2
After the replace function, values: 1 7 3 4 3 6 1 2
After the replace_if function, intList: 1 7 3 4 3 6 1 2
After the replace_copy function, newValues: 88 4 88 9 9 9 9 9
After the replace_copy_if function, newIntList: 1 88 3 88 3 88 1 2

command>
```

replace_copy 函数（第 25 行）将数组 values 中的 3 替换为 88，并将部分数组复制到数组 newValues。原始数组的内容不会更改。在替换之前，数组 values 为 {1, 7, 3, 4, 3, 6, 1, 2}，newValues 为 {9, 9, 9, 9, 9, 9, 9, 9}。替换后，数组 newValues 变为 {88, 4, 88, 9, 9, 9, 9, 9}。注意，只有从位置 2 到 4 的部分数组被复制到从位置 0 开始的目标数组。

replace_copy_if 函数（第 26 行）替换列表中所有使 greaterThan3(element) 为 true 的元素，并将其余元素复制到列表 newIntList 中。原始列表的内容没有更改。在替换之前，列表 intList 是 {1, 7, 3, 4, 3, 6, 1, 2}，newIntList 是 {1, 7, 3, 4, 3, 6, 1, 2}。替换后，列表 newIntList 变为 {1, 88, 3, 88, 3, 88, 1, 2}。

23.9 find、find_if、find_end 和 find_first_of

要点提示：STL 提供了四个查找容器元素的函数：find、find_if、find_end 和 find_first_of。

find 函数使用以下语法查找元素：

```
template<typename InputIterator, typename T>
InputIterator find(InputIterator begin, InputIterator end, T& value)
```

find_if 函数使用以下语法查找使 boolFunction(element) 为 true 的元素：

```
template<typename InputIterator, typename boolFunction>
InputIterator find_if(InputIterator begin, InputIterator end,
   boolFunction f)
```

如果找到第一个匹配元素，这两个函数都返回指向该元素的迭代器；否则，返回 end。

LiveExample 23.11 演示了如何使用 find 和 find_if 函数。

LiveExample 23.11 的互动程序请访问 https://liangcpp.pearsoncmg.com/LiveRunCpp5e/faces/LiveExample.xhtml?header=off&programName=FindDemo&fileType=.cpp&programHeight=630&resultHeight=230。

LiveExample 23.11 FindDemo.cpp

```cpp
#include <iostream>
#include <algorithm>
#include <vector>
#include <iterator>
using namespace std;

int main()
{
  int values[] = {1, 7, 3, 4, 3, 6, 1, 2};
  vector<int> intVector(values, values + 8);

  ostream_iterator<int> output(cout, " ");
  cout << "values: ";
  copy(values, values + 8, output);
  cout << "\nintVector: ";
  copy(intVector.begin(), intVector.end(), output);

  int key;
  cout << "\nEnter a key: ";
```

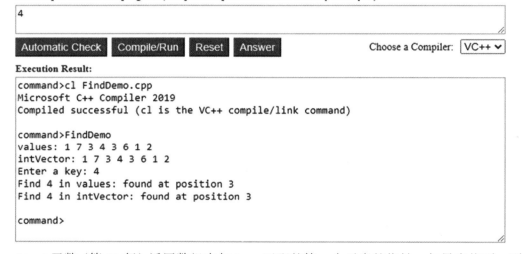

find 函数（第 22 行）返回数组中与 key 匹配的第一个元素的指针。如果未找到，则 p 为 values+8（第 23 行）。如果找到，(p-values) 是数组中匹配元素的索引。

find 函数（第 29 行）返回向量中与 key 匹配的第一个元素的指针。如果未找到，则 itr 为 intVector.end()（第 30 行）。如果找到，(itr-intVector.begin()) 是向量中匹配元素的索引。

find_end 函数用于查找子序列。它有两个版本：

```
template<typename ForwardIterator1, typename ForwardIterator2>
ForwardIterator find_end(ForwardIterator1 begin1, ForwardIterator1 end1,
    ForwardIterator2 begin2, ForwardIterator2 end2)

template<typename ForwardIterator1, typename ForwardIterator2,
    typename boolFunction>
ForwardIterator find_end(ForwardIterator1 begin1, ForwardIterator1 end1,
    ForwardIterator2 begin2, ForwardIterator2 end2, boolFunction f)
```

两个函数在序列范围 begin1 到 end1-1 中查找匹配的 begin2 到 end2-1 整个序列。如果查找成功，返回最后匹配发生的位置；否则返回 end1。在第一个版本中，对元素进行相等比较；在第二个版本中，boolFunction(elementInFirstSequence, elementInSecondSequence) 必须为 true。

LiveExample 23.12 演示了如何使用 find_end 函数的两个版本。

LiveExample 23.12 的互动程序请访问 https://liangcpp.pearsoncmg.com/LiveRunCpp5e/faces/LiveExample.xhtml?header=off&programName=FindEndDemo&fileType=.cpp&programHeight=600&resultHeight=210。

LiveExample 23.12 FindEndDemo.cpp

```cpp
#include <iostream>
#include <algorithm>
#include <vector>
#include <iterator>
using namespace std;

int main()
{
  int array1[] = {1, 7, 3, 4, 3, 6, 1, 2};
  int array2[] = {3, 6, 1};
  vector<int> intVector(array1, array1 + 8);

  ostream_iterator<int> output(cout, " ");
  cout << "array1: ";
  copy(array1, array1 + 8, output);
  cout << "\nintVector: ";
  copy(intVector.begin(), intVector.end(), output);

  int* p = find_end(array1, array1 + 8, array2, array2 + 1);
  if (p != array1 + 8)
    cout << "\nfind {3} in array1 at position " << (p - array1);
  else
    cout << "\nnot found";

  auto itr =
    find_end(intVector.begin(), intVector.end(), array2 + 1, array2 + 3);
  if (itr != intVector.end())
    cout << "\nfind {6, 1} in intVector at position " <<
      (itr - intVector.begin());
  else
    cout << "\nnot found";

  return 0;
}
```

Execution Result:

```
command>cl FindEndDemo.cpp
Microsoft C++ Compiler 2019
Compiled successful (cl is the VC++ compile/link command)

command>FindEndDemo
array1: 1 7 3 4 3 6 1 2
intVector: 1 7 3 4 3 6 1 2
find {3} in array1 at position 4
find {6, 1} in intVector at position 5

command>
```

该程序创建两个数组和一个向量（第 9 ～ 11 行）。这三个容器的内容是

```
array1: {1, 7, 3, 4, 3, 6, 1, 2}
array2: {3, 6, 1}
intVector: {1, 7, 3, 4, 3, 6, 1, 2}
```

调用 `find_end(array1, array1+8, array2, array2+1)` 查找 array1 以匹配 {3}。最后一个成功匹配的位置是 4。

调用 `find_end(intVector.begin(), intVector.end(), array2+1, array2+3)` 查找 intVector 以匹配 {6, 1}。最后一个成功匹配的位置是 5。

函数 `find_first_of` 查找两个序列中的第一个公共元素。它有两个版本：

```cpp
template<typename ForwardIterator1, typename ForwardIterator2>
ForwardIterator find_first_of(ForwardIterator1 begin1,
  ForwardIterator1 end1, ForwardIterator2 begin2,
  ForwardIterator2 end2)
```

```cpp
template<typename ForwardIterator1, typename ForwardIterator2,
  typename boolFunction>
ForwardIterator find_first_of(ForwardIterator1 begin1,
  ForwardIterator1 end1, ForwardIterator2 begin2,
  ForwardIterator2 end2, boolFunction)
```

如果存在匹配，则两个函数都返回第一个序列中的位置；否则返回 end1。在第一个版本中，对元素进行相等比较；在第二个版本中，boolFunction(elementInFirstSequence, elementInSecondSequence) 必须为 true。

LiveExample 23.13 演示了如何使用 `find_first_of` 函数的两个版本。

LiveExample 23.13 的互动程序请访问 https://liangcpp.pearsoncmg.com/LiveRunCpp5e/faces/LiveExample.xhtml?header=off&programName=FindFirstOfDemo&fileType=.cpp&programHeight=720&resultHeight=210。

LiveExample 23.13　FindFirstOfDemo.cpp

```cpp
#include <iostream>
#include <algorithm>
#include <vector>
#include <iterator>
using namespace std;

bool greaterThan(int e1, int e2)
{
   return e1 > e2;
}

int main()
{
   int array1[] = {1, 7, 3, 4, 3, 6, 1, 2};
   int array2[] = {9, 96, 21, 3, 2, 3, 1};
   vector<int> intVector(array1, array1 + 8);

   ostream_iterator<int> output(cout, " ");
   cout << "array1: ";
   copy(array1, array1 + 8, output);
   cout << "\nintVector: ";
   copy(intVector.begin(), intVector.end(), output);

   int* p = find_first_of(array1, array1 + 8, array2 + 2, array2 + 4);
```

```
25      if (p != array1 + 8)
26        cout << "\nfind first of {21, 3} in array1 at position "
27          << (p - array1);
28      else
29        cout << "\nnot found";
30
31      vector<int>::iterator itr =
32        find_first_of(intVector.begin(), intVector.end(),
33          array2 + 2, array2 + 4, greaterThan);
34      if (itr != intVector.end())
35        cout << "\nfind {21, 3} > an element in intVector at position " <<
36          (itr - intVector.begin());
37      else
38        cout << "\nnot found";
39
40      return 0;
41    }
```

Automatic Check | Compile/Run | Reset | Answer   Choose a Compiler: VC++

Execution Result:
```
command>cl FindFirstOfDemo.cpp
Microsoft C++ Compiler 2019
Compiled successful (cl is the VC++ compile/link command)

command>FindFirstOfDemo
array1: 1 7 3 4 3 6 1 2
intVector: 1 7 3 4 3 6 1 2
find first of {21, 3} in array1 at position 2
find first of {21, 3} > an element in intVector at position 1

command>
```

该程序创建两个数组和一个向量（第 14 ~ 16 行）。这三个容器的内容是

```
array1: {1, 7, 3, 4, 3, 6, 1, 2}
array2: {9, 96, 21, 3, 2, 3, 1}
intVector: {1, 7, 3, 4, 3, 6, 1, 2}
```

调用 find_first_of(array1, array1+8, array2+2, array2+4) 查找 array1 以找到 {21, 3} 中的第一个匹配，即 3。3 在 array1 中的位置是 2。

调用 find_first_of(intVector.begin(), intVector.end(), array2+2, array2+4, greaterThan) 在 intVector 中查找第一个大于 {21, 3} 中元素的元素。intVector 中的元素 7 满足条件。intVector 中 7 的位置为 1。

23.10 search 和 search_n

要点提示：STL 提供了 search 和 search_n 函数，用于查找容器中的元素。

函数 search 类似于函数 find_end。两者都查找子序列。find_end 查找最后一个匹配项，但 search 查找第一个匹配项。search 函数有两个版本：

```
template<typename ForwardIterator1, typename ForwardIterator2>
ForwardIterator search(ForwardIterator1 begin1, ForwardIterator1 end1,
  ForwardIterator2 begin2, ForwardIterator2 end2)
```

```
template<typename ForwardIterator1, typename ForwardIterator2,
  typename boolFunction>
ForwardIterator search(ForwardIterator1 begin1, ForwardIterator1 end1,
  ForwardIterator2 begin2, ForwardIterator2 end2, boolFunction)
```

如果存在匹配，则两个函数都返回在第一个序列中的位置；否则，它们返回 end1。在第一个版本中，对元素进行相等比较；在第二个版本中，boolFunction(elementInFirstSequence, elementInSecondSequence) 必须为 true。

search_n 函数用于查找序列中连续出现的某个值。search_n 函数有两个版本：

```
template<typename ForwardIterator, typename size, typename T>
ForwardIterator search_n(ForwardIterator begin,
  ForwardIterator end, size count, const T& value)
```

```
template<typename ForwardIterator, typename size, typename boolFunction>
ForwardIterator search_n(ForwardIterator1 begin,
  ForwardIterator1 end, size count, boolFunction f)
```

如果存在匹配，则这两个函数都返回匹配元素在序列中的位置；否则，它们返回 end。在第一个版本中，对元素进行相等比较；在第二个版本中，boolFunction(element) 必须为 true。

LiveExample 23.14 演示了如何使用 search 和 search_n 函数。

LiveExample 23.14 的互动程序请访问 https://liangcpp.pearsoncmg.com/LiveRunCpp5e/faces/LiveExample.xhtml?header=off&programName=SearchDemo&fileType=.cpp&programHeight=610&resultHeight=230。

LiveExample 23.14　SearchDemo.cpp

```
#include <iostream>
#include <algorithm>
#include <vector>
#include <iterator>
using namespace std;

int main()
{
  int array1[] = {1, 7, 3, 4, 3, 3, 1, 2};
  int array2[] = {9, 96, 4, 3, 2, 3, 1};
  vector<int> intVector(array1, array1 + 8);

  ostream_iterator<int> output(cout, " ");
  cout << "array1: ";
  copy(array1, array1 + 8, output);
  cout << "\nintVector: ";
  copy(intVector.begin(), intVector.end(), output);

  int* p = search(array1, array1 + 8, array2 + 2, array2 + 4);
  if (p != array1 + 8)
    cout << "\nSearch {4, 3} in array1 at position "
      << (p - array1) << endl;
  else
    cout << "\nnot found" << endl;

  vector<int>::iterator itr =
    search_n(intVector.begin(), intVector.end(), 2, 3);
  if (itr != intVector.end())
    cout << "\nSearch two consecutive occurrence of 3 in intVector "
      << "at position " << (itr - intVector.begin()) << endl;
  else
    cout << "\nnot found" << endl;

  return 0;
```

```
 35   }
```

Automatic Check | Compile/Run | Reset | Answer   Choose a Compiler: VC++

Execution Result:
```
command>cl SearchDemo.cpp
Microsoft C++ Compiler 2019
Compiled successful (cl is the VC++ compile/link command)

command>SearchDemo
array1: 1 7 3 4 3 3 1 2
intVector: 1 7 3 4 3 3 1 2
Search {4, 3} in array1 at position 3

Search two occurrence of 3 in intVector at position 4

command>
```

该程序创建两个数组和一个向量（第 9 ~ 11 行）。这三个容器的内容是

```
array1: {1, 7, 3, 4, 3, 3, 1, 2}
array2: {9, 96, 4, 3, 2, 3, 1}
intVector: {1, 7, 3, 4, 3, 3, 1, 2}
```

调用 search(array1, array1+8, array2+2, array2+4) 查找 array1 以找到序列 {4, 3}。匹配项的位置在 array1 中是 3。

调用 search_n(intVector.begin(), intVector.end(), 2, 3) 查找两个连续的 3。匹配项的位置在 array1 中是 4。

23.11 sort 和 binary_search

要点提示：STL 提供了对容器元素进行排序的 sort 函数，以及二分查找元素的 binary_search 函数。

sort 函数需要随机访问迭代器。可以用以下两种版本将函数应用于数组、向量或双端队列的排序：

```
template<typename randomAccessIterator>
void sort(randomAccessIterator begin, randomAccessIterator end)

template<typename randomAccessIterator, typename relationalOperator>
void sort(randomAccessIterator begin, randomAccessIterator end,
    relationalOperator op)
```

可以用以下两种版本的 binary_search 函数查找有序序列中的值：

```
template<typename ForwardIterator, typename T>
bool binary_search(ForwardIterator begin,
    ForwardIterator end, const T& value)

template<typename ForwardIterator, typename T,
    typename strictWeakOrdering>
bool binary_search(ForwardIterator begin,
    ForwardIterator end, const T& value, strictWeakOrdering op)
```

注意：某些 STL 算法支持传递函数运算符。它实际上是一个**函数对象**，对象的指针被传递来调用 STL 函数。函数对象有三种：关系型、逻辑型和算术型，如表 23.5 所示。sort 和 binary_search 算法需要关系运算符。使用函数对象，需要包含 <functional> 头文件。

表 23.5 函数对象

STL 函数对象	类型
equal_to<T>	关系型
not_equal_to<T>	关系型
greater<T>	关系型
greater_equal<T>	关系型
less<T>	关系型
less_equal<T>	关系型
logical_and<T>	逻辑型
logical_not<T>	逻辑型
logical_or<T>	逻辑型
plus<T>	算术型
minus<T>	算术型
multiplies<T>	算术型
divides<T>	算术型
modulus<T>	算术型
negate<T>	算术型

注意：严格弱序运算符是 less<T> 和 greater(T)。binary_search 算法的第一个版本使用 less<T> 运算符进行比较，第二个版本指定 less<T> 或 greater(T)。

LiveExample 23.15 演示了如何使用函数 sort 和 binary_search。

LiveExample 23.15 的互动程序请访问 https://liangcpp.pearsoncmg.com/LiveRunCpp5e/faces/LiveExample.xhtml?header=off&programName=SortDemo&fileType=.cpp&programHeight=580&resultHeight=260。

LiveExample 23.15　SortDemo.cpp

Source Code Editor:

```cpp
#include <iostream>
#include <algorithm>
#include <iterator>
#include <functional>
using namespace std;

int main()
{
  int array1[] = {1, 7, 3, 4, 3, 3, 1, 2};

  ostream_iterator<int> output(cout, " ");
  cout << "Before sort, array1: ";
  copy(array1, array1 + 8, output);

  sort(array1, array1 + 8);// Sort array1

  cout << "\nAfter sort, array1: ";
  copy(array1, array1 + 8, output);

  cout << (binary_search(array1, array1 + 8, 4) ?
    "\n4 is in array1" : "\n4 is not in array1") << endl;

  sort(array1, array1 + 8, greater<int>());

  cout << "\nAfter sort with function operator(>), array1: \n";
```

```
26       copy(array1, array1 + 8, output);
27
28       cout << (binary_search(array1, array1 + 8, 4,
29         greater<int>()) ?
30         "\n4 is in array1" : "\n4 is not in array1") << endl;
31
32       return 0;
33   }
```

[Automatic Check] [Compile/Run] [Reset] [Answer]   Choose a Compiler: VC++ ▼

Execution Result:
```
command>cl SortDemo.cpp
Microsoft C++ Compiler 2019
Compiled successful (cl is the VC++ compile/link command)

command>SortDemo
Before sort, array1: 1 7 3 4 3 3 1 2
After sort, array1: 1 1 2 3 3 3 4 7
4 is in array1

After sort with function operator(>), array1:
7 4 3 3 3 2 1 1
4 is in array1

command>
```

sort 和 binary_search 的默认函数运算符是 less_equal<T>()。调用 sort(array1, array1+8, greater<int>())（第 23 行），使用 greater<int>() 函数对象对数组进行排序。由于元素是降序排列的，因此你可以调用 binary_search(array1, array1+8, 4, greater<int>())（第 28～29 行）在 array1 中二分查找到元素 4。

23.12 adjacent_find、merge 和 inplace_merge

要点提示：STL 提供了查找值相同的相邻元素的 adjacent_find 函数，以及归并两个有序序列的 merge 和 inplace_merge 函数。

adjacent_find 函数用以下语法查找值相等或满足 boolFunction(element) 的首次出现的相邻元素：

```
template<typename ForwardIterator>
ForwardIterator adjacent_find(ForwardIterator begin,
  ForwardIterator end)

template<typename ForwardIterator, typename boolFunction>
ForwardIterator adjacent_find(ForwardIterator begin,
  ForwardIterator end, boolFunction f)
```

adjacent_find 函数返回指向匹配序列中第一个元素的迭代器。如果未找到，则返回 end。

merge 函数使用以下语法将两个有序序列归并为一个新序列：

```
template<typename InputIterator1, typename InputIterator2,
    typename OutputIterator>
OutputIterator merge(InputIterator1 begin1,
  InputIterator1 end1, InputIterator2 begin2,
  InputIterator2 end2, OutputIterator targetPosition)
```

```cpp
template<typename InputIterator1, typename InputIterator2,
    typename OutputIterator, typename relationalOperator>
OutputIterator merge(InputIterator1 begin1,
    InputIterator1 end1, InputIterator2 begin2,
    InputIterator2 end2, OutputIterator targetPosition,
    relationalOperator)
```

inplace_merge 函数将序列的第一部分与第二部分归并；假设这两个部分包含排序后的连续元素。语法为：

```cpp
template<typename BidirectionalIterator>
void inplace_merge(bidirectionalIterator begin,
    bidirectionalIterator middle, bidirectionalIterator end)

template<typename BidirectionalIterator, typename relationalOperator>
void inplace_merge(bidirectionalIterator begin,
    bidirectionalIterator middle, bidirectionalIterator end,
    relationalOperator)
```

函数归并已排序的从 begin 到 middle-1 和从 middle 到 end-1 的连续序列，并且排序后的序列被存储在原始序列中。因此，这也被称为就地归并。

LiveExample 23.16 演示了如何使用函数 adjacent_find、merge 和 inplace_merge。

LiveExample 23.16 的互动程序请访问 https://liangcpp.pearsoncmg.com/LiveRunCpp5e/faces/LiveExample.xhtml?header=off&programName=MergeDemo&fileType=.cpp&programHeight=540&resultHeight=230。

LiveExample 23.16 MergeDemo.cpp

Source Code Editor:

```cpp
 1  #include <iostream>
 2  #include <algorithm>
 3  #include <list>
 4  #include <iterator>
 5  using namespace std;
 6
 7  int main()
 8  {
 9    int array1[] = {1, 7, 3, 4, 3, 3, 1, 2};
10    list<int> intList(8);
11    auto p = adjacent_find(array1, array1 + 8); // adjacent_find
12    cout << "value " << *p << " found at index " << (p - array1) << endl;
13
14    ostream_iterator<int> output(cout, " ");
15    cout << "array1: ";
16    copy(array1, array1 + 8, output);
17
18    sort(array1, array1 + 3);
19    sort(array1 + 3, array1 + 8);
20    cout << "\nafter sort partial arrays, array1: ";
21    copy(array1, array1 + 8, output);
22
23    merge(array1, array1 + 3, array1 + 3, array1 + 8, intList.begin());
24    cout << "\nafter merger, intList: ";
25    copy(intList.begin(), intList.end(), output);
26
27    inplace_merge(array1, array1 + 3, array1 + 8);// Inplace merger
28    cout << "\nafter inplace merger, array1: ";
29    copy(array1, array1 + 8, output);
30    return 0;
```

```
31    }
```

Execution Result:

```
command>cl MergeDemo.cpp
Microsoft C++ Compiler 2019
Compiled successful (cl is the VC++ compile/link command)

command>MergeDemo
Value 3 found at index 4
array1: 1 7 3 4 3 3 1 2
after sort partial arrays, array1: 1 3 7 1 2 3 3 4
after merger, intList: 1 1 2 3 3 3 4 7
after inplace merger, array1: 1 1 2 3 3 3 4 7

command>
```

该程序创建一个数组和一个列表（第 9～10 行）。数组的内容为

```
array1: {1, 7, 3, 4, 3, 3, 1, 2}
```

排序 {1, 7, 3} 和 {4, 3, 3, 1, 2}（第 18～19 行）后，数组变为

```
array1: {1, 3, 7, 1, 2, 3, 3, 4}
```

在将 {1, 3, 7} 和 {1, 2, 3, 3, 4} 归并到 intList 中（第 23 行）后，intList 变为

```
intList: {1, 1, 2, 3, 3, 3, 4, 7}
```

在就地归并 {1, 3, 7} 和 {1, 2, 3, 3, 4}（第 27 行）之后，array1 变为

```
array1: {1, 1, 2, 3, 3, 3, 4, 7}
```

23.13 reverse 和 reverse_copy

要点提示：STL 提供 reverse 和 reverse_copy 函数来反转序列中的元素。

reverse 函数用于反转序列中的元素。reverse_copy 函数将一个序列中的元素以相反的顺序复制到另一个序列。reverse_copy 函数不会更改源容器中的内容。这些函数的语法为：

```
template<typename BidirectionalIterator>
void reverse(BidirectionalIterator begin, BidirectionalIterator end)
```

```
template<typename BidirectionalIterator, typename OutputIterator>
OutputIterator reverse_copy(BidirectionalIterator begin,
  BidirectionalIterator end, OutputIterator targetPosition)
```

LiveExample 23.17 演示了如何使用函数 reverse 和 reverse_copy。

LiveExample 23.17 的互动程序请访问 https://liangcpp.pearsoncmg.com/LiveRunCpp5e/faces/LiveExample.xhtml?header=off&programName=ReverseDemo&fileType=.cpp&programHeight=480&resultHeight=210。

LiveExample 23.17　ReverseDemo.cpp

```cpp
#include <iostream>
#include <algorithm>
#include <list>
#include <iterator>
using namespace std;

int main()
{
    int array1[] = {1, 7, 3, 4, 3, 3, 1, 2};
    list<int> intList(8);

    ostream_iterator<int> output(cout, " ");
    cout << "array1: ";
    copy(array1, array1 + 8, output);

    reverse(array1, array1 + 8);// Reverse array1
    cout << "\nafter reverse arrays, array1: ";
    copy(array1, array1 + 8, output);

    reverse_copy(array1,array1+8,intList.begin());// Reverse array1 copy to intList
    cout << "\nafter reverse_copy, array1: ";
    copy(array1, array1 + 8, output);
    cout << "\nafter reverse_copy, intList: ";
    copy(intList.begin(), intList.end(), output);

    return 0;
}
```

Execution Result:

```
command>cl ReverseDemo.cpp
Microsoft C++ Compiler 2019
Compiled successful (cl is the VC++ compile/link command)

command>ReverseDemo
array1: 1 7 3 4 3 3 1 2
after reverse arrays, array1: 2 1 3 3 4 3 7 1
after reverse_copy, array1: 2 1 3 3 4 3 7 1
after reverse_copy, intList: 1 7 3 4 3 3 1 2

command>
```

23.14　`rotate` 和 `rotate_copy`

要点提示：STL 提供了 `rotate` 和 `rotate_copy` 函数来轮转序列中的元素。

`rotate` 函数使用以下语法轮转序列元素：

```
template<typename ForwardIterator>
void rotate(ForwardIterator begin, ForwardIterator newBegin,
    ForwardIterator end)
```

`newBegin` 指定的元素将成为轮转后序列中的第一个元素。

rotate_copy 函数类似于 rotate，只是它使用以下语法将结果复制到目标序列：

```
template<typename ForwardIterator, typename OutputIterator>
OutputIterator rotate_copy(ForwardIterator begin, ForwardIterator
    newBeg, ForwardIterator end, OutputIterator targetPosition)
```

LiveExample 23.18 演示了如何使用函数 rotate 和 rotate_copy。

LiveExample 23.18 的互动程序请访问 https://liangcpp.pearsoncmg.com/LiveRunCpp5e/faces/LiveExample.xhtml?header=off&programName=RotateDemo&fileType=.cpp&programHeight=480&resultHeight=210。

LiveExample 23.18　RotateDemo.cpp

Source Code Editor:

```cpp
#include <iostream>
#include <algorithm>
#include <list>
#include <iterator>
using namespace std;

int main()
{
  int array1[] = {1, 2, 3, 4, 5, 6, 7, 8};
  list<int> intList(8);

  ostream_iterator<int> output(cout, " ");
  cout << "array1: ";
  copy(array1, array1 + 8, output);

  rotate(array1, array1 + 3, array1 + 8); // Rotate
  cout << "\nafter rotate arrays, array1: ";
  copy(array1, array1 + 8, output);

  // Rotate copy
  rotate_copy(array1, array1 + 1, array1 + 8, intList.begin());
  cout << "\nafter rotate_copy, array1: ";
  copy(array1, array1 + 8, output);
  cout << "\nafter rotate_copy, intList: ";
  copy(intList.begin(), intList.end(), output);

  return 0;
}
```

[Automatic Check] [Compile/Run] [Reset] [Answer]　　Choose a Compiler: VC++

Execution Result:

```
command>cl RotateDemo.cpp
Microsoft C++ Compiler 2019
Compiled successful (cl is the VC++ compile/link command)

command>RotateDemo
array1: 1 2 3 4 5 6 7 8
after rotate arrays, array1: 4 5 6 7 8 1 2 3
after rotate_copy, array1: 4 5 6 7 8 1 2 3
after rotate_copy, intList: 5 6 7 8 1 2 3 4

command>
```

该程序创建一个数组和一个列表（第 9 ～ 10 行）。数组的内容为

```
array1: {1, 2, 3, 4, 5, 6, 7, 8}
```

指针 array1+3 指向 4，因此在调用 rotate(array1, array1+3, array1+8)（第 16 行）后，array1 变为

```
array1: {4, 5, 6, 7, 8, 1, 2, 3}
```

现在指针 array1+1 指向 5，调用 rotate_copy(array1, array1+1, array1+8, intList.begin())（第 21 行）后，intList 变为

```
intList: {5, 6, 7, 8, 1, 2, 3, 4}
```

23.15 swap、iter_swap 和 swap_ranges

要点提示：STL 提供了 swap、iter_swap 和 swap_ranges 函数来交换序列中的元素。swap 函数交换两个变量中的值。iter_swap 函数交换迭代器所指向的值。swap_ranges 函数交换两个序列。这些函数的语法定义如下：

```cpp
template<typename T>
void swap(T& value1, T& value2)
```

```cpp
template<typename ForwardIterator1, typename ForwardIterator2>
void iter_swap(ForwardIterator p1, ForwardIterator p2)
```

```cpp
template<typename ForwardIterator1, typename ForwardIterator2>
ForwardIterator swap_ranges(ForwardIterator1 begin1,
    ForwardIterator1 end1, ForwardIterator2 begin2)
```

LiveExample 23.19 演示了如何使用这三个函数。

LiveExample 23.19 的互动程序请访问 https://liangcpp.pearsoncmg.com/LiveRunCpp5e/faces/LiveExample.xhtml?header=off&programName=SwapDemo&fileType=.cpp&programHeight=460&resultHeight=210。

LiveExample 23.19　SwapDemo.cpp

Source Code Editor:

```cpp
#include <iostream>
#include <algorithm>
#include <iterator>
using namespace std;

int main()
{
    int array1[] = {1, 2, 3, 4, 5, 6, 7, 8};
    ostream_iterator<int> output(cout, " ");
    cout << "array1: ";
    copy(array1, array1 + 8, output);

    cout << "\nafter swap variables, array1: ";
    swap(array1[0], array1[1]); // Swap array1[0] with array1[1]
    copy(array1, array1 + 8, output);

    cout << "\nafter swap via pointers, array1: ";
    iter_swap(array1 + 2, array1 + 3);
    copy(array1, array1 + 8, output);

    cout << "\nafter swap ranges, array1: ";
    swap_ranges(array1, array1 + 4, array1 + 4);
```

```
23      copy(array1, array1 + 8, output);
24
25      return 0;
26  }
```

Execution Result:

```
command>cl SwapDemo.cpp
Microsoft C++ Compiler 2019
Compiled successful (cl is the VC++ compile/link command)

command>SwapDemo
array1: 1 2 3 4 5 6 7 8
after swap variables, array1: 2 1 3 4 5 6 7 8
after swap via pointers, array1: 2 1 4 3 5 6 7 8
after swap ranges, array1: 5 6 7 8 2 1 4 3

command>
```

调用 swap(array1[0], array1[1]) 将 array1[0] 与 array1[1] 交换（第 14 行）。

调用 iter_swap(array1+2, array1+3) 交换 array1+2 和 array1+3 所指向的元素（第 18 行）。

调用 swap_ranges(array1, array1+4, array1+4) 将从 array1 到 array1+3 中的元素与从 array1+4 到 array1+7 中的元素交换（第 22 行）。

23.16 count 和 count_if

要点提示：STL 提供了 count 和 count_if 函数，用于对序列元素的出现次数进行计数。count 函数用以下语法对序列中出现的给定值进行计数：

```
template<typename InputIterator, typename T>
int count(InputIterator begin, InputIterator end, const T& value)
```

count_if 函数用以下语法统计使得 boolFunction(element) 为 true 的元素的出现次数：

```
template<typename InputIterator, typename boolFunction>
int count_if(InputIterator begin, InputIterator end, boolFunction f)
```

LiveExample 23.20 演示了如何使用这些函数。

LiveExample 23.20 的互动程序请访问 https://liangcpp.pearsoncmg.com/LiveRunCpp5e/faces/LiveExample.xhtml?header=off&programName=CountDemo&fileType=.cpp&programHeight=380&resultHeight=170。

LiveExample 23.20   CountDemo.cpp

```
1  #include <iostream>
2  #include <algorithm>
3  using namespace std;
4
5  bool greaterThan1(int value)
6  {
```

```cpp
 7      return value > 1;
 8    }
 9
10    int main()
11    {
12      int array1[] = {1, 2, 3, 4, 5, 6, 7, 8};
13
14      cout << "The number of 1's in array1: " <<
15        count(array1, array1 + 8, 1) << endl;
16
17      cout << "The number of elements > 1 in array1: " <<
18        count_if(array1, array1 + 8, greaterThan1) << endl;
19
20      return 0;
21    }
```

Execution Result:

```
command>cl CountDemo.cpp
Microsoft C++ Compiler 2019
Compiled successful (cl is the VC++ compile/link command)

command>CountDemo
The number of 1's in array1: 1
The number of elements > 1 in array1: 7

command>
```

23.17 `max_element` 和 `min_element`

要点提示：STL 提供了 `max_element` 和 `min_element` 函数，以返回序列中的最大和最小元素。

我们已经熟悉了 max 和 min 函数。`max_element` 和 `min_element` 可获得序列的最大元素和最小元素。函数定义如下：

```cpp
template<typename ForwardIterator>
ForwardIterator max_element(ForwardIterator begin,
  ForwardIterator end)
```

```cpp
template<typename ForwardIterator>
ForwardIterator min_element(ForwardIterator begin,
  ForwardIterator end)
```

LiveExample 23.21 给出了如何使用这些函数的示例。

LiveExample 23.21 的互动程序请访问 https://liangcpp.pearsoncmg.com/LiveRunCpp5e/faces/LiveExample.xhtml?header=off&programName=MaxMinDemo&fileType=.cpp&programHeight=290&resultHeight=180。

LiveExample 23.21　MaxMinDemo.cpp

Source Code Editor:
```cpp
1  #include <iostream>
2  #include <algorithm>
3  using namespace std;
4
```

```
 5  int main()
 6  {
 7      int array1[] = {1, 2, 3, 4, 5, 6, 7, 8};
 8
 9      cout << "The max element in array1: " <<
10          *max_element(array1, array1 + 8) << endl;
11
12      cout << "The min element in array1: " <<
13          *min_element(array1, array1 + 8) << endl;
14
15      return 0;
16  }
```

Execution Result:

```
command>cl MaxMinDemo.cpp
Microsoft C++ Compiler 2019
Compiled successful (cl is the VC++ compile/link command)
command>MaxMinDemo
The max element in array1: 8
The min element in array1: 1

command>
```

23.18 random_shuffle

要点提示: random_shuffle 函数对序列元素进行随机重新排序。

random_shuffle 函数有两个版本:

```
template<typename randomAccessIterator>
void random_shuffle(randomAccessIterator begin,
    randomAccessIterator end)
```

```
template<typename randomAccessIterator>
void random_shuffle(randomAccessIterator begin,
    randomAccessIterator end, RandomNumberGenerator& rand)
```

LiveExample 23.22 给出了如何使用此函数的示例。

LiveExample 23.22 的互动程序请访问 https://liangcpp.pearsoncmg.com/LiveRunCpp5e/faces/LiveExample.xhtml?header=off&programName=ShuffleDemo&fileType=.cpp&programHeight=480&resultHeight=170。

LiveExample 23.22　ShuffleDemo.cpp

Source Code Editor:

```
 1  #include <iostream>
 2  #include <algorithm>
 3  #include <iterator>
 4  #include <ctime>
 5  using namespace std;
 6
 7  int randGenerator(int aRange)
 8  {
 9      srand(time(0));
10      return rand() % aRange;
```

```
11  }
12
13  int main()
14  {
15      int array1[] = {1, 2, 3, 4, 5, 6, 7, 8};
16      random_shuffle(array1, array1 + 8);// Shuffle array1
17      cout << "After random shuffle, array1: ";
18      ostream_iterator<int> output(cout, " ");
19      copy(array1, array1 + 8, output);
20
21      int array2[] = {1, 2, 3, 4, 5, 6, 7, 8};
22      random_shuffle(array2, array2 + 8, randGenerator);
23      cout << "\nAfter random shuffle, array2: ";
24      copy(array2, array2 + 8, output);
25
26      return 0;
27  }
```

Execution Result:
```
command>cl ShuffleDemo.cpp
Microsoft C++ Compiler 2019
Compiled successful (cl is the VC++ compile/link command)

command>ShuffleDemo
After random shuffle, array1: 5 2 7 3 1 6 8 4
After random shuffle, array2: 8 3 1 7 6 5 2 4

command>
```

该程序创建 array1（第 15 行），并调用 random_shuffle 函数的第一个版本来随机混洗数组元素。random_shuffle 函数的第一个版本使用内部默认随机数生成器，该生成器按固定序列生成随机数。每次运行代码时，混洗后都会得到相同的 array1 元素序列。

该程序创建 array2（第 21 行），并调用 random_shuffle 函数的第二个版本来随机混洗数组元素。random_shuffle 函数的第二个版本使用自定义随机数生成器，该生成器按随机序列生成随机数，因为随机数种子是随机设置的（第 9 行）。每次运行代码时，混洗后都会得到不同的 array2 元素序列。

23.19　for_each 和 transform

要点提示：STL 提供了 for_each 和 transform 函数，它们将函数应用于序列中的每个元素。

for_each 函数通过以下语法用某个函数处理序列中的每个元素：

```
template<typename InputIterator, typename function>
void for_each(InputIterator begin, InputIterator end, function f)
```

可以使用 transform 函数将某个函数应用于序列中的每个元素，并将结果复制到目标序列。函数定义如下：

```
template<typename InputIterator, typename OutputIterator,
    typename function>
OutputIteration transform(InputIterator begin,
    InputIterator end, OutputIterator targetPosition, function f)
```

LiveExample 23.23 演示了如何使用这些函数。

LiveExample 23.23 的互动程序请访问 https://liangcpp.pearsoncmg.com/LiveRunCpp5e/faces/LiveExample.xhtml?header=off&programName=ForEachDemo&fileType=.cpp&programHeight=510&resultHeight=170。

LiveExample 23.23　ForEachDemo.cpp

```cpp
#include <iostream>
#include <algorithm>
#include <vector>
#include <iterator>
using namespace std;

void display(int& value)
{
  cout << value << " ";
}

int square(int& value)
{
  return value * value;
}

int main()
{
  int array1[] = {1, 2, 3, 4, 5, 6, 7, 8};
  cout << "array1: ";
  for_each(array1, array1 + 8, display); // Display array1

  vector<int> intVector(8);
  transform(array1, array1 + 8, intVector.begin(), square);
  cout << "\nintVector: ";
  for_each(intVector.begin(), intVector.end(), display);

  return 0;
}
```

```
command>cl ForEachDemo.cpp
Microsoft C++ Compiler 2019
Compiled successful (cl is the VC++ compile/link command)

command>ForEachDemo
array1: 1 2 3 4 5 6 7 8
intVector: 1 4 9 16 25 36 49 64

command>
```

display 函数（第 7 ～ 10 行）向控制台显示一个数字。调用 for_each(array1, array1+8, display)（第 21 行），将 display 函数应用于序列中的每个元素。因此，显示 array1 中的所有元素。

square 函数（第 12 ～ 15 行）返回一个数字的平方。调用 transform(array1, array1+8, intVector.begin(), square)（第 24 行），将 square 函数应用于序列中的每个元素，并将新序列复制到 intVector。

23.20 includes、set_union、set_difference、set_intersection 和 set_symmetric_difference

要点提示：STL 为集合操作提供了 includes、set_union、set_difference、set_intersection 和 set_symmetric_difference 函数。

STL 支持测试子集、并、差、交和对称差的集合运算。所有这些函数都要求序列元素已排序。

如果第一个序列中的元素包含第二个序列的元素，includes 函数将返回 true。

```
template<typename InputIterator1, typename InputIterator2>
bool includes(InputIterator1 begin1, InputIterator1 end1,
    InputIterator2 begin2, InputIterator2 end2)
```

set_union 函数获取属于每个序列的元素。

```
template<typename InputIterator1, typename InputIterator2,
    typename OutputIterator>
OutputIterator set_union(InputIterator1 begin1, InputIterator1 end1,
    InputIterator2 begin2, InputIterator2 end2,
    OutputIterator targetPosition)
```

set_difference 函数获取属于第一个序列但不属于第二个序列的元素。

```
template<typename InputIterator1, typename InputIterator2,
    typename OutputIterator>
OutputIterator set_difference(InputIterator begin1,
    InputIterator end1, InputIterator begin2, InputIterator end2,
    OutputIterator targetPosition)
```

set_intersection 函数获取同时出现在两个序列中的元素。

```
template<typename InputIterator1, typename InputIterator2,
    typename OutputIterator>
OutputIterator set_intersection(InputIterator begin1,
    InputIterator end1, InputIterator begin2, InputIterator end2,
    OutputIterator targetPosition)
```

set_symmetric_difference 函数获取出现在一个序列但不同时出现在两个序列中的元素。

```
template<typename InputIterator1, typename InputIterator2,
    typename OutputIterator>
OutputIterator set_symmetric_difference(InputIterator begin1,
    InputIterator end1, InputIterator begin2, InputIterator end2,
    OutputIterator targetPosition)
```

假设 array1 和 array2 如下所示。

array1	array2
{1, 2, 3, 4, 5, 6, 7, 8}	{1, 3, 6, 9, 12}

它们的集合操作如下所示。

操作	结果
array1 union array2	{1, 2, 3, 4, 5, 6, 7, 8, 9, 12}
array1 difference array2	{2, 4, 5, 7, 8}
array1 intersection array2	{1, 3, 6}
array1 symmetric_diff array2	{2, 4, 5, 7, 8, 9, 12}

注意：集合函数返回指向目标中最后一个元素之后位置的迭代器。

LiveExample 23.24 演示了如何使用集合函数。

LiveExample 23.24 的互动程序请访问 https://liangcpp.pearsoncmg.com/LiveRunCpp5e/faces/LiveExample.xhtml?header=off&programName=SetOperationDemo&fileType=.cpp&programHeight=780&resultHeight=260。

LiveExample 23.24   SetOperationDemo.cpp

Source Code Editor:

```cpp
#include <iostream>
#include <algorithm>
#include <vector>
#include <iterator>
using namespace std;

int main()
{
  int array1[] = {1, 2, 3, 4, 5, 6, 7, 8};
  int array2[] = {1, 3, 6, 9, 12};
  vector<int> intVector(15);

  ostream_iterator<int> output(cout, " ");
  cout << "array1: ";
  copy(array1, array1 + 8, output);
  cout << "\narray2: ";
  copy(array2, array2 + 5, output);

  bool isContained =
    includes(array1, array1 + 8, array2, array2 + 3);
  cout << (isContained ? "\n{1, 3, 6} is a subset of array1" :
    "\n{1, 3, 6} is not a subset of array1");

  vector<int>::iterator last = set_union(array1, array1 + 8,
    array2, array2 + 5, intVector.begin()); // Union
  cout << "\nAfter union, intVector: ";
  copy(intVector.begin(), last, output);

  last = set_difference(array1, array1 + 8,
    array2, array2 + 5, intVector.begin()); // Difference
  cout << "\nAfter difference, intVector: ";
  copy(intVector.begin(), last, output);

  last = set_intersection(array1, array1 + 8,
    array2, array2 + 5, intVector.begin()); // Intersection
  cout << "\nAfter intersection, intVector: ";
  copy(intVector.begin(), last, output);

  last = set_symmetric_difference(array1, array1 + 8,
    array2, array2 + 5, intVector.begin()); // Symmetric difference
  cout << "\nAfter symmetric difference, intVector: ";
  copy(intVector.begin(), last, output);

  return 0;
}
```

Execution Result:

```
command>cl SetOperationDemo.cpp
Microsoft C++ Compiler 2019
```

```
Compiled successful (cl is the VC++ compile/link command)

command>SetOperationDemo
array1: 1 2 3 4 5 6 7 8
array2: 1 3 6 9 12
{1, 3, 6} is a subset of array1
After union, intVector: 1 2 3 4 5 6 7 8 9 12
After difference, intVector: 2 4 5 7 8
After intersection, intVector: 1 3 6
After symmetric difference, intVector: 2 4 5 7 8 9 12

command>
```

该程序创建两个数组和一个向量（第 9～11 行）。

```
array1: {1, 2, 3, 4, 5, 6, 7, 8}
array2: {1, 3, 6, 9, 12}
```

调用 includes(array1, array1+8, array2, array2+3)（第 20 行）返回 true，因为 {1, 3, 6} 是 array1 的子集。

调用 set_union(array1, array1+8, array2, array2+5, intVector.begin())（第 24～25 行）得到 array1 和 array2 的并集，并放在 intVector 中。intVector 变为

```
intVector: {1, 2, 3, 4, 5, 6, 7, 8, 9, 12}
```

调用 set_difference(array1, array1+8, array2, array2+5, intVector.begin())（第 29～30 行）得到 array1 和 array2 的差，并放在 intVector 中。intVector 变为

```
intVector: {2, 4, 5, 7, 8}
```

调用 set_intersection(array1, array1+8, array2, array2+5, intVector.begin())（第 34～35 行）得到 array1 和 array2 的交集，并放在 intVector 中。intVector 变为

```
intVector: {1, 3, 6}
```

调用 set_symmetric_difference(array1, array1+8, array2, array2+5, intVector.begin())（第 39～40 行）得到 array1 和 array2 的对称差，并放在 intVector 中。intVector 变为

```
intVector: {2, 4, 5, 7, 8, 9, 12}
```

23.21 accumulate、adjacent_difference、inner_product 和 partial_sum

要点提示：STL 支持数学函数 accumulate、adjacent_difference、inner_product 和 partial_sum。它们在 <numeric> 头文件中定义。

accumulate 函数有两个版本：

```
template<typename InputIterator, typename T>
T accumulate(InputIterator begin, InputIterator end, T initValue)

template<typename InputIterator, typename T,
  typename arithmeticOperator>
T accumulate(InputIterator begin, InputIterator end, T initValue,
  arithmeticOperator op)
```

第一个版本返回所有元素和 initValue 的总和。第二个版本对 initValue 和所有元素应用算术运算符（例如乘法）并返回结果。例如

array1	accumulate(array1, array1+5, 0) 的结果
{1, 2, 3, 4, 5}	15

array1	accumulate(array1, array1+5, 1, multiplies<int>()) 的结果
{1, 2, 3, 4, 5}	120

adjacent_difference 函数有两个版本：

```
template<typename InputIterator, typename T>
OutputIterator adjacent_difference(InputIterator begin,
  InputIterator end, OutputIterator targetPosition)

template<typename InputIterator, typename T,
  typename arithmeticOperator>
OutputIterator adjacent_difference(InputIterator begin,
  InputIterator end, OutputIterator targetPosition,
  arithmeticOperator op)
```

第一个版本创建了一个元素序列，它的第一个元素与输入序列中的第一个元素相同，每个后续元素都是输入序列当前元素与前一个元素之间的差。第二个版本与第一个版本相同，只是应用指定的算术运算符来代替减法运算符。例如

array1	adjacent_difference(array1, array1+5, intVector.begin()) 的结果
{1, 2, 3, 4, 5}	{1, 1, 1, 1, 1}

inner_product 函数有两个版本：

```
template<typename InputIterator1, typename InputIterator2,
  typename T>
T inner_product(InputIterator1 begin1,
  InputIterator1 end1, InputIterator2 begin2, T initValue)

template<typename InputIterator1, typename InputIterator2,
  typename T, typename arithmeticOperator1,
  typename arithmeticOperator2>
T inner_product(InputIterator1 begin1,
  InputIterator1 end1, InputIterator2 begin2, T initValue,
  arithmeticOperator1 op1, arithmeticOperator2 op2)
```

两个序列 {a1, a2, …, ai} 和 {b1, b2, …, bi} 的内积定义为 a1*b1+a2*b2+…+ai*bi。

第一个版本返回 initValue 和序列内积的总和。第二个版本与第一个版本相同，只是默认的加法运算符被 op1 取代，乘法运算符被 op2 取代。例如

array1	inner_product(array1, array1+5,array1, 0) 的结果
{1, 2, 3, 4, 5}	55

partial_sum 函数有两个版本：

```
template<typename InputIterator1, typename InputIterator2,
  typename OutputIterator>
OutputIterator partial_sum(InputIterator1 begin1,
  InputIterator1 end1, OutputIterator2 begin2)

template<typename InputIterator1, typename InputIterator2,
  typename OutputIterator, typename arithmeticOperator>
OutputIterator partial_sum(InputIterator begin1,
  InputIterator end1, OutputIterator targetPosition
  arithmeticOperator op)
```

第一个版本创建了一个序列，它的每个元素都是包括该元素在内的所有先前元素的总和。第二个版本与第一个版本相同，只是将默认的加法运算符替换为 op。例如，

array1	partial_sum(array1, array1+5, intVector.begin()) 的结果
{1, 2, 3, 4, 5}	{1, 3, 6, 10, 15}

LiveExample 23.25 演示了如何使用这些数学函数。

LiveExample 23.25 的互动程序请访问 https://liangcpp.pearsoncmg.com/LiveRunCpp5e/faces/LiveExample.xhtml?header=off&programName=MathOperationDemo&fileType=.cpp&programHeight=640&resultHeight=240。

LiveExample 23.25　MathOperationDemo.cpp

Source Code Editor:

```cpp
#include <iostream>
#include <algorithm>
#include <numeric>
#include <vector>
#include <iterator>
#include <functional>
using namespace std;

int main()
{
  int array1[] = {1, 2, 3, 4, 5};
  vector<int> intVector(5);

  ostream_iterator<int> output(cout, " ");
  cout << "array1: ";
  copy(array1, array1 + 5, output);

  cout << "\nSum of array1: " <<
    accumulate(array1, array1 + 5, 0) << endl;

  cout << "Product of array1: " <<
    accumulate(array1, array1 + 5, 1, multiplies<int>()) << endl;

  vector<int>::iterator last =
    adjacent_difference(array1, array1 + 5, intVector.begin());
  cout << "After adjacent difference, intVector: ";
  copy(intVector.begin(), last, output);

  cout << "\nInner product of array1 * array1 is " <<
    inner_product(array1, array1 + 5, array1, 0);

  last = partial_sum(array1, array1 + 5, intVector.begin());
  cout << "\nAfter partial sum, intVector: ";
  copy(intVector.begin(), last, output);
```

```
36      return 0;
37  }
```

Execution Result:

```
command>cl MathOperationDemo.cpp
Microsoft C++ Compiler 2019
Compiled successful (cl is the VC++ compile/link command)

command>MathOperationDemo
array1: 1 2 3 4 5
Sum of array1: 15
Product of array1: 120
After adjacent difference, intVector: 1 1 1 1 1
Inner product of array1 * array1 is 55
After partial sum, intVector: 1 3 6 10 15

command>
```

该程序创建一个数组和一个向量（第 11 ~ 12 行）。

```
array1: {1, 2, 3, 4, 5}
```

调用 `accumulate(array1, array1+5, 0)`（第 19 行）返回 `array1` 中所有元素的总和。

调用 `accumulate(array1, array1+5, 1, multiplies<int>())`（第 22 行）返回 `array1` 中所有元素的乘积。

调用 `adjacent_difference(array1, array1+5, intVector.begin())`（第 25 行）得到 `array1` 的相邻差的序列，并放在 `intVector` 中。

调用 `inner_product(array1, array1+5, array1, 0)`（第 30 行）得到 `array1` 和 `array1` 的内积。

调用 `partial_sum(array1, array1+5, intVector.begin())`（第 32 行）得到 `array1` 的部分和的序列，并放在 `intVector` 中。

23.22 lambda 表达式

要点提示：在 C++11 中，可使用 lambda 表达式在一个表达式中定义和传递函数。

许多 STL 算法支持将函数作为参数传递，但必须先创建一个函数，然后传递函数参数。C+11 的 lambda 表达式使我们能够定义一个匿名函数并传递它。lambda 表达式可以极大地简化编码。lambda 表达式也称为 lambda 函数，因为表达式本质上是定义一个函数。

要了解使用 lambda 函数的必要性，考虑 LiveExample 23.23。该程序使用 STL `for_each` 算法显示 `array1`（第 21 行）和 `intVector`（第 26 行）中的每个元素。第 7 ~ 10 行定义了 `display` 函数。

该程序还使用 STL `transform` 算法将 `square` 函数应用于 `intVector`（第 24 行）中的每个元素。第 12 ~ 15 行定义了 `square` 函数。

`display` 和 `square` 都是短函数，`square` 函数只在程序中的一个位置被调用。lambda 函数可以通过定义调用它的函数来简化编码。

LiveExample 23.26 使用 lambda 函数重写 LiveExample 23.23。

LiveExample 23.26 的互动程序请访问 https://liangcpp.pearsoncmg.com/LiveRunCpp5e/faces/LiveExample.xhtml?header=off&programName=ForEachDemoUsingLambdaFunctions&fileType=.cpp&programHeight=360&resultHeight=160。

LiveExample 23.26　ForEachDemoUsingLambdaFunctions.cpp

```cpp
#include <iostream>
#include <algorithm>
#include <vector>
using namespace std;

int main()
{
  int array1[] = {1, 2, 3, 4, 5, 6, 7, 8};
  cout << "array1: ";
  for_each(array1, array1 + 8, [](int v) { cout << v << " "; });

  vector<int> intVector(8);
  transform(array1, array1 + 8, intVector.begin(),
    [](int v) { return v * v; });

  cout << "\nintVector: ";
  for_each(intVector.begin(), intVector.end(),
    [](int v) { cout << v << " "; });

  return 0;
}
```

Execution Result:

```
command>cl ForEachDemoUsingLambdaFunctions.cpp
Microsoft C++ Compiler 2019
Compiled successful (cl is the VC++ compile/link command)

command>ForEachDemoUsingLambdaFunctions
array1: 1 2 3 4 5 6 7 8
intVector: 1 4 9 16 25 36 49 64

command>
```

该程序使用 `for_each` 算法将 lambda 函数应用于 `array1` 中的每个元素（第 10 行）。lambda 函数定义为

```cpp
[](int v) { cout << v << " "; }
```

此函数没有名称。因此，lambda 函数被称为匿名函数。参数 v 是指数组中的元素。

lambda 函数也可以返回一个值。第 14 行中的函数

```cpp
[](int v) { return v * v; }
```

返回 v*v。

lambda 函数中的参数可以通过值或引用传递。LiveExample 23.26 中的 lambda 函数是通过值传递的。可以通过以下方式传递引用来重写它：

```cpp
[](int& v) { cout << v << " "; }
[](int& v) { return v * v; }
```

此处示例中的 [] 为空。也可以在 lambda 函数中使用其他变量。这些变量可以在括号 [] 中定义。这称为捕获变量。

LiveExample 23.27 给出了一个将给定值乘以向量元素的例子。

LiveExample 23.27 的互动程序请访问 https://liangcpp.pearsoncmg.com/LiveRunCpp5e/faces/LiveExample.xhtml?header=off&programName=CaptureVariable&fileType=.cpp&programHeight=410&resultHeight=140。

LiveExample 23.27　CaptureVariable.cpp

```cpp
#include <iostream>
#include <algorithm>
#include <vector>
using namespace std;

void multiply(vector<int>& v, int k)
{
  transform(v.begin(), v.end(), v.begin(),
    [k](int& value) { return value * k; });
}

int main()
{
  vector<int> v {0, 1, 2, 3};

  multiply(v, 3);

  for_each(v.begin(), v.end(),
    [](int value) { cout << value << " "; });

  return 0;
}
```

```
command>cl CaptureVariable.cpp
Microsoft C++ Compiler 2019
Compiled successful (cl is the VC++ compile/link command)

command>CaptureVariable
0 3 6 9

command>
```

变量 k 用以下语法在 lambda 函数中被捕获：

[k] (int& v) { return v * k; }

如果未在括号中捕获，则无法在 lambda 函数内部引用它。可以在 lambda 函数中捕获任意数量的变量。

通常，lambda 函数可以用以下语法定义：

[捕获变量] (参数) {函数体}

23.23 新的 C++11 STL 算法

要点提示：C++11 为频繁使用的操作提供了新的算法。

C++11 提供了几种有用的新 STL 算法：`all_of`、`any_of`、`copy_if`、`copy_n`、`find_if_not` 和 `none_of`。

如果 `begin` 迭代器和 `end` 迭代器指定范围内的所有元素都满足某个谓词函数，则 `all_of` 函数返回 `true`。例如

```cpp
int array1[] = {1, 2, 3, 4, 5, 6, 7, 8};
cout << all_of(array1, array1 + 8,
    [](int v){ return v >= 1; }) << endl; // Return true
cout << all_of(array1, array1 + 8,
    [](int v){ return v >= 2; }) << endl; // Return false
```

如果 `begin` 迭代器和 `end` 迭代器指定范围内的一个元素满足谓词函数，则 `any_of` 函数返回 `true`。例如

```cpp
int array1[] = {1, 2, 3, 4, 5, 6, 7, 8};
cout << any_of(array1, array1 + 8,
    [](int v){ return v >= 3; }) << endl; // Return true
cout << any_of(array1, array1 + 8,
    [](int v){ return v >= 12; }) << endl; // Return false
```

`copy_if` 函数将 `begin` 迭代器和 `end` 迭代器指定范围内的满足谓词函数的元素复制到输出迭代器。例如

```cpp
int array1[] = { 1, 2, 3, 4, 5, 6, 7, 8 };
ostream_iterator<int> output(cout, " ");
copy_if(array1, array1 + 8, output,
    [](int v){return v > 4; }); // Displays 5 6 7 8
```

`copy_n` 函数将指定数量的元素从容器的 `begin` 迭代器复制到输出迭代器。例如

```cpp
int array1[] = { 1, 2, 3, 4, 5, 6, 7, 8 };
ostream_iterator<int> output(cout, " ");
copy_n(array1 + 5, 3, output); // Displays 6 7 8
```

`find_if_not` 函数返回 `begin` 迭代器和 `end` 迭代器指定范围内不满足谓词函数的第一个元素的迭代器。例如

```cpp
int array1[] = { 1, 2, 3, 4, 5, 6, 7, 8 };
int* p = find_if_not(array1, array1 + 8,
    [](int v){return v == 1;});
if (p != array1 + 8)
    cout << "found at position " << (p - array1);
else
    cout << "not found";
```

前面的代码显示"found at position 1"。

如果 `begin` 迭代器和 `end` 迭代器指定范围内的元素都不满足谓词函数，则 `none_of` 函数返回 `true`。例如

```cpp
int array1[] = { 1, 2, 3, 4, 5, 6, 7, 8 };
cout << none_of(array1, array1 + 8,
    [](int v){ return v >= 3; }) << endl; // Return false
cout << none_of(array1, array1 + 8,
    [](int v){ return v >= 20; }) << endl; // Return true
```

关键术语

`accumulate` algorithm（`accumulate`算法）
`adjacent_difference` algorithm（`adjacent_difference`算法）
`adjacent_find` algorithm（`adjacent_find`算法）
`binary_search` algorithm（`binary_search`算法）
`copy` algorithm（`copy`算法）
`count` algorithm（`count`算法）
`count_if` algorithm（`count_if`算法）
`fill` algorithm（`fill`算法）
`fill_n` algorithm（`fill_n`算法）
`find` algorithm（`find`算法）
`find_end` algorithm（`find_end`算法）
`find_first_of` algorithm（`find_first_of`算法）
`find_if` algorithm（`find_if`算法）
`for_each` algorithm（`for_each`算法）
function object（函数对象）
`generate` algorithm（`generate`算法）
`generate_n` algorithm（`generate_n`算法）
heap algorithms（堆算法）
`includes` algorithm（`includes`算法）
`inplace_merge` algorithm（`inplace_merge`算法）
`inner_product` algorithm（`inner_product`算法）
`iter_swap` algorithm iter（`iter_swap`算法）
`max_element` algorithm（`max_element`算法）
`merge` algorithm（`merge`算法）
`min_element` algorithm（`min_element`算法）
modifying STL algorithms（修改性STL算法）

nonmodifying STL algorithms（非修改性STL算法）
numeric STL algorithms（数值STL算法）
`partial_sum` algorithm（`partial_sum`算法）
`random_shuffle` algorithm（`random_shuffle`算法）
`remove` algorithm（`remove`算法）
`remove_copy` algorithm（`remove_copy`算法）
`remove_copy_if` algorithm（`remove_copy_if`算法）
`remove_if` algorithm（`remove_if`算法）
`replace` algorithm（`replace`算法）
`replace_if` algorithm（`replace_if`算法）
`replace_copy` algorithm（`replace_copy`算法）
`replace_copy_if` algorithm（`replace_copy_if`算法）
`reverse` algorithm（`reverse`算法）
`reverse_copy` algorithm（`reverse_copy`算法）
`rotate` algorithm（`rotate`算法）
`rotate_copy` algorithm（`rotate_copy`算法）
`search` algorithm（`search`算法）
`search_n` algorithm（`search_n`算法）
`sort` algorithm（`sort`算法）
`swap` algorithm（`swap`算法）
`swap_range` algorithm（`swap_range`算法）
`set_difference` algorithm（`set_difference`算法）
`set_intersection` algorithm（`set_intersection`算法）
`set_symmetric_difference` algorithm（`set_symmetric_difference`算法）
`set_union` algorithm（`set_union`算法）
`transform` algorithm（`transform`算法）

章节总结

1. STL将算法与容器分离。这使得算法能够通过迭代器通用地应用于所有容器。STL使算法和容器易于维护。
2. STL提供了大约80种算法。它们可以分为四组：非修改性算法、修改性算法、数值算法和堆算法。
3. 所有的算法都是通过迭代器运行的。许多算法在由两个迭代器指向的元素序列上进行操作。第一个迭代器指向序列的第一个元素，第二个迭代器指向序列最后一个元素之后的元素。
4. `copy`函数将序列中的元素从一个容器复制到另一个容器。
5. 函数`fill`和`fill_n`用指定值填充容器。
6. 函数`generate`和`generate_n`用某个函数返回的值填充容器。

7. 函数 `remove` 和 `remove_if` 从序列中删除符合某些条件的元素。函数 `remove_copy` 和 `remove_copy_if` 分别类似于 `remove` 和 `remove_if`，只是它们将结果复制到目标序列中。
8. 函数 `replace` 和 `replace_if` 将序列中所有出现的给定值替换为新值。函数 `replace_copy` 和 `replace_copy_if` 分别类似于 `replace` 和 `replace_if`，只是它们将结果复制到目标序列中。
9. 函数 `find`、`find_if`、`find_end` 和 `find_first_of` 用于在序列中查找元素。
10. 函数 `search` 和 `search_n` 查找子序列。
11. `sort` 函数需要随机访问迭代器。可将其应用于数组、向量或双端队列的排序。
12. `adjacent_find` 函数查找值相等或满足 `boolFunction(element)` 的第一个出现的相邻元素。
13. `merge` 函数将两个已排序的序列归并为一个新序列。
14. `inplace_merge` 函数将序列的第一部分与第二部分归并；假定这两个部分包含排序后的连续元素。
15. `reverse` 函数反转序列中的元素。`reverse_copy` 函数将一个序列中的元素以相反的顺序复制到另一个序列。
16. `rotate` 函数轮转序列元素。`rotate_copy` 函数类似于 `rotate`，只是它将结果复制到目标序列中。
17. `swap` 函数交换两个变量中的值。`iter_swap` 函数交换迭代器所指向的值。`swap_ranges` 函数交换两个序列。
18. `count` 函数对序列中出现的给定值进行计数。`count_if` 函数统计使 `boolFunction(element)` 为 `true` 的元素的出现次数。
19. 函数 `max_element` 和 `min_element` 获取序列中的最大元素和最小元素。
20. `random_shuffle` 函数对序列中的元素进行随机重新排序。
21. `for_each` 函数通过应用某个函数处理序列中的每个元素。`transform` 函数将某个函数应用于序列中的每个元素，并将结果复制到目标序列中。
22. STL 支持集合操作 `includes`、`set_union`、`set_difference`、`set_intersection`，以及 `set_symmetric_difference`。所有这些函数都要求序列元素已排序。
23. STL 支持数学函数 `accumulate`、`adjacent_difference`、`inner_product` 和 `partial_sum`。它们在 `<numeric>` 头文件中定义。

编程练习

23.1　创建一个包含五个数字的 `double` 数组：1.3、2.4、4.5、6.7、9.0。用 `fill` 函数将前三个元素填充为 5.5。用 `fill_n` 函数将前四个元素填充为 6.9。

23.2　创建一个包含五个数字的双端队列：1.3、2.4、4.5、6.7、9.0。用 `generate` 函数在双端队列中填充随机数。用 `generate_n` 函数在双端队列中填充随机数。

23.3　用以下数字创建数组：1.3、2.4、4.5、6.7、4.5、9.0。用 `remove` 函数删除所有值为 4.5 的元素。用 `remove_if` 函数删除所有小于 2.0 的元素。用 `remove_copy` 函数将除 6.7 以外的所有元素复制到列表中。用 `remove_copy_if` 函数将除大于 4.0 以外的所有元素复制到列表中。

23.4　用以下数字创建数组：2.4、1.3、2.4、4.5、6.7、4.5、9.0。用 `replace` 函数将所有出现的 2.4 替换为 9.9。用 `replace_if` 函数将小于 2.0 的所有元素替换为 12.5。用 `replace_copy` 函数将所有出现的 6.7 替换为 9.7，并将所有序列复制到向量中。用 `replace_copy_if` 函数将所有大于或等于 1.3 的元素替换为 747，并将序列复制到向量中。

23.5　用以下数字创建数组：2.4、1.3、2.4、4.5、6.7、4.5、9.0。用 `find` 函数查找 4.5 在数组中的位置。用 `find_if` 函数查找第一个小于 2 的元素的位置。用 `find_end` 函数查找序列 `{2.4, 4.5}` 在数组中的位置。用 `find_first_of` 函数查找数组和列表 `{34, 55, 2.4, 4.5}` 中第一个公共元素的位置。

23.6 用以下数字创建数组：2.4、1.3、2.4、2.4、4.5、6.7、4.5、9.0。用 search 函数查找序列 {2.4, 4.5} 在数组中的位置。用 search_n 函数查找值为 2.4 的两个连续元素。

23.7 用以下数字创建数组：2.4、1.3、2.4、2.4、4.5、6.7、4.5、9.0。用 sort 函数对数组进行排序。用 binary_search 函数分别查找值 6.7 和 4.3。

23.8 实现 fill 和 fill_n 函数。

```
template<typename ForwardIterator, typename T>
void fill(ForwardIterator beg, ForwardIterator end, const T& value)
template<typename ForwardIterator, typename size, typename T>
void fill_n(ForwardIterator beg, size n, const T& value)
```

23.9 实现 generate 和 generate_n 函数。

```
template<typename ForwardIterator, typename function>
void generate(ForwardIterator beg, ForwardIterator end, function gen)
template<typename ForwardIterator, typename size, typename function>
void generate_n(ForwardIterator beg, size n, function gen)
```

用 https://liangcpp.pearsoncmg.com/test/Exercise23_09.txt 中的代码完成你的程序。

23.10 实现 reverse 和 reverse_copy 函数。

```
template<typename BidirectionalIterator>
void reverse(BidirectionalIterator beg,
  BidirectionalIterator end)
template<typename BidirectionalIterator, typename OutputIterator>
OutputIterator reverse_copy(BidirectionalIterator beg,
  BidirectionalIterator end, OutputIterator targetPosition)
```

23.11 实现 replace 和 replace_if 函数。

```
template<typename ForwardIterator, typename T>
void replace(ForwardIterator beg, ForwardIterator end,
  const T& oldValue, const T& newValue)
template<typename ForwardIterator, typename boolFunction, typename T>
void replace_if(ForwardIterator beg, ForwardIterator end,
  boolFunction f, const T& newValue)
```

用 https://liangcpp.pearsoncmg.com/test/Exercise23_11.txt 中的模板写你的代码。

23.12 实现 find 和 find_if 函数。

```
template<typename InputIterator, typename T>
InputIterator find(InputIterator beg, InputIterator end, T& value)
template<typename InputIterator, typename boolFunction>
InputIterator find_if(InputIterator beg, InputIterator end,
  boolFunction f)
```

第 24 章

Introduction to C++ Programming and Data Structures, Fifth Edition

散 列

学习目标
1. 了解散列是什么以及散列的用途（24.2 节）。
2. 获取对象的散列码，设计散列函数，将键映射到索引（24.3 节）。
3. 用开放寻址法处理冲突（24.4 节）。
4. 了解线性探测法、平方探测法和双重散列法之间的差异（24.4 节）。
5. 用独立链处理冲突（24.5 节）。
6. 了解负载因子和再散列的必要性（24.6 节）。
7. 用散列实现映射（24.7 节）。
8. 用散列实现集合（24.8 节）。

24.1 简介

要点提示：散列是非常高效的。用散列查找、插入和删除元素需要 $O(1)$ 时间。

前一章介绍了二分查找树。在平衡良好的查找树中，可以在 $O(\log n)$ 时间内找到一个元素。有没有更高效的方法查找容器中的元素？本章介绍一种称为散列的技术。可以用散列来实现映射或集合，以在 $O(1)$ 时间内查找、插入和删除元素。

24.2 散列是什么

要点提示：散列使用散列函数将键映射到索引。

在介绍散列之前，让我们回顾一下映射，它是一种使用散列实现的数据结构。回想一下，映射（在 22.6 节中介绍）是一个存储项的容器对象。每个项包含两个部分：一个键和一个值。键，也称为搜索键，用于查找相应的值。例如，字典可以存储在映射中，其中单词是键，单词的定义是值。

注意：映射也称为**字典**、**散列表**或**关联数组**。

STL 定义用于建模映射的 `map` 和 `multimap` 类。在本章中，我们将学习散列的概念，并使用它来实现映射。

如果知道数组中某个元素的索引，则可以在 $O(1)$ 时间内用该索引检索该元素。那么，这是否意味着我们可以将值存储在数组中，并用键作为索引来查找值？答案是肯定的——如果你能把一个键映射到一个索引。存储这些值的数组称为**散列表**。将键映射到散列表中的索引的函数称为**散列函数**。如图 24.1 所示，散列函数从键中获取索引，并用该索引来检索键的值。散列是一种用从键中获得的索引检索值，而不执行搜索的技术。

如何设计一个从键生成索引的散列函数？理想情况下，我们希望设计一个函数，将每个搜索键映射到散列表中的不同索引。这样的函数被称为**完美散列函数**。然而，很难找到一个完美散列函数。当两个或多个键映射到同一个散列值时，我们认为发生了冲突。尽管本章稍后将讨论处理冲突的方法，但最好首先避免冲突。因此，我们应该设计一个快速、易于计算的散列函数，以最大限度地减少冲突。

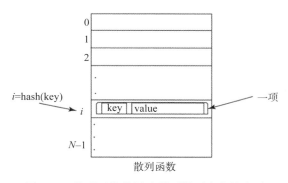

图24.1 散列函数将键映射到散列表中的索引

24.3 散列函数和散列码

要点提示：典型的散列函数首先将搜索键转换为称为散列码的整数值，然后将散列码压缩为散列表的索引。

在C++11中，可以用hash<KeyType>()(key)为基元类型或对象类型的元素返回无符号的整数**散列码**，具体取决于元素的类型和值。例如，hash<string>()("abc")返回字符串"abc"的散列码，hash<int>()(454)返回454的散列码。散列函数的约定如下：

1. 如果两个元素相等，则函数返回相同的散列码。
2. 在程序执行过程中，如果对象的数据没有更改，多次调用散列函数会返回相同的整数。
3. 两个不相等的对象可能具有相同的散列码，但C++11中的函数将避免太多这样的情况。

以下部分讨论数字和字符串的散列码的可能实现。

24.3.1 基元类型的散列码

对于short、int和char类型的搜索键，只需将它们强制转换为int即可。因此，这些类型的两个不同搜索键都具有不同的散列码。

对于float类型的搜索键，用floatToIntBits(key)作为散列码，其中floatToIntBits(float f)返回一个int值，该值的位表示与浮点数f的位表示相同。因此，两个不同的float类型搜索键具有不同的散列码。有关floatToIntBits的实现，参见编程练习24.6。

对于long long类型的搜索键（在C++11中long long是64位），简单地将其强制转换为int不是一个好的选择，因为所有仅在前32位不同的键都将具有相同的散列码。为了考虑前32位，将64位分成两部分，并执行异或运算以组合两部分。这个过程被称为折叠。long long键的散列码是

```
int hashCode = static_cast<int>(key ^ (key >> 32));
```

注意到>>是右移运算符，这里将每个位向右移动32个位置。例如，1010110>>2产生0010101。^是按位异或运算符。它对二进制操作数的两个相应位进行异或运算。例如，1010110^0110111产生1100001。有关按位运算的更多信息，参见附录E。

对于 double 类型的搜索键，首先用 doubleToLongLongBits 函数将其转换为 long long 值，然后执行如下折叠：

```
long long bits = *reinterpret_cast<long long*>(&key);
int hashCode = static_cast<int>(bits ^ (bits >> 32));
```

注意 13.7 节中引入的 reinterpret_cast。

24.3.2 字符串的散列码

搜索键通常是字符串，因此为字符串设计一个好的散列函数很重要。一种直观的方法是将所有字符的数字码相加作为字符串的散列码。如果一个应用程序的两个搜索键不包含相同的字符，这种方法可能会奏效，但如果搜索键包含相同的字符，如 tod 和 dot，则会产生很多冲突。更好的方法是生成一个考虑字符位置的散列码。具体来说，让散列码为

$$s_0 \times b^{(n-1)} + s_1 \times b^{(n-2)} + \cdots + s_{n-1}$$

这里 s_i 是 s[i]（字符串的第 i 个字符）。这个表达式是某个正数 b 的多项式，所以这被称为**多项式散列码**。如下所示，用 Horner 规则进行多项式计算（见 6.14 节），可以高效地计算散列码：

$$(\cdots((s_0 \times b + s_1) \times b + s_2) \times b + \cdots + s_{n-2}) \times b + s_{n-1}$$

这种计算可能会导致长字符串溢出，但算术溢出在 C++ 中被忽略。应该选择一个适当的值 b 来最小化冲突。实验表明，较好的 b 的选择是 31、33、37、39 和 41。

24.3.3 压缩散列码

键的散列码可能是超出散列表索引范围的大整数，因此需要将其缩小以适应索引的范围。假设散列表的索引在 0 到 N-1 之间。将整数缩放到 0 到 N-1 之间最常见的方法是用

```
h(hashCode) = hashCode % N
```

理想情况下，应该为 N 选择一个质数，以确保索引分布均匀。然而，找到一个大质数是很耗时的。我们选择 N 作为 2 的幂，做出这个选择是有充分理由的。当 N 是 2 的幂时，

```
h(hashCode) = hashCode % N
```

与

```
h(hashCode) = hashCode & (N - 1)
```

相同。

& 是一个按位与运算符（请参阅附录 E）。如果两个对应的位都是 1，则两个对应位的与结果为 1。例如，假设 N=4，hashCode=11，11%4=3，这与 01011&00011=11 相同。& 运算符的执行速度比 % 运算符快得多。

为了确保散列分布均匀，通常也会把一个补充散列函数和主散列函数一起使用。补充函数可以定义为：

```
int supplementalHash(int h)
{
    h ^= (h >> 20) ^ (h >> 12);
    return h ^ (h >> 7) ^ (h >> 4);
}
```

^ 和 >> 分别是按位异或和无符号右移运算（也在附录 E 中介绍）。按位运算比乘法、除法和取余运算快得多。应该尽可能用按位运算替换这些运算。

完整的散列函数定义为：

```
h(hashCode) = supplementalHash(hashCode) % N
```

这与下面式子相同：

```
h(hashCode) = supplementalHash(hashCode) & (N - 1)
```

因为 N 是 2 的幂。

24.4 用开放寻址处理冲突

要点提示：当两个键映射到散列表中的同一索引时，会发生冲突。通常有两种处理冲突的方法：开放寻址和独立链法。

开放寻址是在发生冲突时，在散列表中查找开放位置的过程。开放寻址有几种变体：**线性探测**、**平方探测**和**双重散列**。

24.4.1 线性探测

当在向散列表插入项的过程中发生冲突时，线性探测会依次查找下一个可用位置。例如，如果在 `hashTable[k%N]` 处发生冲突，则检查 `hashTable[(k+1)%N]` 是否可用。如果不可用，检查 `hashTable[(k+2)%N]` 是否可用，以此类推，直到找到可用的单元格，如图 24.2 所示。

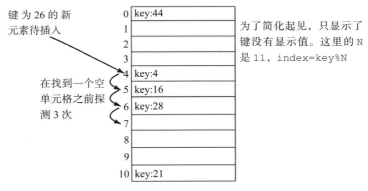

图 24.2 线性探测按顺序查找下一个可用位置

注意：当探测到表的末尾时，它会返回到表的开头。因此，将散列表视为循环表。

在散列表中搜索项时，先从键的散列函数中获得索引，例如 k。检查 `hashTable[k%N]` 是否包含该项。如果不包含，检查 `hashTable[(k+1)%N]` 是否包含该项，以此类推，直到找到该项，或者到达一个空单元格。

要从散列表中删除项，先搜索与键匹配的项。如果找到项，放置一个特殊标记，表示该项可用。散列表中的每个单元格都有三种可能的状态：已占用、已标记或为空。注意，标记的单元格也可用于插入。

线性探测往往会导致散列表中的连续单元格组被占用。每个组称为一个**集群**（簇）。每个集群实际上都是一个探测序列，在检索、添加或删除项时必须进行搜索。随着集群规模的增长，它们可能会合并成更大的集群，从而进一步减慢搜索时间。这是线性探测的一大缺点。

24.4.2 平方探测

平方探测可以避免线性探测中可能出现的聚集问题。线性探测着眼于从索引 k 开始的连续单元格。而平方探测则着眼于索引 $(k+j^2)\%N(j \geqslant 0)$ 处的单元格，即 $k\%N, (k+1)\%N$, $(k+4)\%n, (k+9)\%N, \cdots$ 如图 24.3 所示。

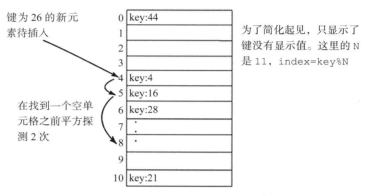

图 24.3 平方探测将序列中的下一个索引增加 j^2，j=1，2，3，…

平方探测的工作方式与线性探测相同，只是搜索序列发生了变化。平方探测避免了线性探测的聚集问题，但它有自己的聚集问题——**二次聚集**；也就是说，与被占用的项冲突的那些项使用相同的探测序列。

线性探测保证只要表未满，就可以找到可用的单元格进行插入。但平方探测不能保证。

24.4.3 双重散列

另一种避免聚集问题的开放寻址方案被称为**双重散列**。从初始索引 k 开始，线性探测和平方探测都向 k 添加增量来定义搜索序列。线性探测的增量是 1，平方探测的增量为 j^2。这些增量与键无关。双重散列在键上使用第二散列函数 $h'(key)$ 来确定增量，以避免聚集问题。具体来说，双重散列会查看索引 $(k+j*h'(key))\%N(j \geqslant 0)$ 处的单元格，即，$k\%N$, $(k+h'(key))\%N$, $(k+2*h'(key))\%N$, $(k+3*h'(key))\%N$, ⋯ 。

例如，如下定义大小为 11 的散列表上的主散列函数 h 和第二散列函数 h':

```
h(key) = key % 11;
h'(key) = 7 - key % 7;
```

对于搜索键 12，我们有

```
h(12) = 12 % 11 = 1;
h'(12) = 7 - 12 % 7 = 2;
```

假设有键 45、58、4、28 和 21 的元素已经被放置在散列表中。我们现在插入键为 12 的元素。键为 12 的探测序列从索引 1 开始。由于索引 1 处的单元格已被占用，则搜索索引 3(1+1*2) 处的下一个单元格。由于索引 3 处的单元格已被占用，则搜索索引 5(1+2*2) 处的下一个单元格。由于索引 5 处的单元格为空，因此现在将键为 12 的元素插入该单元格。

搜索过程如图 24.4 所示。

探测序列的索引如下：1、3、5、7、9、0、2、4、6、8、10。此序列覆盖整个表。我们应该设计自己的函数来生成一个能到达整个表的探测序列。注意，第二函数永远不应该有零值，因为零不是增量。

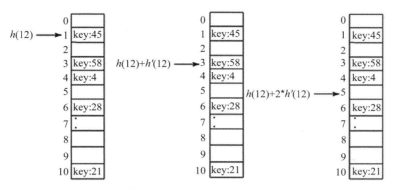

图 24.4　双重散列中的第二散列函数确定探测序列中下一个索引的增量

24.5　用独立链处理冲突

要点提示：独立链方案将具有相同散列索引的所有项放置在同一位置，而不是查找新位置。独立链方案中的每个位置都使用一个桶来保存多个项。

前一节介绍了使用开放寻址处理冲突。当发生冲突时，开放寻址方案会找到一个新的位置。本节介绍使用**独立链**处理冲突。独立链方案将具有相同散列索引的所有项放置在同一位置，而不是查找新位置。独立链方案中的每个位置都被称为一个桶。桶是一个包含多个项的容器。

桶可以用数组、向量或链表实现。我们用链表进行演示。可以将散列表中的每个单元格视为链表头的引用，链表中的元素从头开始链接，如图 24.5 所示。

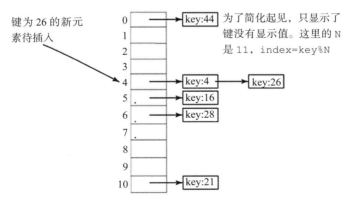

图 24.5　独立链方案将桶中具有相同散列索引的项链接起来

24.6　负载因子和再散列

要点提示：负载因子测量散列表的装载程度。如果超过负载因子，要增加散列表的大小，并将项重新装载到新的较大散列表中。这就是所谓的再散列。

负载因子 λ（lambda）衡量散列表的装载程度。它是元素数量与散列表大小的比率，即 $\lambda = \dfrac{n}{N}$，n 表示元素数量，N 表示散列表中的位置数量。

注意，如果映射为空，则 λ 为零。对于开放寻址方案，λ 在 0 和 1 之间；如果散列表已满，则 λ 为 1。对于独立链方案，λ 可以是任意值。随着 λ 的增加，冲突的概率也会增加。研究表明，对于开放寻址方案，应将负载因子保持在 0.5 以下，对于独立链方案，应保持在 0.9 以下。

将负载因子保持在某个阈值以下对于散列的性能很重要。每当负载因子超过阈值时，就需要增加散列表的大小，并将映射中的所有项**再散列**到新的更大的散列表中。注意，需要更改散列函数，因为散列表的大小已经更改。由于代价高昂，为了降低再散列的可能性，至少应该将散列表大小增加一倍。即使有周期性的再散列，散列也是映射的高效实现。

24.7 用散列实现映射

要点提示：映射可以用散列来实现。

现在我们了解了散列的概念，知道如何设计一个好的散列函数来将键映射到散列表中的索引，如何用负载因子衡量性能，以及如何增加表大小和再散列以保持性能。本节演示如何用独立链实现映射。

STL 定义用于建模映射的 `map` 和 `multimap` 类。在本节中，我们将实现 `map` 类。可以很容易修改该代码以实现 `multimap` 类。我们设计了自定义 `map` 类，并将其命名为 `MyMap`，如图 24.6 所示。

图 24.6 MyMap 用散列实现映射

如何实现 MyMap？我们将为散列表使用一个向量，散列表中的每个元素都是一个桶。桶也是一个向量。LiveExample 24.1 使用独立链实现 MyMap。

LiveExample 24.1 的互动程序请访问 https://liangcpp.pearsoncmg.com/LiveRunCpp5e/faces/LiveExample.xhtml?header=off&programName=MyMap&programHeight=5190&fileType=.h&resultVisible=false。

LiveExample 24.1　MyMap.h

Source Code Editor:

```cpp
#ifndef MYMAP_H
#define MYMAP_H

#include <vector>
#include <string>
#include <sstream>
#include <stdexcept>
using namespace std;

int DEFAULT_INITIAL_CAPACITY = 4;
float DEFAULT_MAX_LOAD_FACTOR = 0.75f;
unsigned MAXIMUM_CAPACITY = 1 << 30;

template<typename K, typename V>
class Entry // Represent an entry with key and value
{
public:
  Entry(const K& key, const V& value)
  {
    this->key = key;
    this->value = value;
  }

  string toString() const
  {
    stringstream ss;
    ss << "[" << key << ", " << value << "]";
    return ss.str();
  }

  K key;
  V value;
};

template<typename K, typename V>
class MyMap
{
public:
  MyMap();
  MyMap(int initialCapacity);
  MyMap(int initialCapacity, float loadFactorThreshold);

  V put(const K& key, const V& value);
  V& get(const K& key) const;
  int getSize() const;
  bool isEmpty() const;
  vector<Entry<K, V>> getEntries() const;
  vector<K> getKeys() const;
  vector<V> getValues() const;
  string toString() const;
  bool containsKey(const K& key) const;
  bool containsValue(const V& value) const;
  void remove(const K& key);
```

```cpp
 54      void clear();
 55
 56    private:
 57      int size;
 58      float loadFactorThreshold;
 59      int capacity;
 60
 61      // Hash table is an array with each cell as a vector
 62      vector<Entry<K, V>>* table; // table is a pointer to the array
 63
 64      int hash(int hashCode) const;
 65      unsigned hashCode(const K& key) const;
 66      int supplementalHash(int h) const;
 67      int trimToPowerOf2(int initialCapacity);
 68      void rehash(); // rehash() function
 69      void removeEntries();
 70    };
 71
 72    template<typename K, typename V>
 73    MyMap<K, V>::MyMap()
 74    {
 75      capacity = DEFAULT_INITIAL_CAPACITY;
 76      table = new vector<Entry<K, V>>[capacity]; // Create an array of vector<Entry<K, V>
 77      loadFactorThreshold = DEFAULT_MAX_LOAD_FACTOR;
 78      size = 0; // Set size to 0
 79    }
 80
 81    template<typename K, typename V>
 82    MyMap<K, V>::MyMap(int initialCapacity)
 83    {
 84      capacity = initialCapacity;
 85      table = new vector<Entry<K, V>>[capacity]; // Create an array of vector<Entry<K, V>
 86      loadFactorThreshold = DEFAULT_MAX_LOAD_FACTOR;
 87      size = 0;
 88    }
 89
 90    template<typename K, typename V>
 91    MyMap<K, V>::MyMap(int initialCapacity, float loadFactorThreshold)
 92    {
 93      if (initialCapacity > MAXIMUM_CAPACITY)
 94        capacity = MAXIMUM_CAPACITY;
 95      else
 96        capacity = trimToPowerOf2(initialCapacity);
 97
 98      this->loadFactorThreshold = loadFactorThreshold;
 99      table = new vector<Entry<K, V>>[capacity];
100      size = 0;
101    }
102
103    template<typename K, typename V>
104    V MyMap<K, V>::put(const K& key, const V& value)
105    {
106      if (containsKey(key)) // Test if key is in the map
107      { // The key is already in the map. Update the value for the key.
108        int bucketIndex = hash(hashCode(key)); // Hash key to an index
109        for (Entry<K, V>& entry : table[bucketIndex])
110        {
111          if (entry.key == key) // Test if key is already in the map
112          {
113            V oldValue = entry.value;
114            // Replace old value with new value
115            entry.value = value;
116            // Return the old value for the key
117            return oldValue;
```

```cpp
118          }
119        }
120      }
121
122      // Check load factor
123      if (size >= capacity * loadFactorThreshold)
124      {
125        if (capacity == MAXIMUM_CAPACITY)
126          throw runtime_error("Exceeding maximum capacity");
127
128        rehash();
129      }
130
131      int bucketIndex = hash(hashCode(key)); // Hash key to an index
132
133      // Add a new entry (key, value) to hashTable[index]
134      table[bucketIndex].push_back(Entry<K, V>(key, value));
135
136      size++; // Increase size
137
138      return value;
139    }
140
141    template<typename K, typename V>
142    V& MyMap<K, V>::get(const K& key) const
143    {
144      int bucketIndex = hash(hashCode(key));
145
146      for (Entry<K, V>& entry : table[bucketIndex])
147        if (entry.key == key)
148          return entry.value; // Return the value for the key
149
150      throw runtime_error("Key not found");
151    }
152
153    template<typename K, typename V>
154    bool MyMap<K, V>::isEmpty() const
155    {
156      return size == 0;
157    }
158
159    template<typename K, typename V>
160    vector<Entry<K, V>> MyMap<K, V>::getEntries() const
161    {
162      vector<Entry<K, V>> v;
163
164      for (int i = 0; i < capacity; i++)
165      {
166        for (Entry<K, V>& entry : table[i])
167          v.push_back(entry);
168      }
169
170      return v;
171    }
172
173    template<typename K, typename V>
174    bool MyMap<K, V>::containsKey(const K& key) const
175    {
176      try
177      {
178        get(key);
179        return true;
180      }
181      catch (runtime_error& ex)
```

```cpp
182     {
183       return false; // Key not found
184     }
185  }
186
187  template<typename K, typename V>
188  bool MyMap<K, V>::containsValue(const V& value) const
189  {
190    for (int i = 0; i < capacity; i++)
191    {
192      for (Entry<K, V> entry : table[i])
193        if (entry.value == value)
194          return true;
195    }
196
197    return false;
198  }
199
200  template<typename K, typename V>
201  void MyMap<K, V>::remove(const K& key)
202  {
203    int bucketIndex = hash(hashCode(key));
204
205    // Remove the entry that matches the key from a bucket
206    if (table[bucketIndex].size() > 0)
207    {
208      for (auto p = table[bucketIndex].begin();
209        p != table[bucketIndex].end(); p++)
210        if (p->key == key) // Test if p->key is key
211        {
212          table[bucketIndex].erase(p);
213          size--; // Decrease size
214          break; // No need to continue in the loop
215        }
216    }
217  }
218
219  template<typename K, typename V>
220  void MyMap<K, V>::clear()
221  {
222    size = 0;
223    removeEntries();
224  }
225
226  template<typename K, typename V>
227  void MyMap<K, V>::removeEntries()
228  {
229    for (int i = 0; i < capacity; i++)
230    {
231      table[i].clear();
232    }
233  }
234
235  template<typename K, typename V>
236  vector<K> MyMap<K, V>::getKeys() const
237  {
238    // Left as exercise
239  }
240
241  template<typename K, typename V>
242  vector<V> MyMap<K, V>::getValues() const
243  {
244    // Left as exercise
245  }
```

```cpp
246
247  template<typename K, typename V>
248  string MyMap<K, V>::toString() const
249  {
250    stringstream ss;
251    ss << "[";
252
253    for (int i = 0; i < capacity; i++)
254    {
255      for (Entry<K, V>& entry : table[i])
256        ss << entry.toString();
257    }
258
259    ss << "]";
260    return ss.str();
261  }
262
263  template<typename K, typename V>
264  unsigned MyMap<K, V>::hashCode(const K& key) const
265  {
266    return std::hash<K>()(key);
267  }
268
269  template<typename K, typename V>
270  int MyMap<K, V>::hash(int hashCode) const
271  {
272    return supplementalHash(hashCode) & (capacity - 1);
273  }
274
275  template<typename K, typename V>
276  int MyMap<K, V>::supplementalHash(int h) const
277  {
278    h ^= (h >> 20) ^ (h >> 12);
279    return h ^ (h >> 7) ^ (h >> 4);
280  }
281
282  template<typename K, typename V>
283  int MyMap<K, V>::trimToPowerOf2(int initialCapacity)
284  {
285    int capacity = 1;
286    while (capacity < initialCapacity) {
287      capacity <<= 1;
288    }
289
290    return capacity;
291  }
292
293  template<typename K, typename V>
294  void MyMap<K, V>::rehash()
295  {
296    vector<Entry<K, V>> set = getEntries(); // Get entries
297    capacity <<= 1; // Double capacity
298    delete[] table; // Delete old hash table
299    table = new vector<Entry<K, V>>[capacity]; // Create a new hash table
300    size = 0; // Reset size to 0
301
302    for (Entry<K, V>& entry : set)
303      put(entry.key, entry.value); // Store to new table
304  }
305
306  template<typename K, typename V>
307  int MyMap<K, V>::getSize() const
308  {
309    return size; // Return size
310  }
```

```
311
312  #endif
```

MyMap 类用独立链实现。在类中定义了决定负载因子（第 58 行）和散列表容量（第 59 行）的参数。默认初始容量为 4（第 10 行），最大容量为 2^30（第 12 行）。当前散列表容量被设计为 2 的幂的整数。默认负载因子阈值为 0.75f（第 11 行）。构建映射时，可以指定自定义负载因子阈值。自定义负载因子阈值存储在 loadFactorThreshold 中（第 58 行）。数据字段 size 表示映射中的项的数量（第 57 行）。散列表是一个数组。数组中的每个单元都是一个向量（第 62 行）。

提供了三个构造函数构造映射。可以用无参数构造函数（第 72 ~ 79 行）构造具有默认容量与负载因子阈值的默认映射，可以构造具有指定容量和默认负载因子阈值的映射（第 81 ~ 88 行），还可以构造具有指定容量与负载因子阈值的映射（第 90 ~ 101 行）。

put(key, value) 函数将一个新项添加到映射中（第 103 ~ 139 行）。该函数首先测试该键是否已经在映射中（第 106 行），如果在，则定位该项并用新值替换该键的项中的旧值（第 115 行），并且返回旧值（第 117 行）。如果键在映射中是新的，则在映射中创建新项（第 134 行）。在插入新项之前，该函数检查 size 是否超过负载因子阈值（第 123 行）。如果超过，程序调用 rehash()（第 128 行）增加容量并将项存储到新的较大散列表中。

rehash() 函数首先将所有项复制到一个向量中（第 296 行），将容量加倍（第 297 行），删除当前散列表（第 298 行），创建新的散列表（第 299 行），并将 size 重置为 0（第 300 行）。然后，该函数将项复制到新的散列表中（第 302 ~ 303 行）。rehash 函数需要 $O(capacity)$ 时间。如果不执行 rehash，put 函数花费 $O(1)$ 时间添加新项。

get(key) 函数返回具有指定键的第一个项的值（第 141 ~ 151 行）。此函数需要 $O(1)$ 时间。此函数调用 hashCode(key)（第 144 行）返回 key 的散列码。

hash() 函数调用 supplementalHash 确保散列分布均匀，从而为散列表生成索引（第 269 ~ 273 行）。此函数需要 $O(1)$ 时间。

remove(key) 函数删除映射中具有指定键的项（第 200 ~ 217 行）。此函数需要 $O(1)$ 时间。

getSize() 函数只返回映射的大小（第 306 ~ 310 行）。此函数需要 $O(1)$ 时间。

getKeys() 函数将映射中的所有键作为一个集合返回。该函数从每个存储桶中找到键，并将它们添加到一个集合中（第 235 ~ 239 行）。此函数需要 $O(capacity)$ 时间。有关此函数的实现，参见编程练习 24.9。

getValues() 函数返回映射中的所有值。该函数检查所有桶中的每个项，并将其添加到一个集合中（第 241 ~ 245 行）。此函数需要 $O(capacity)$ 时间。有关此函数的实现，参见编程练习 24.10。

clear 函数从映射中删除所有项（第 219 ~ 224 行）。它调用 removeEntries() 删除桶中的所有项（第 223 行）。removeEntries() 函数（第 226 ~ 233 行）需要 $O(capacity)$ 时间清除表中的所有项。

containsKey(key) 函数通过调用 get 函数（第 173 ~ 185 行）检查指定的键是否在映射中。由于 get 函数需要 $O(1)$ 时间，所以 containsKey(key) 函数需要 $O(1)$ 时间。

containsValue(value) 函数检查值是否在映射中（第 187 ～ 198 行）。此函数需要 O(capacity+size) 时间。它实际上是 O(capacity)，因为 capacity>size。

getEntries() 函数返回一个向量，该向量包含映射中的所有项（第 159 ～ 171 行）。此函数需要 O(capacity) 时间。

如果映射为空，则 isEmpty() 函数简单地返回 true（第 153 ～ 157 行）。此函数需要 O(1) 时间。

表 24.1 总结了 MyMap 中函数的时间复杂度。

表 24.1 MyMap 中函数的时间复杂度

函数	时间
clear()	O(capacity)
containsKey(key: Key)	O(1)
containsValue(value: V)	O(capacity)
getEntries()	O(capacity)
get(key: K)	O(1)
isEmpty()	O(1)
getKeys()	O(capacity)
put(key: K, value: V)	O(1)
remove(key: K)	O(1)
getSize()	O(1)
getValues()	O(capacity)
rehash()	O(capacity)

由于再散列并不经常发生，所以 put 函数的时间复杂度为 O(1)。注意，clear、getEntries、getKeys、getValues 和 rehash 函数的复杂度取决于 capacity，因此为了避免这些函数的性能不佳，应该仔细选择初始容量。

LiveExample 24.2 给出了一个使用 MyMap 的测试程序。

LiveExample 24.2 的互动程序请访问 https://liangcpp.pearsoncmg.com/LiveRunCpp5e/faces/LiveExample.xhtml?header=off&programName=TestMyMap&programHeight=560&resultHeight=250。

LiveExample 24.2 TestMyMap.cpp

Source Code Editor:

```cpp
#include <iostream>
#include <string>
#include "MyMap.h"
using namespace std;

int main()
{
  // Create a map
  MyMap<string, int> map;
  map.put("Smith", 30);
  map.put("Anderson", 31);
  map.put("Lewis", 29);
  map.put("Cook", 29);
  map.put("Smith", 65); // Add ("smith", 65) to the map

  cout << "Entries in map: " << map.toString() << endl;
```

```
17      cout << "The age for " << "Lewis is " <<
18        map.get("Lewis") << endl;
19      cout << "Is Smith in the map? " <<
20        (map.containsKey("Smith") ? "true" : "false") << endl;
21      cout << "Is age 33 in the map? " <<
22        (map.containsValue(33) ? "true" : "false") << endl;
23
24      map.remove("Smith"); // Remove the entry with key Smith
25      cout << "Entries in map: " << map.toString() << endl;
26
27      map.clear(); // Clear the map
28      cout << "Entries in map: " << map.toString() << endl;
29
30      return 0;
31    }
```

[Automatic Check] [Compile/Run] [Reset] [Answer]   Choose a Compiler: VC++

Execution Result:

```
command>cl TestMyMap.cpp
Microsoft C++ Compiler 2019
Compiled successful (cl is the VC++ compile/link command)

command>TestMyMap
Entries in map: [[Smith, 65][Anderson, 31][Lewis, 29][Cook, 29]]
The age for Lewis is 29
Is Smith in the map? true
Is age 33 in the map? false
Entries in map: [[Anderson, 31][Lewis, 29][Cook, 29]]
Entries in map: []

command>
```

该程序用 MyMap 创建一个映射（第 9 行），并在映射中添加五个项（第 10 ~ 14 行）。第 10 行添加值为 30 的键 Smith，第 14 行添加值为 65 的 Smith。后一个值覆盖前一个值。映射实际上只有四项。程序显示映射中的项（第 16 行），获取某个键的值（第 18 行），检查映射是否包含键（第 20 行）和值（第 22 行），删除带有 Smith 键的项（第 24 行），并重新显示映射中的项（第 25 行）。最后，程序清空映射（第 27 行）并显示空映射（第 28 行）。

24.8　用散列实现集合

要点提示：集合可以用散列来实现。

C++ STL 定义了用于建模集合的 set 和 multiset。可以用与实现 MyMap 相同的方法来实现它们。唯一的区别是映射中存储的是键/值对，而集合中存储的是元素。

本节我们将实现 set 类。可以很容易修改该代码来实现 multiset 类。我们将设计自定义 set 类来镜像 set，并将其命名为 MySet，如图 24.7 所示。

LiveExample 24.3 用独立链实现 MySet。

LiveExample 24.3 的互动程序请访问 https://liangcpp.pearsoncmg.com/LiveRunCpp5e/faces/LiveExample.xhtml?header=off&programName=MySet&programHeight=3920&fileType=.h&resultVisible=false。

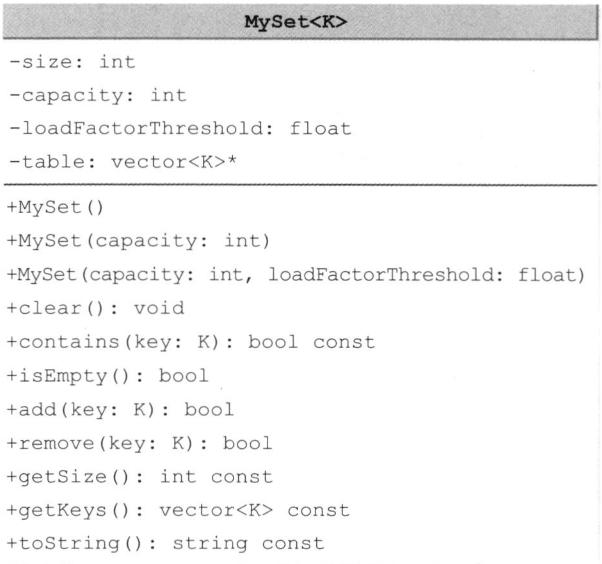

```
                    MySet<K>
    -size: int
    -capacity: int
    -loadFactorThreshold: float
    -table: vector<K>*
    +MySet()
    +MySet(capacity: int)
    +MySet(capacity: int, loadFactorThreshold: float)
    +clear(): void
    +contains(key: K): bool const
    +isEmpty(): bool
    +add(key: K): bool
    +remove(key: K): bool
    +getSize(): int const
    +getKeys(): vector<K> const
    +toString(): string const
```

图 24.7 MySet 实现了 set 类

LiveExample 24.3   MySet.h

Source Code Editor:

```cpp
#ifndef MYSet_H
#define MYSet_H

#include <vector>
#include <string>
#include <sstream>
#include <stdexcept>
using namespace std;

int DEFAULT_INITIAL_CAPACITY = 4;
float DEFAULT_MAX_LOAD_FACTOR = 0.75f;
unsigned MAXIMUM_CAPACITY = 1 << 30;

template<typename K>
class MySet
{
public:
  MySet();
  MySet(int initialCapacity);
  MySet(int initialCapacity, float loadFactorThreshold);

  int getSize() const;
  bool isEmpty() const;
  bool contains(const K& key) const;
  bool add(const K& key);
  bool remove(const K& key);
  void clear();
  vector<K> getKeys() const;
  string toString() const;

private:
  int size;
  float loadFactorThreshold;
  int capacity;
```

```cpp
36      // Hash table is an array with each cell as a vector
37      vector<K>* table; // table is a pointer to the array
38
39      int hash(int hashCode) const;
40      unsigned hashCode(const K& key) const;
41      int supplementalHash(int h) const;
42      int trimToPowerOf2(int initialCapacity);
43      void rehash(); // rehash() function
44      void removeKeys();
45    };
46
47    template<typename K>
48    MySet<K>::MySet()
49    {
50      capacity = DEFAULT_INITIAL_CAPACITY;
51      table = new vector<K>[capacity]; // Create the array for hash table
52      loadFactorThreshold = DEFAULT_MAX_LOAD_FACTOR;
53      size = 0;
54    }
55
56    template<typename K>
57    MySet<K>::MySet(int initialCapacity)
58    {
59      capacity = initialCapacity;
60      table = new vector<K>[capacity]; // Create the array for hash table
61      loadFactorThreshold = DEFAULT_MAX_LOAD_FACTOR;
62      size = 0;
63    }
64
65    template<typename K>
66    MySet<K>::MySet(int initialCapacity, float loadFactorThreshold)
67    {
68      if (initialCapacity > MAXIMUM_CAPACITY)
69        capacity = MAXIMUM_CAPACITY;
70      else
71        capacity = trimToPowerOf2(initialCapacity);
72
73      this->loadFactorThreshold = loadFactorThreshold;
74      table = new vector<K>[capacity]; // Create the array for hash table
75      size = 0;
76    }
77
78    template<typename K>
79    bool MySet<K>::add(const K& key)
80    {
81      if (contains(key)) return false; // key is already in the set
82
83      // Check load factor
84      if (size >= capacity * loadFactorThreshold)
85      {
86        if (capacity == MAXIMUM_CAPACITY)
87          throw runtime_error("Exceeding maximum capacity");
88
89        rehash();
90      }
91
92      int bucketIndex = hash(hashCode(key)); // Map the key to an index
93
94      // Add a new entry (key, value) to hashTable[index]
95      table[bucketIndex].push_back(key);
96
97      size++; // Increase size
98
99      return true;
100   }
101
```

```cpp
102  template<typename K>
103  bool MySet<K>::isEmpty() const
104  {
105    return size == 0; // Return true if size is 0
106  }
107
108  template<typename K>
109  bool MySet<K>::contains(const K& key) const
110  {
111    int bucketIndex = hash(hashCode(key));
112
113    for (K& e: table[bucketIndex]) // Test every element in table[bucket]
114      if (e == key)
115        return true;
116
117    return false;
118  }
119
120  template<typename K>
121  bool MySet<K>::remove(const K& key)
122  {
123    int bucketIndex = hash(hashCode(key));
124
125    // Remove the first entry that matches the key from a bucket
126    if (table[bucketIndex].size() > 0)
127    {
128      for (auto p = table[bucketIndex].begin();
129           p != table[bucketIndex].end(); p++)
130        if (*p == key) // Test if *p is key
131        {
132          table[bucketIndex].erase(p);
133          size--; // Decrease size
134          return true; // Remove just one entry that matches the key
135        }
136    }
137
138    return false;
139  }
140
141  template<typename K>
142  void MySet<K>::clear()
143  {
144    size = 0;
145    removeKeys();
146  }
147
148  template<typename K>
149  vector<K> MySet<K>::getKeys() const
150  {
151    vector<K> v;
152
153    for (int i = 0; i < capacity; i++)
154    {
155      for (K& e: table[i])
156        v.push_back(e); // Add e to v
157    }
158
159    return v;
160  }
161
162  template<typename K>
163  void MySet<K>::removeKeys()
164  {
165    for (int i = 0; i < capacity; i++)
166    {
```

```cpp
167            table[i].clear();
168        }
169    }
170
171    template<typename K>
172    string MySet<K>::toString() const
173    {
174        stringstream ss;
175        ss << "[";
176
177        for (int i = 0; i < capacity; i++)
178        {
179            for (K& e: table[i])
180                ss << e << " ";
181        }
182
183        ss << "]";
184        return ss.str();
185    }
186
187    template<typename K>
188    unsigned MySet<K>::hashCode(const K& key) const
189    {
190        return std::hash<K>()(key); // Use hash<K>()(key)
191    }
192
193    template<typename K>
194    int MySet<K>::hash(int hashCode) const
195    {
196        return supplementalHash(hashCode) & (capacity - 1);
197    }
198
199    template<typename K>
200    int MySet<K>::supplementalHash(int h) const
201    {
202        h ^= (h >> 20) ^ (h >> 12);
203        return h ^ (h >> 7) ^ (h >> 4);
204    }
205
206    template<typename K>
207    int MySet<K>::trimToPowerOf2(int initialCapacity)
208    {
209        int capacity = 1;
210        while (capacity < initialCapacity)
211            capacity <<= 1;
212
213        return capacity;
214    }
215
216    template<typename K>
217    void MySet<K>::rehash()
218    {
219        vector<K> set = getKeys(); // Get entries
220        capacity <<= 1; // Double capacity
221        delete[] table; // Delete old hash table
222        table = new vector<K>[capacity]; // Create a new hash table
223        size = 0; // Reset size to 0
224
225        for (K& e: set)
226            add(e); // Store to new table
227    }
228
229    template<typename K>
230    int MySet<K>::getSize() const
```

```
231 ▸ {
232     return size; // Return size
233 }
234
235 #endif
```

实现 `MySet` 与实现 `MyMap` 非常相似，不同之处在于，`MySet` 的散列表中存储的是键，但 `MyMap` 的散列表中存储的是项（键/值对）。

提供了三个构造函数来构造集合。可以用无参数构造函数构造具有默认容量和负载因子的默认集合（第 47 ~ 54 行），可以构造具有指定容量和默认负载因子的集合（第 56 ~ 63 行），还可以构造具有指定容量和负载因子的集合（第 65 ~ 76 行）。

`add(key)` 函数将一个新键添加到集合中。该函数首先检查键是否已经在集合中（第 81 行）。如果在，函数将返回 `false`。然后，该函数检查 size 是否超过负载因子阈值（第 84 行）。如果超过，程序调用 `rehash()`（第 89 行）增加容量并将键存储到新的更大的散列表中。

`rehash()` 函数首先复制列表中的所有键（第 219 行），将容量加倍（第 220 行），删除旧表（第 221 行），创建新的散列表（第 222 行），并将 size 重置为 0（第 223 行）。然后，该函数将键复制到新的较大散列表中（第 225 ~ 226 行）。rehash 函数需要 $O(capacity)$ 时间。如果不执行 rehash，则 add 函数花费 $O(1)$ 时间添加新键。

`clear` 函数可删除集合中的所有键（第 141 ~ 146 行）。它调用 `removeKeys()` 删除所有表单元格（第 167 行）。每个表单元格都是一个向量，用来存储具有相同散列码的键。`removeKeys()` 函数需要 $O(capacity)$ 时间。

`contains(key)` 函数通过检查指定的存储桶是否包含键来检查指定的键是否在集合中（第 108 ~ 118 行）。此函数需要 $O(1)$ 时间。

`remove(key)` 函数删除集合中指定的键（第 120 ~ 139 行）。此函数需要 $O(1)$ 时间。

`getSize()` 函数只返回集合中的键的数目（第 229 ~ 233 行）。此函数需要 $O(1)$ 时间。

`hashCode`、`hash`、`supplementalHash` 和 `trimToPowerOf2` 与 `MyMap` 类中的相同。

表 24.2 总结了 `MySet` 中函数的时间复杂度。

表 24.2 `MySet` 中函数的时间复杂度

函数	时间
clear()	$O(capacity)$
contains(k: K)	$O(1)$
add(k: K)	$O(1)$
remove(k: K)	$O(1)$
isEmpty()	$O(1)$
getSize()	$O(1)$
getKeys()	$O(size)$
rehash()	$O(cupacity)$

LiveExample 24.4 给出了一个使用 `MySet` 的测试程序。

LiveExample 24.4 的互动程序请访问 https://liangcpp.pearsoncmg.com/LiveRunCpp5e/faces/LiveExample.xhtml?header=off&programName=TestMySet&programHeight=510&resultHeight=230。

LiveExample 24.4   TestMySet.cpp

Source Code Editor:

```cpp
#include <iostream>
#include <string>
#include "MySet.h"
using namespace std;

int main()
{
  // Create a MySet
  MySet<string> set;
  set.add("Smith"); // Add Smith to set
  set.add("Anderson");
  set.add("Lewis");
  set.add("Cook");
  set.add("Smith"); // Add Smith to set

  cout << "Keys in set: " << set.toString() << endl;
  cout << "Number of keys in set: " << set.getSize() << endl;
  cout << "Is Smith in set? " << set.contains("Smith") << endl;

  set.remove("Smith");
  cout << "Names in set are ";
  for (string s: set.getKeys())
    cout << s << " ";

  set.clear();
  cout << "\nKeys in set: " << set.toString() << endl;

  return 0;
}
```

Execution Result:

```
command>cl TestMySet.cpp
Microsoft C++ Compiler 2019
Compiled successful (cl is the VC++ compile/link command)

command>TestMySet
Keys in set: [Smith Anderson Lewis Cook ]
Number of keys in set: 4
Is Smith in set? 1
Names in set are Anderson Lewis Cook
Keys in set: []

command>
```

该程序使用 `MySet` 创建一个集合（第 9 行），并向该集合添加五个键（第 10 ~ 14 行）。第 10 行添加 `Smith`，第 14 行再次添加 `Smith`。由于集合中只存储非重复键，因此 `Smith` 在集合中只出现一次。集合中实际上有四个键。程序显示键（第 16 行），获取其大小（第 17 行），检查集合是否包含指定的键（第 18 行），并删除一个键（第 20 行）。由于集合中的键是可迭代的，因此使用 `foreach` 循环遍历集合中的所有键（第 22 ~ 23 行）。最后，程序清空该集合（第 25 行）并显示空集合（第 26 行）。

关键术语

associative array（关联数组）
cluster（集群）
dictionary（字典）
double hashing（双重散列）
hash code（散列码）
hash function（散列函数）
hash map（散列映射）
hash set（散列集合）
hash table（散列表）
linear probing（线性探测）
load factor（负载因子）
open addressing（开放寻址）
perfect hash function（完美散列函数）
polynomial hash code（多项式散列码）
quadratic probing（平方探测）
rehashing（再散列）
secondary clustering（二次聚集）
separate chaining（独立链）

章节总结

1. 映射是一种存储项的数据结构。每个项包含两个部分：一个键和一个值。该键也称为搜索键，用于搜索相应的值。可以使用散列技术实现映射，以实现 $O(1)$ 时间复杂度的查找、检索、插入和删除。
2. 集合是一种存储元素的数据结构。可以使用散列技术实现集合，以实现 $O(1)$ 时间复杂度的集合查找、插入和删除。
3. 散列是一种不执行搜索，而是通过键获得索引来检索值的技术。典型的散列函数首先将搜索键转换为称为散列码的整数值，然后将散列码压缩为散列表的索引。
4. 当两个键映射到散列表中的同一索引时，会发生冲突。通常有两种处理冲突的方法：开放寻址和独立链。
5. 开放寻址是冲突时在散列表中查找开放位置的过程。开放寻址有几种变体：线性探测、平方探测和双重散列。
6. 独立链方案将具有相同散列索引的所有项放置在同一位置，而不是查找新位置。独立链方案中的每个位置都被称为一个桶。桶是一个包含多个项的容器。

编程练习

**24.1 （用线性探测的开放寻址实现 `MyMap`）创建一个新的具体类，该类用线性探测的开放寻址实现 `MyMap`。简单起见，用 `f(key)=key%size` 作为散列函数，其中 `size` 是散列表的大小。最初，散列表的大小是 4。只要负载因子超过阈值（0.5），表的大小就会加倍。

**24.2 （用平方探测的开放寻址实现 `MyMap`）创建一个新的具体类，该类用平方探测的开放寻址实现 `MyMap`。简单起见，用 `f(key)=key%size` 作为散列函数，其中 `size` 是散列表的大小。最初，散列表的大小是 4。只要负载因子超过阈值（0.5），表的大小就会加倍。

**24.3 （用双重散列的开放寻址实现 `MyMap`）创建一个新的具体类，该类用双重散列的开放寻址实现 `MyMap`。简单起见，用 `f(key)=key%size` 作为散列函数，其中 `size` 是散列表的大小。最初，散列表的大小是 4。只要负载因子超过阈值（0.5），表的大小就会加倍。

**24.4 （修改 `MyMap` 支持重复键）修改 `MyMap` 以支持项使用重复键。需要修改 `put(key, value)` 函数的实现。再添加一个名为 `getAll(key)` 的新函数，该函数返回一组与映射中的键匹配的值。

**24.5 （用 `MyMap` 实现 `MySet`）用 `MyMap` 实现 `MySet`。注意，可以用 `(key, key)` 而不是 `(key, value)` 创建项。

**24.6 （实现 `floatToIntBits`）编写以下函数，将 32 位 `float` 值返回为 `int`。注意，`float` 值和 `int` 值具有相同的二进制表示形式。

```
long long floatToIntBits(float value)
```

*24.7 （实现 doubleToLongLongBits）编写以下函数，将 64 位 double 值返回为 long long。注意，long long 值和 double 值具有相同的二进制表示形式。

    int doubleToLongLongBits(double value)

**24.8 （实现字符串的散列码）用 24.3.2 节中描述的方法编写一个函数，返回字符串的散列码，其中 b 值为 31。函数头如下：

    int hashCodeForString(string& s)

*24.9 （实现 getKeys）实现 LiveExample 24.1 中定义的 getKeys 函数。用 https://liangcpp.pearsoncmg.com/test/Exercise24_09.txt 完成代码。

*24.10 （实现 getValues）实现 LiveExample 24.1 中定义的 getValues 函数。

*24.11 （用迭代器实现 MyMap）修改 LiveExample 24.1 以定义迭代器类，并添加函数 begin() 和 end()，这两个函数返回用于遍历映射中的项的迭代器。

*24.12 （实现 MyMultiMap）修改 LiveExample 24.1 实现 MyMultiMap 类来存储允许重复键的键/值项。

*24.13 （用迭代器实现 MySet）修改 LiveExample 24.3 以定义迭代器类，并添加函数 begin() 和 end()，这两个函数返回用于遍历集合中的键的迭代器。

*24.14 （实现 MyMultiSet）修改 LiveExample 24.3 实现 MyMultiSet 类来存储允许重复的键。

*24.15 （比较 MySet 和 vector）MySet 在 LiveExample 24.3 中定义。编写一个程序，生成 1000000 个 0 到 999999 之间的随机整数，对它们进行混洗，并将它们存储在 vector 和 MySet 中。生成一个包含 1000000 个介于 0 和 1999999 之间的随机整数的列表。对于列表中的每个数字，测试它是否在数组列表和散列集合中。运行程序以显示数组列表和散列集合的总测试时间。

第 25 章

Introduction to C++ Programming and Data Structures, Fifth Edition

AVL 树

学习目标

1. 描述什么是 AVL 树（25.1 节）。
2. 用 LL 旋转、LR 旋转、RR 旋转和 RL 旋转再平衡树（25.2 节）。
3. 设计 AVLTree 类（25.3 节）。
4. 将元素插入 AVL 树（25.4 节）。
5. 实现节点再平衡（25.5 节）。
6. 删除 AVL 树中的元素（25.6 节）。
7. 实现 AVLTree 类（25.7 节）。
8. 测试 AVLTree 类（25.8 节）。
9. 分析 AVL 树查找、插入和删除操作的复杂度（25.9 节）。

25.1 简介

要点提示：AVL 树是一个平衡的二叉查找树。

第 21 章介绍了二叉查找树。二叉查找树的查找、插入和删除时间取决于树的高度。最坏情况下，高度为 $O(n)$。如果一棵树是**完全平衡**的，即它是一棵完全二叉树，它的高度是 $O(\log n)$。我们能保持一棵完全平衡的树吗？答案是可以的，但这样做代价高昂。折衷方案是保持一棵平衡性良好的树，即每个节点的两个子树的高度大致相同。

AVL 树具有很好的平衡性。AVL 树是由两位俄罗斯计算机科学家 G.M.Adelson-Velsky 和 E.M.Landis 于 1962 年发明的。在 AVL 树中，每个节点的两个子树的高度之差为 0 或 1。可以看出，AVL 树的最大高度是 $O(\log n)$。

在 AVL 树中插入或删除元素的过程与常规二叉查找树的过程相同。不同之处在于，在插入或删除操作之后，可能需要再平衡树。节点的**平衡因子**是其右子树高度减去其左子树高度。例如，图 25.1a 中节点 87 的平衡因子为 0，节点 67 的为 1，节点 55 的为 -1。如果一个节点的平衡因子为 -1、0 或 1，则称其为**平衡节点**。如果一个节点的平衡因子为 -1，则称其为**左重节点**。如果一个节点的平衡因子为 1，则称其为**右重节点**。

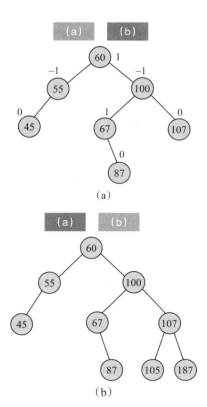

图 25.1 平衡因子决定节点是否平衡

25.2 再平衡树

要点提示：在 AVL 树中插入或删除元素后，如果树变得不平衡，则执行旋转操作来再平衡树。

如果节点在插入或删除操作后不平衡，则需要再平衡它。再平衡节点的过程称为**旋转**。有四种可能的旋转。

LL 旋转：如图 25.2a 所示，LL 不平衡发生在节点 A，A 的平衡因子为 -2，而左子节点 B 的平衡因子则为 -1 或 0。这种类型的不平衡可以通过在 A 处执行一次右旋转来解决，如图 25.2b 所示。

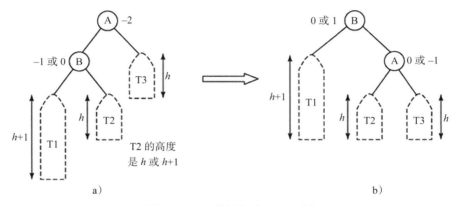

图 25.2　LL 旋转解决 LL 不平衡

RR 旋转：如图 25.3a 所示，RR 不平衡发生在节点 A，A 的平衡因子为 2，而右子节点 B 的平衡因子则为 1 或 0。这种类型的不平衡可以通过在 A 处执行一次左旋转来解决，如图 25.3b 所示。

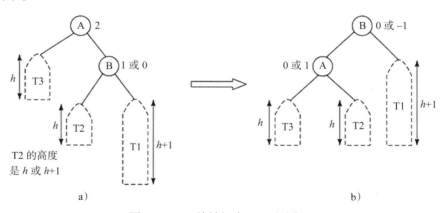

图 25.3　RR 旋转解决 RR 不平衡

LR 旋转：如图 25.4a 所示，LR 不平衡发生在节点 A，A 的平衡因子为 -2，左子节点 B 的平衡因子为 1。假设 B 的右子节点为 C。这种类型的不平衡可以通过在 A 处执行两次旋转（首先在 B 处执行一次左旋转，然后在 A 处执行一次右旋转）来解决，如图 25.4b 所示。

RL 旋转：如图 25.5a 所示，RL 不平衡发生在节点 A，A 的平衡因子为 2，右子节点 B 的平衡因子是 -1。假设 B 的左子节点为 C。这种类型的不平衡可以通过在 A 处执行两次旋转（首先在 B 处执行一次右旋转，然后在 A 处执行一次左旋转）来解决，如图 25.5b 所示。

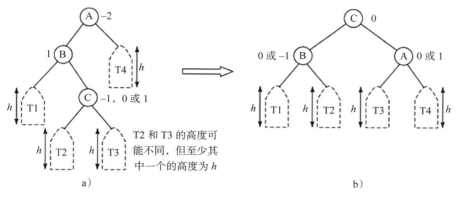

图 25.4 LR 旋转解决 LR 不平衡

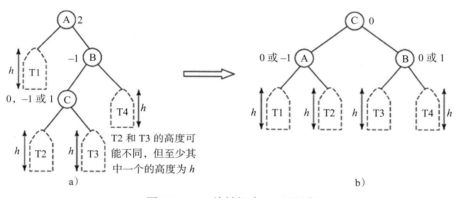

图 25.5 RL 旋转解决 RL 不平衡

25.3 设计 AVL 树类

要点提示：由于 AVL 树是二叉查找树，因此 AVLTree 被设计为 BST 的子类。

AVL 树是一种二叉树。因此，可以定义 AVLTree 类去扩展 BST 类，如图 25.6 所示。BST 和 TreeNode 类在 21.8 节中进行了定义。

为了平衡树，我们需要知道每个节点的高度。方便起见，将每个节点的高度存储在 AVLTreeNode 中，并将 AVLTreeNode 定义为 TreeNode 的子类，如 LiveExample 21.1 中的第 8～22 行所定义。注意，TreeNode 包含数据字段 element、left 和 right，这些字段被 AVLTreeNode 继承。因此，AVLTreeNode 包含四个数据字段，如图 25.7 所示。

在 BST 类中，createNewNode() 函数创建一个 TreeNode 对象。此函数在 AVLTree 类中被重写以创建 AVLTreeNode。注意，BST 类中 createNewNode() 函数的返回类型为 TreeNode，但 AVLTree 类中 createNewNode() 函数的返回类型是 AVLTreeNode。这是可以的，因为 AVLTreeNode 是 TreeNode 的一个子类型。

在 AVL 树中查找元素与在常规二叉树中查找相同。因此，BST 类中定义的 search 函数也适用于 AVLTree。

insert 和 remove 函数被重写以插入和删除元素，并在必要时执行再平衡操作，以确保树是平衡的。

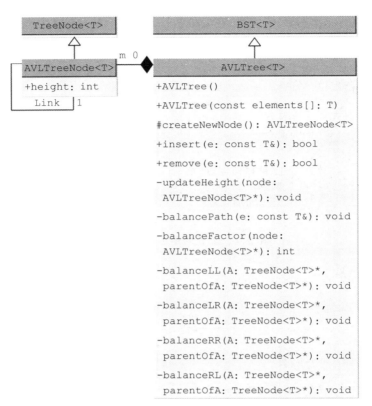

图 25.6 AVLTree 类通过 createNewNode、insert 和 remove 函数的新实现扩展 BST 类

图 25.7 AVLTreeNode 包含受保护的数据字段 element、height、left 和 right

25.4 重写 insert 函数

要点提示：将元素插入 AVL 树与将其插入 BST 相同，只是可能需要再平衡树。

新元素总是作为叶节点插入。作为添加新节点的结果，新节点的祖先的高度可能增加。插入后，检查从新叶节点到根节点的路径上的节点。如果发现节点不平衡，用 Listing 25.1 执行适当的旋转。

Listing 25.1　平衡路径上的节点

```
void balancePath(T& e)
{
    获取从包含元素 e 的节点到根的路径,
        如图 25.8 所示;
    for 通往根的路径上的节点 A
    {
        更新 A 的高度;
```

令 parentOfA 表示 A 的父节点，它是路径中的下一个节点。如果 A 为根，
则 parentOfA 为空；
switch (balanceFactor(A))
{
  case -2: **if** **balanceFactor**(A.left) = -1 or 0
         执行 LL 旋转；// 见图 25.2
       else
         执行 LR 旋转；// 见图 25.4
       break;
  case +2: **if** **balanceFactor**(A.right) = +1 or 0
         执行 RR 旋转；// 见图 25.3
       else
         执行 RL 旋转；// 见图 25.5
} // switch 结束
} // for 结束
} // 函数结束

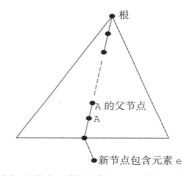

图 25.8　从新叶节点开始的路径上的节点可能变得不平衡

该算法考虑从新叶节点到根节点的路径中的每个节点。更新路径上节点的高度。如果一个节点是平衡的，则不需要执行任何操作。如果节点不平衡，则执行适当的旋转。

25.5　实现旋转

要点提示：不平衡的树通过执行适当的旋转操作变得平衡。

25.2 节说明了如何在节点处执行旋转。Listing 25.2 给出了 LL 旋转的算法，如图 25.2 所示。

Listing 25.2　LL 旋转算法

```
void balanceLL(TreeNode* A, TreeNode* parentOfA)
{
  令 B 为 A 的左子节点；

  if (A 是根)
    令 B 是新根；
  else
  {
    if (A 是 parentOfA 的左子节点)
      令 B 是 parentOfA 的左子节点；
    else
      令 B 是 parentOfA 的右子节点；
  }

  将 B.right 赋值给 A.left，使 T2 成为 A 的左子树；
  将 A 赋值给 B.right，使 A 成为 B 的右子节点；
  更新节点 A 和节点 B 的高度；
} // 函数结束
```

注意，节点 A 和 B 的高度可能会改变，但树中其他节点的高度不会改变。类似地，可以实现 RR 旋转、LR 旋转和 RL 旋转。

25.6 实现 remove 函数

要点提示：从 AVL 树中删除元素与从 BST 中删除元素相同，只是该树可能需要再平衡。

如 21.9 节所述，要从二叉树中删除元素，算法首先定位包含该元素的节点。让 current 指向二叉树中包含元素的节点，parent 指向 current 节点的父节点。current 节点可以是 parent 节点的左子节点或右子节点。删除元素时会出现两种情况。

情况 1：current 节点没有左子节点，如图 21.8a 所示。要删除 current 节点，只需将 parent 节点与 current 节点的右子节点连接即可，如图 21.8b 所示。

沿着从 parent 节点到根节点的路径，节点的高度可能已经降低。为确保树是平衡的，调用

```
balancePath(parent.element); // Defined in Listing 25.1
```

情况 2：current 节点有一个左子节点。如图 21.10a 所示，让 rightMost 指向 current 节点的左子树中包含最大元素的节点，parentOfRightMost 则指向 rightMost 节点的父节点。rightMost 节点不能有右子节点，但可以有左子节点。将 current 节点中的元素值替换为 rightMost 节点中的值，将 parentOfRightMost 节点与 rightMost 节点的左子节点连接，并删除 rightMost 节点，如图 21.10b 所示。

沿着从 parentOfRightMost 到根的路径，节点的高度可能已经降低。为确保树是平衡的，调用

```
balancePath(parentOfRightMost); // Defined in Listing 25.1
```

25.7 AVLTree 类

要点提示：AVLTree 类扩展 BST 类以重写 insert 和 remove 函数，并在必要时再平衡树。

LiveExample 25.1 给出了 AVLTree 类的完整源代码。

LiveExample 25.1 的互动程序请访问 https://liangcpp.pearsoncmg.com/LiveRunCpp5e/faces/LiveExample.xhtml?header=off&programName=AVLTree&programHeight=5370&fileType=.h&resultVisible=150。

LiveExample25.1　AVLTree.h

Source Code Editor:

```
1  #ifndef AVLTREE_H
2  #define AVLTREE_H
3
4  #include "BST.h"
5  #include <vector>
6  #include <algorithm>
7  #include <stdexcept>
8  using namespace std;
9
10 template<typename T>
11 class AVLTreeNode : public TreeNode<T>
12 {
```

```cpp
13    public:
14      int height; // height of the node
15
16      AVLTreeNode(const T& e) : TreeNode<T>(e) // Constructor
17      {
18        height = 0; // Set height to 0
19      }
20    };
21
22    template <typename T>
23    class AVLTree : public BST<T>
24    {
25    public:
26      AVLTree();
27      AVLTree(const T elements[], int arraySize);
28      // AVLTree(BST<T> &tree); left as exercise
29      // ~AVLTree(); left as exercise
30      // The = operator is also left as an exercise
31      bool insert(const T& e); // Redefine insert defined in BST
32      bool remove(const T& e); // Redefine remove defined in BST
33
34    protected:
35      // Redefine createNewNode defined in BST
36      AVLTreeNode<T>* createNewNode(const T& e);
37
38    private:
39      // Balance the nodes in the path from the specified
40      // node to the root if necessary
41      void balancePath(const T& e);
42
43      // Update the height of a specified node
44      void updateHeight(AVLTreeNode<T>* node);
45
46      // Return the balance factor of the node
47      int balanceFactor(AVLTreeNode<T>* node);
48
49      // Balance LL (see Figure 25.2 )
50      void balanceLL(TreeNode<T>* A, TreeNode<T>* parentOfA);
51
52      // Balance LR (see Figure 25.4 )
53      void balanceLR(TreeNode<T>* A, TreeNode<T>* parentOfA);
54
55      // Balance RR (see Figure 25.3 )
56      void balanceRR(TreeNode<T>* A, TreeNode<T>* parentOfA);
57
58      // Balance RL (see Figure 25.5 )
59      void balanceRL(TreeNode<T>* A, TreeNode<T>* parentOfA);
60    };
61
62    template <typename T>
63    AVLTree<T>::AVLTree()
64    {
65
66    }
67
68    template <typename T>
69    AVLTree<T>::AVLTree(const T elements[], int arraySize)
70    {
71      root = nullptr;
72      size = 0;
73
74      for (int i = 0; i < arraySize; i++)
75      {
76        insert(elements[i]);
77      }
```

```cpp
78    }
79
80    template <typename T>
81    AVLTreeNode<T>* AVLTree<T>::createNewNode(const T& e)
82    {
83        return new AVLTreeNode<T>(e);
84    }
85
86    template <typename T>
87    bool AVLTree<T>::insert(const T& e)
88    {
89        bool successful = BST<T>::insert(e);
90        if (!successful)
91            return false; // element is already in the tree
92        else
93            // Balance from element to the root if necessary
94            balancePath(e);
95
96        return true; // element is inserted
97    }
98
99    template <typename T>
100   void AVLTree<T>::balancePath(const T& e)
101   {
102       vector<TreeNode<T>*>* p = path(e);
103       for (int i = (*p).size() - 1; i >= 0; i--)
104       {
105           AVLTreeNode<T>* A = static_cast<AVLTreeNode<T>*>((*p)[i]);
106           updateHeight(A);
107           AVLTreeNode<T>* parentOfA = (A == root) ? nullptr :
108               static_cast<AVLTreeNode<T>*>((*p)[i - 1]);
109
110           switch (balanceFactor(A)) // Test balance factor
111           {
112               case -2:
113                   if (balanceFactor(
114                       static_cast<AVLTreeNode<T>*>(((*A).left))) <= 0)
115                       balanceLL(A, parentOfA); // Perform LL rotation
116                   else
117                       balanceLR(A, parentOfA); // Perform LR rotation
118                   break;
119               case +2:
120                   if (balanceFactor(
121                       static_cast<AVLTreeNode<T>*>(((*A).right))) >= 0)
122                       balanceRR(A, parentOfA); // Perform RR rotation
123                   else
124                       balanceRL(A, parentOfA); // Perform RL rotation
125           }
126       }
127   }
128
129   template <typename T>
130   void AVLTree<T>::updateHeight(AVLTreeNode<T>* node)
131   {
132       if (node->left == nullptr && node->right == nullptr) // node is a leaf
133           node->height = 0;
134       else if (node->left == nullptr) // node has no left subtree
135           node->height =
136               1 + (*static_cast<AVLTreeNode<T>*>((node->right))).height;
137       else if (node->right == nullptr) // node has no right subtree
138           node->height =
139               1 + (*static_cast<AVLTreeNode<T>*>((node->left))).height;
140       else
141           node->height = 1 +
142               max((*static_cast<AVLTreeNode<T>*>((node->right))).height,
```

```cpp
143           (*static_cast<AVLTreeNode<T>*>((node->left))).height);
144  }
145
146  template <typename T>
147  int AVLTree<T>::balanceFactor(AVLTreeNode<T>* node)
148  {
149    if (node->right == nullptr) // node has no right subtree
150      return -node->height;
151    else if (node->left == nullptr) // node has no left subtree
152      return +node->height;
153    else
154      return (*static_cast<AVLTreeNode<T>*>((node->right))).height -
155        (*static_cast<AVLTreeNode<T>*>((node->left))).height;
156  }
157
158  template <typename T>
159  void AVLTree<T>::balanceLL(TreeNode<T> * A, TreeNode<T> * parentOfA)
160  {
161    TreeNode<T> * B = (*A).left; // // A and B are left-heavy
162
163    if (A == root) // Test if A is root
164      root = B;
165    else
166      if (parentOfA->left == A)
167        parentOfA->left = B;
168      else
169        parentOfA->right = B;
170
171    A->left = B->right; // Make T2 the left subtree of A
172    B->right = A; // Make A the left child of B
173    updateHeight(static_cast<AVLTreeNode<T>*>(A));
174    updateHeight(static_cast<AVLTreeNode<T>*>(B));
175  }
176
177  template <typename T>
178  void AVLTree<T>::balanceLR(TreeNode<T>* A, TreeNode<T>* parentOfA)
179  {
180    TreeNode<T>* B = A->left; // A is left-heavy
181    TreeNode<T>* C = B->right; // B is right-heavy
182
183    if (A == root)
184      root = C;
185    else
186      if (parentOfA->left == A)
187        parentOfA->left = C;
188      else
189        parentOfA->right = C;
190
191    A->left = C->right; // Make T3 the left subtree of A
192    B->right = C->left; // Make T2 the right subtree of B
193    C->left = B;
194    C->right = A;
195
196    // Adjust heights
197    updateHeight(static_cast<AVLTreeNode<T>*>(A));
198    updateHeight(static_cast<AVLTreeNode<T>*>(B));
199    updateHeight(static_cast<AVLTreeNode<T>*>(C));
200  }
201
202  template <typename T>
203  void AVLTree<T>::balanceRR(TreeNode<T>* A, TreeNode<T>* parentOfA)
204  {
205    // A is right-heavy and B is right-heavy
206    TreeNode<T>* B = A->right;
207
208    if (A == root)
```

```cpp
      root = B;
    else
      if (parentOfA->left == A)
        parentOfA->left = B;
      else
        parentOfA->right = B;

    A->right = B->left; // Make T2 the right subtree of A
    B->left = A;
    updateHeight(static_cast<AVLTreeNode<T> *>(A));
    updateHeight(static_cast<AVLTreeNode<T>*>(B));
}

template <typename T>
void AVLTree<T>::balanceRL(TreeNode<T> * A, TreeNode<T> * parentOfA)
{
    TreeNode<T> * B = A->right; // A is right-heavy
    TreeNode<T> * C = B->left; // B is left-heavy

    if (A == root)
      root = C;
    else
      if (parentOfA->left == A)
        parentOfA->left = C;
      else
        parentOfA->right = C;

    A->right = C->left; // Make T2 the right subtree of A
    B->left = C->right; // Make T3 the left subtree of B
    C->left = A;
    C->right = B;

    // Adjust heights
    updateHeight(static_cast<AVLTreeNode<T>*>(A));
    updateHeight(static_cast<AVLTreeNode<T>*>(B));
    updateHeight(static_cast<AVLTreeNode<T>*>(C));
}

template <typename T>
bool AVLTree<T>::remove(const T& e)
{
    if (root == nullptr)
      return false; // Element e is not in the tree

    // Locate the node to be deleted and also locate its parent node
    TreeNode<T>* parent = nullptr;
    TreeNode<T>* current = root;
    while (current != nullptr)
    {
      if (e < current->element) // Test if e < current->element
      {
        parent = current;
        current = current->left;
      }
      else if (e > current->element)
      {
        parent = current;
        current = current->right;
      }
      else
        break; // Element e is in the tree pointed by current
    }

    if (current == nullptr)
      return false; // Element e is not in the tree
```

```
274
275      // Case 1: current has no left children
276      if (current->left == nullptr)
277      {
278        // Connect the parent with the right child of the current node
279        if (parent == nullptr)
280          root = current->right;
281        else
282        {
283          if (e < parent->element)
284            parent->left = current->right; // Connect to parent's left
285          else
286            parent->right = current->right; // Connect to parent's right
287
288          // Balance the tree if necessary
289          balancePath(parent->element);
290        }
291      }
292      else
293      {
294        // Case 2: The current node has a left child
295        // Locate the rightmost node in the left subtree of
296        // the current node and also its parent
297        TreeNode<T>* parentOfRightMost = current;
298        TreeNode<T>* rightMost = current->left;
299
300        while (rightMost->right != nullptr)
301        {
302          parentOfRightMost = rightMost;
303          rightMost = rightMost->right; // Keep going to the right
304        }
305
306        // Replace the element in current by the element in rightMost
307        current->element = rightMost->element;
308
309        // Eliminate rightmost node
310        if (parentOfRightMost->right == rightMost)
311          parentOfRightMost->right = rightMost->left;
312        else
313          // Special case: parentOfRightMost is current
314          parentOfRightMost->left = rightMost->left;
315
316        // Balance the tree if necessary
317        balancePath(parentOfRightMost->element);
318      }
319
320      size--; // Decrease size
321      return true; // Element inserted
322    }
323
324    #endif
```

Answer　Reset

AVLTree 类扩展了 BST 类（第 23 行）。与 BST 类一样，AVLTree 类有一个构造空 AVLTree 的无参数构造函数（第 62～66 行）和一个从元素数组创建初始 AVLTree（第 68～78 行）的构造函数。

BST 类中定义的 createNewNode() 函数创建一个 TreeNode。此函数被重写以返回 AVLTreeNode（第 80～84 行）。注意，此函数是从 BST 定义的 insert 函数动态调用的（参见 LiveExample 21.1 中的第 199、221、223 行）。

AVLTree 中的 insert 函数在第 86～97 行被重写。该函数首先调用 BST 中的 insert

函数，然后调用 balancePath(element)（第 94 行）以确保树是平衡的。

balancePath 函数首先获取从包含元素的节点到根的路径上的节点（第 102 行）。对于路径中的每个节点，更新其高度（第 106 行），检查其平衡因子（第 110 行），并在必要时执行适当的旋转（第 112 ~ 124 行）。

第 158 ~ 245 行定义了执行旋转的四个函数。每个函数都用两个 TreeNode<T> 参数 A 和 parentOfA 调用，以在节点 A 执行适当的旋转。旋转如何执行如图 25.2 ~ 25.5 所示。旋转后，节点的高度也会更新。

AVLTree 中的 remove 函数在第 247 ~ 322 行被重写。该函数与 BST 类中实现的函数相同，只是在第 289 行和第 317 行删除节点后必须重新平衡节点。

25.8 测试 **AVLTree** 类

要点提示： 本节给出一个使用 AVLTree 类的示例。

LiveExample 25.2 给出一个测试程序。该程序创建一个 AVLTree，该 AVLTree 用包含整数 25、20 和 5 的数组初始化（第 22 ~ 23 行），在第 28 ~ 39 行插入元素，并在第 41 ~ 49 行删除元素。

LiveExample 25.2 的互动程序请访问 https://liangcpp.pearsoncmg.com/LiveRunCpp5e/faces/LiveExample.xhtml?header=off&programName=TestAVLTree&programHeight=900&resultHeight=1180。

LiveExample25.2　TestAVLTree.cpp

Source Code Editor:

```cpp
#include <iostream>
#include "AVLTree.h"
using namespace std;

template <typename T>
void printTree(const AVLTree<T>& tree)
{
  // Traverse tree
  cout << "\nInorder (sorted): " << endl;
  tree.inorder();
  cout << "\nPostorder: " << endl;
  tree.postorder();
  cout << "\nPreorder: " << endl;
  tree.preorder();
  cout << "\nThe number of nodes is " << tree.getSize();
  cout << endl;
}

int main()
{
  // Create an AVL tree
  int numbers[] = {25, 20, 5};
  AVLTree<int> tree(numbers, 3);

  cout << "After inserting 25, 20, 5:" << endl;
  printTree<int>(tree);

  tree.insert(34);
  tree.insert(50); // Insert 50 to tree
  cout << "\nAfter inserting 34, 50:" << endl;
  printTree<int>(tree);
```

```
32
33      tree.insert(30);
34      cout << "\nAfter inserting 30" << endl;
35      printTree<int>(tree);
36
37      tree.insert(10);
38      cout << "\nAfter inserting 10" << endl;
39      printTree(tree);
40
41      tree.remove(34);
42      tree.remove(30);
43      tree.remove(50);
44      cout << "\nAfter removing 34, 30, 50:" << endl;
45      printTree<int>(tree);
46
47      tree.remove(5); // Delete 5 from tree
48      cout << "\nAfter removing 5:" << endl;
49      printTree<int>(tree);
50
51      return 0;
52  }
```

Execution Result:

```
command>cl TestAVLTree.cpp
Microsoft C++ Compiler 2019
Compiled successful (cl is the VC++ compile/link command)

command>TestAVLTree
After inserting 25, 20, 5:

Inorder (sorted):
5 20 25
Postorder:
5 25 20
Preorder:
20 5 25
The number of nodes is 3

After inserting 34, 50:

Inorder (sorted):
5 20 25 34 50
Postorder:
5 25 50 34 20
Preorder:
20 5 34 25 50
The number of nodes is 5

After inserting 30

Inorder (sorted):
5 20 25 30 34 50
Postorder:
5 20 30 50 34 25
Preorder:
25 20 5 34 30 50
The number of nodes is 6

After inserting 10
```

```
Inorder (sorted):
5 10 20 25 30 34 50
Postorder:
5 20 10 30 50 34 25
Preorder:
25 10 5 20 34 30 50
The number of nodes is 7

After removing 34, 30, 50:
Inorder (sorted):
5 10 20 25
Postorder:
5 20 25 10
Preorder:
10 5 25 20
The number of nodes is 4

After removing 5:
Inorder (sorted):
10 20 25
Postorder:
10 25 20
Preorder:
20 10 25
The number of nodes is 3

command>
```

图 25.9 显示了把元素添加到树中时树是如何演变的。添加 25 和 20 后，树如图 25.9a 所示。5 被插入为 20 的左子节点，如图 25.9b 所示。树是不平衡的。它在节点 25 处是左重的。执行 LL 旋转以生成 AVL 树，如图 25.9c 所示。

插入 34 后，树如图 25.9d 所示。插入 50 后，树如图 25.9e。该树不平衡。它在节点 25 处是右重的。执行 RR 旋转生成 AVL 树，如图 25.9f 所示。

插入 30 后，树如图 25.9g 所示。该树不平衡。执行 RL 旋转生成 AVL 树，如图 25.9h 所示。

插入 10 后，树如图 25.9i 所示。该树不平衡。执行 LR 旋转生成 AVL 树，如图 25.9j 所示。

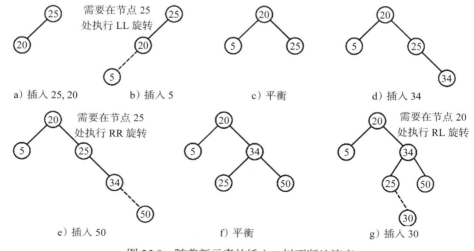

图 25.9 随着新元素的插入，树不断地演变

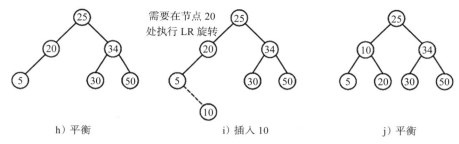

图 25.9 随着新元素的插入，树不断地演变（续）

图 25.10 显示了元素被删除时树是如何演变的。删除 34、30 和 50 后，树如图 25.10b 所示。该树不平衡。执行 LL 旋转生成 AVL 树，如图 25.10c 所示。

删除 5 后，树如图 25.10d 所示。该树不平衡。执行 RL 旋转生成 AVL 树，如图 25.10e 所示。

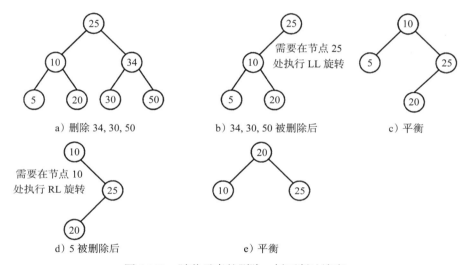

图 25.10 随着元素的删除，树不断地演变

25.9 AVL 树的时间复杂度分析

要点提示：由于 AVL 树的高度是 $O(\log n)$，因此 AVLTree 中 search、insert 和 remove 函数的时间复杂度是 $O(\log n)$。

AVLTree 中 search、insert 和 remove 函数的时间复杂度取决于树的高度。我们可以证明树的高度是 $O(\log n)$。

设 $G(h)$ 表示高度为 h 的 AVL 树中节点的最小数量。显然，$G(0)$ 是 1，$G(1)$ 是 2。具有高度 $h \geq 2$ 的 AVL 树必须至少有两个子树：一个子树具有高度 $h-1$，另一个子树具有高度 $h-2$。因此，$G(h) = G(h-1) + G(h-2) + 1$。回想一下，索引 i 处的斐波那契数可以使用递归关系 $F(i) = F(i-1) + F(i-2)$ 来描述。因此，函数 $G(h)$ 本质上与 $F(i)$ 相同。可以证明，$h < 1.4405 \log(n+2) - 1.3277$，其中 n 为树中节点的数量。因此，AVL 树的高度是 $O(\log n)$。

search、insert 和 remove 函数只涉及树中路径上的节点。路径中的每个节点在常数时间内执行 updateHeight 和 balanceFactor 函数。路径中的某个节点在常数时间内执行balancePath函数。因此，search、insert和remove函数的时间复杂度为$O(\log n)$。

关键术语

AVL tree（AVL 树）
balance factor（平衡因子）
left-heavy（左重）
LL rotation（LL 旋转）
LR rotation（LR 旋转）
perfectly balanced（完全平衡）

right-heavy（右重）
RL rotation（RL 旋转）
rotation（旋转）
RR rotation（RR 旋转）
well balanced（平衡性良好）

章节总结

1. AVL 树是一棵平衡性良好的二叉树。
2. 在 AVL 树中，每个节点的两个子树的高度之差为 0 或 1。
3. 在 AVL 树中插入或删除元素的过程与常规二叉查找树的过程相同。不同之处在于，在插入或删除操作之后，可能需要再平衡树。
4. 插入和删除导致的树的不平衡通过在不平衡节点处旋转子树进行再平衡。
5. 再平衡节点的过程称为旋转。有四种可能的旋转：LL 旋转、LR 旋转、RR 旋转和 RL 旋转。
6. AVL 树的高度为 $O(\log n)$。因此，查找、插入和删除函数的时间复杂度为 $O(\log n)$。

编程练习

*25.1 （存储字符）编写一个程序，将从 a 到 z 的 26 个小写字母按顺序插入 BST 和 AVLTree，并分别按中序、前序和后序显示树中的字符。

25.2 （比较性能）编写一个测试程序，随机生成 500000 个数字，并将它们插入 BST，重新混洗 500000 个数字并执行查找，然后在将它们从树中删除之前重新混洗这些数字。编写另一个测试程序，除了将数字插入到 AVLTree，其余操作相同。比较这两个程序的执行时间。

25.3 （修改 AVLTree）通过添加复制构造函数和析构函数来修改 AVLTee 类。

**25.4 （BST 的父引用）假设 BST 中定义的 TreeNode 类包含对节点的父引用，如编程练习 21.7 所示。实现 AVLTree 类以支持此更改。编写一个测试程序，将数字 1，2，⋯，100 添加到树并显示所有叶节点的路径。

**25.5 （第 k 个最小元素）可以在 $O(n)$ 时间内从中序迭代器中找到 BST 中的第 k 个最小元素。对于 AVL 树，可以在 $O(\log n)$ 时间内找到它。要实现这一点，请在 AVLTreeNode 中添加一个名为 size 的新数据字段，用来存储以此节点为根的子树的节点数。注意，节点 v 的 size 比其两个子节点的 size 之和多 1。图 25.11 显示了一个 AVL 树和树中每个节点的 size 值。

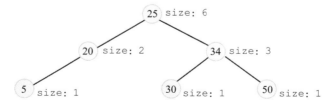

图 25.11 AVLTreeNode 中的 size 数据字段存储以该节点为根的子树中的节点数

在 AVLTree 类中，添加以下函数以返回树中第 k 个最小的元素。

```
T find(int k)
```

如果 k<1 或 k> 树的 size，则函数返回 nullptr。这个函数可以用递归函数 find(k, root) 来实现，该函数返回具有指定根的树中第 k 个最小元素。设 A 和 B 分别是根的左子节点和右子节点。假设树不是空的，并且 k ≤ root.size，find(k, root) 可以递归定义如下：

$$\text{find}(k, \text{root}) = \begin{cases} \text{root.element}, & A \text{ 为 null 且 } k \text{ 为 } 1; \\ \text{B.element}, & A \text{ 为 null 且 } k \text{ 为 } 2; \\ \text{find}(k, A), & k \leq \text{A.size}; \\ \text{root.element}, & k = \text{A.size} + 1; \\ \text{find}(k - \text{A.size} - 1, B), & k > \text{A.size} + 1; \end{cases}$$

修改 AVLTree 中的 insert 和 delete 函数，为每个节点中的 size 属性设置正确的值。insert 和 delete 函数仍需要 $O(\log n)$ 时间。find(k) 函数可以在 $O(\log n)$ 时间内实现。因此，可以在 $O(\log n)$ 时间内找到 AVL 树中的第 k 个最小元素。

**25.6（最近点对）18.8 节介绍了一种使用分治法在 $O(n \log n)$ 时间内找到最近点对的算法。该算法用递归实现的开销很大。使用普通扫描方法和 AVL 树，可以在 $O(n \log n)$ 时间内解决相同的问题。使用 AVLTree 实现该算法。

第 26 章

Introduction to C++ Programming and Data Structures, Fifth Edition

图及其应用

学习目标

1. 用图对真实世界的问题进行建模，并解释柯尼斯堡七桥问题（26.1 节）。
2. 描述图术语：顶点、边、简单图、加权/无权图和有向/无向图（26.2 节）。
3. 用列表、邻接矩阵和邻接列表表示顶点和边（26.3 节）。
4. 用 `Graph` 类对图进行建模（26.4 节）。
5. 用 `Tree` 类表示图的遍历（26.5 节）。
6. 设计并实现深度优先搜索（26.6 节）。
7. 设计并实现广度优先搜索（26.7 节）。
8. 用广度优先搜索来解决九枚硬币翻转问题（26.8 节）。

26.1 简介

要点提示：许多现实世界中的问题都可以使用图算法来解决。

图在建模真实世界的问题中发挥着重要作用。例如，找到两个城市之间最少航班数的问题可以用图建模，其中顶点表示城市，边表示两个邻接城市之间的航班，如图 26.1 所示。找到两个城市之间最少航班数的问题简化为找到图中两个顶点之间的最短路径。

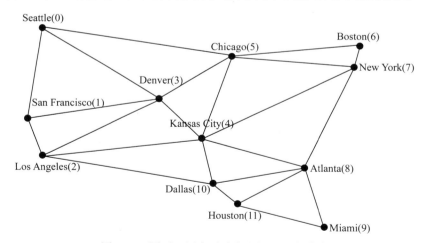

图 26.1 图可以用来对城市之间的距离建模

对图问题的研究称为图论。图论由莱昂哈德·欧拉于 1736 年创立，当时他引入了图论术语来解决著名的**柯尼斯堡七桥问题**。普鲁士的柯尼斯堡市（今俄罗斯加里宁格勒）被普雷格尔河分割。河上有两个岛屿。城市和岛屿由七座桥连接，如图 26.2a 所示。问题是：一个人能穿过每座桥一次然后回到起点吗？欧拉证明了这是不可能的。

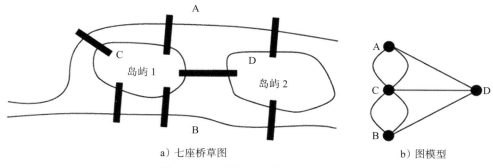

图 26.2 连接岛屿和陆地的七座桥

为了证明这个问题，欧拉首先通过消除所有街道，将柯尼斯堡城市地图抽象为图 26.2a 所示的草图。然后，他用一个点（称为顶点或节点）代替每个陆地，用一条线（称为边）代替每座桥，如图 26.2b 所示。这种有顶点和边的结构被称为图。

在图中，我们的问题变为：是否存在一条路径，它从任意顶点开始，正好遍历所有边一次，然后返回到起始顶点。欧拉证明了如果这样的路径存在，则每个顶点必须有偶数条边。因此，柯尼斯堡七桥问题没有解决办法。

图问题通常使用算法来解决。图算法在各个领域都有许多应用，如计算机科学、数学、生物学、工程、经济学、遗传学和社会科学。本章介绍深度优先搜索和广度优先搜索的算法及其应用。下一章介绍加权图中求最小生成树和最短路径的算法及其应用。

26.2 基本图术语

要点提示：图由顶点和连接顶点的边组成。

本章并不假设读者具有图论或离散数学的知识背景。我们使用简单明了的术语来定义图。什么是图？**图**是一种数学结构，表示现实世界中实体之间的关系。例如，图 26.1 中的图表示城市之间的航班，图 26.2b 中的图表示陆地之间的桥梁。

图由一组顶点、节点或点，以及连接这些顶点的一组边组成。方便起见，我们将图定义为 $G = (V, E)$，其中 V 表示一组顶点，E 表示一组边。例如，图 26.1 中图的 V 和 E 如下所示：

```
V = {"Seattle", "San Francisco", "Los Angeles",
     "Denver", "Kansas City", "Chicago", "Boston", "New York",
     "Atlanta", "Miami", "Dallas", "Houston"};
E = {{"Seattle", "San Francisco"}, {"Seattle", "Chicago"},
     {"Seattle", "Denver"}, {"San Francisco", "Denver"},
     ...
    };
```

图可以是有向的，也可以是无向的。在**有向图**中，每条边都有一个方向，表示可以通过边从一个顶点移动到另一个顶点。我们可以用有向图来建模父子关系，其中从顶点 A 到 B 的边表示 A 是 B 的父亲。

图 26.3a 显示了一个有向图。在**无向图**中，可以在顶点之间双向移动。图 26.1 中的图是无向的。

边可以**加权**也可以**不加权**。例如，可以为图 26.1 的图中的每条边指定一个权重，以表示两个城市之间的飞行时间。

如果图中的两个顶点由同一条边连接，则称它们邻接。类似地，如果两条边连接到同一顶点，则称它们邻接。图中连接两个顶点的边被称为与两个顶点相关联。顶点的**度**是指与顶点相关联的边的数量。

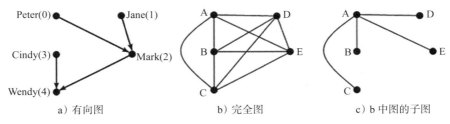

图 26.3 图可以多种形式出现

如果两个顶点邻接，则它们是邻居。类似地，如果两条边邻接，那么它们就是邻居。

环是连接顶点到自身的边。如果两个顶点由两条或多条边连接，则这些边称为**平行边**。**简单图**是指没有环或平行边的图。**完全图**是任意两个顶点都邻接的图，如图 26.3b 所示。

如果图中任意两个顶点之间存在一条路径，则图是连通的。图 G 的子图是一个图，其顶点集是 G 的顶点集的子集，其边集是 G 边集的子集。例如，图 26.3c 中的图是图 26.3b 中的图的子图。

假设图是连通的且无向的。图的**生成树**是连接图中所有顶点且为树的子图。

26.3 图的表示

要点提示：图的表示就是将其顶点和边存储在程序中。用于存储图的数据结构有数组、向量或列表。

要编写一个处理和操纵图的程序，必须在计算机中存储或表示图。

26.3.1 顶点的表示

顶点可以存储在数组中。例如，可以用以下数组存储图 26.1 中的图的所有城市名称：

```
string vertices[] = {"Seattle", "San Francisco", "Los Angeles",
  "Denver", "Kansas City", "Chicago", "Boston", "New York",
  "Atlanta", "Miami", "Dallas", "Houston"};
```

注意：顶点可以是任何类型的对象。例如，可以将城市视为包含名称、人口和市长等信息的对象。因此，可以按如下方式定义顶点：

```
City city0("Seattle", 563374, "Greg Nickels");
...
City city11("Houston", 1000203, "Bill White");
City vertices[] = {city0, city1, ..., city11};

class City
{
public:
  City(const string& cityName, int population, const string& mayor)
  {
    this->cityName = cityName;
    this->population = population;
    this->mayor = mayor;
  }

  string getCityName() const
```

```cpp
  {
    return cityName;
  }

  int getPopulation() const
  {
    return population;
  }

  string getMayor() const
  {
    return mayor;
  }

  void setMayor(const string& mayor)
  {
    this->mayor = mayor;
  }

  void setPopulation(int population)
  {
    this->population = population;
  }

private:
  string cityName;
  int population;
  string mayor;
};
```

对于具有 n 个顶点的图，可以用自然数 0，1，2，…，n–1 方便地标记顶点。因此，`vertices[0]` 表示 `"Seattle"`，`vertices[1]` 表示 `"San Francisco"`，以此类推，如图 26.4 所示。

vertices[0]	Seattle
vertices[1]	San Francisco
vertices[2]	Los Angeles
vertices[3]	Denver
vertices[4]	Kansas City
vertices[5]	Chicago
vertices[6]	Boston
vertices[7]	New York
vertices[8]	Atlanta
vertices[9]	Miami
vertices[10]	Dallas
vertices[11]	Houston

图 26.4　数组存储顶点名称

注意：可通过顶点的名称或索引（以方便的方式为准）引用顶点。显然，通过程序中的索引访问顶点很容易。

26.3.2　边的表示（用于输入）：边数组

边可以用二维数组来表示。例如，可以用以下数组存储图 26.1 中图的所有边：

```
int edges[][2] = {
  {0, 1}, {0, 3}, {0, 5},
  {1, 0}, {1, 2}, {1, 3},
  {2, 1}, {2, 3}, {2, 4}, {2, 10},
  {3, 0}, {3, 1}, {3, 2}, {3, 4}, {3, 5},
  {4, 2}, {4, 3}, {4, 5}, {4, 7}, {4, 8}, {4, 10},
  {5, 0}, {5, 3}, {5, 4}, {5, 6}, {5, 7},
  {6, 5}, {6, 7},
  {7, 4}, {7, 5}, {7, 6}, {7, 8},
  {8, 4}, {8, 7}, {8, 9}, {8, 10}, {8, 11},
  {9, 8}, {9, 11},
  {10, 2}, {10, 4}, {10, 8}, {10, 11},
  {11, 8}, {11, 9}, {11, 10}
};
```

这被称为使用数组的边表示。

26.3.3 边的表示（用于输入）：**Edge** 对象

表示边的另一种方法是将边定义为对象，并将它们存储在向量中。Edge 类的定义见 LiveExample 26.1。

LiveExample 26.1 的互动程序请访问 https://liangcpp.pearsoncmg.com/LiveRunCpp5e/faces/LiveExample.xhtml?header=off&programName=Edge&fileType=.h&programHeight=290&resultVisible=false。

LiveExample 26.1　Edge.h

```cpp
#ifndef EDGE_H
#define EDGE_H

class Edge
{
public:
  int u;
  int v;

  Edge(int u, int v)
  {
    this->u = u;
    this->v = v;
  }
};
#endif
```

例如，你可以用以下向量存储图 26.1 中的图的所有边：

```cpp
vector<Edge> edgeVector;
edgeVector.push_back(Edge(0, 1));
edgeVector.push_back(Edge(0, 3));
edgeVector.push_back(Edge(0, 5));
...
```

如果事先不知道有多少条边，将 Edge 对象存储在向量中非常有用。

26.3.2 节和 26.3.3 节中用边数组或 Edge 对象表示边对于输入来说是直观的，但这对于内部处理来说并不是高效的。接下来的两节将介绍用邻接矩阵和邻接列表来表示图。这两种数据结构对于图的处理是高效的。

26.3.4 边的表示：邻接矩阵

假设图有 n 个顶点。可以用二维 $n \times n$ 矩阵，比如**邻接矩阵** adjacencyMatrix，来表示边。数组中的每个元素都是 0 或 1。如果存在从顶点 i 到顶点 j 的边，则 adjacencyMatrix[i][j] 为 1；否则，adjacencyMatrix[i][j] 为 0。如果图是无向的，则矩阵是对称的，因为 adjacencyMatrix[i][j] 与 adjacencyMatrix[j][i] 相同。例如，图 26.1 中的图的边可以如下用邻接矩阵表示：

```
int adjacencyMatrix[12][12] = {
  {0, 1, 0, 1, 0, 1, 0, 0, 0, 0, 0, 0}, // Seattle
  {1, 0, 1, 1, 0, 0, 0, 0, 0, 0, 0, 0}, // San Francisco
  {0, 1, 0, 1, 1, 1, 0, 0, 0, 0, 0, 0}, // Los Angeles
  {1, 1, 1, 0, 1, 1, 0, 0, 0, 0, 0, 0}, // Denver
  {0, 0, 1, 1, 0, 1, 0, 1, 1, 0, 1, 0}, // Kansas City
  {1, 0, 1, 1, 1, 0, 1, 1, 0, 0, 0, 0}, // Chicago
  {0, 0, 0, 0, 0, 1, 0, 1, 0, 0, 0, 0}, // Boston
  {0, 0, 0, 0, 1, 1, 1, 0, 1, 0, 0, 0}, // New York
  {0, 0, 0, 0, 1, 0, 0, 1, 0, 1, 1, 1}, // Atlanta
  {0, 0, 0, 0, 0, 0, 0, 0, 1, 0, 0, 1}, // Miami
  {0, 0, 0, 0, 1, 0, 0, 0, 1, 0, 0, 1}, // Dallas
  {0, 0, 0, 0, 0, 0, 0, 0, 1, 1, 1, 0}  // Houston
};
```

图 26.3a 中有向图的邻接矩阵可以如下表示：

```
int a[5][5] = {{0, 0, 1, 0, 0}, // Peter
               {0, 0, 1, 0, 0}, // Jane
               {0, 0, 0, 0, 1}, // Mark
               {0, 0, 0, 0, 1}, // Cindy
               {0, 0, 0, 0, 0}  // Wendy
              };
```

如 12.7 节所述，用向量表示数组更为灵活。将数组传递给函数时，还必须传递其大小，但将向量传递给函数，则不必传递其大小，因为向量对象包含大小信息。前面的邻接矩阵可以如下所示用向量来表示，用 C++11 向量初始化语句初始化：

```
vector<vector<int>> a = {{0, 0, 1, 0, 0}, // Peter
                         {0, 0, 1, 0, 0}, // Jane
                         {0, 0, 0, 0, 1}, // Mark
                         {0, 0, 0, 0, 1}, // Cindy
                         {0, 0, 0, 0, 0}  // Wendy
                        };
```

26.3.5 边的表示：邻接列表

可以用邻接顶点列表或邻接边列表来表示边。顶点 i 的邻接顶点列表包含与 i 邻接的顶点，顶点 i 的邻接边列表包含与 i 邻接的边。可以定义列表数组。数组有 n 项，每项都是一个列表。顶点 i 的邻接顶点列表包含所有满足从顶点 i 到顶点 j 有一条边的顶点 j。例如，为了表示图 26.1 中的图的边，可以如下所示创建一个列表数组：

```
vector<int> neighbors[12];
```

neighbors[0] 包含与顶点 0（即 Seattle）邻接的所有顶点，neighbors[1] 包含与

顶点 1（即 San Francisco）邻接的所有顶点，以此类推，如图 26.5 所示。

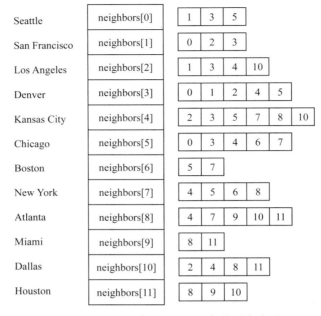

图 26.5　图 26.1 中的图的边用邻接列表表示

图 26.1 中的图的边用链表表示。

为了表示图 26.1 中图的邻接边列表，可以如下所示创建一个列表数组：

```
vector<Edge> neighbors[12]
```

neighbors[0] 包含与顶点 0（即 Seattle）邻接的所有边，neighbors[1] 包含与顶点 1（即 San Francisco）邻接的所有边，以此类推，如图 26.6 所示。

neighbors[0]	Edge(0,1)	Edge(0,3)	Edge(0,5)			
neighbors[1]	Edge(1,0)	Edge(1,2)	Edge(1,3)			
neighbors[2]	Edge(2,1)	Edge(2,3)	Edge(2,4)	Edge(2,10)		
neighbors[3]	Edge(3,0)	Edge(3,1)	Edge(3,2)	Edge(3,4)	Edge(3,5)	
neighbors[4]	Edge(4,2)	Edge(4,3)	Edge(4,5)	Edge(4,7)	Edge(4,8)	Edge(4,10)
neighbors[5]	Edge(5,0)	Edge(5,3)	Edge(5,4)	Edge(5,6)	Edge(5,7)	
neighbors[6]	Edge(6,5)	Edge(6,7)				
neighbors[7]	Edge(7,4)	Edge(7,5)	Edge(7,6)	Edge(7,8)		
neighbors[8]	Edge(8,4)	Edge(8,7)	Edge(8,9)	Edge(8,10)	Edge(8,11)	
neighbors[9]	Edge(9,8)	Edge(9,11)				
neighbors[10]	Edge(10,2)	Edge(10,4)	Edge(10,8)	Edge(10,11)		
neighbors[11]	Edge(11,8)	Edge(11,9)	Edge(11,10)			

图 26.6　图 26.1 中的图的边用邻接边列表表示

注意：可以用**邻接矩阵**或**邻接列表**表示图。哪个更好呢？如果图是密集的（即有很多边），则优选用邻接矩阵。如果图非常稀疏（即有很少的边），则使用邻接列表更好，因为使用邻接矩阵会浪费大量空间。

邻接矩阵和邻接列表都可以在程序中使用，以使算法更加高效。例如，使用邻接矩阵检查两个顶点是否连接需要 $O(1)$ 常数时间，使用邻接列表打印图中的所有边需要 $O(m)$ 线性时间，其中 m 是边的数量。

注意：邻接顶点对于表示无权图更简单。但邻接边列表对于广泛的图应用来说更加灵活。用邻接边列表可以很容易地在边上添加附加约束。因此，本书将使用邻接边列表来表示图。

为了灵活性和简单性，我们使用向量而不是数组，因为向量很容易扩展，可以添加新的顶点。此外，我们使用向量而不是列表，因为我们的算法只需要搜索表中的邻接顶点。用向量可以简化编码。如下所示，用向量可以构建图 26.6 中的邻接列表：

```
vector<vector<Edge*>> neighbors(12);
neighbors[0].push_back(new Edge(0, 1));
neighbors[0].push_back(new Edge(0, 3));
neighbors[0].push_back(new Edge(0, 5));
neighbors[1].push_back(new Edge(1, 0));
neighbors[1].push_back(new Edge(1, 2));
neighbors[1].push_back(new Edge(1, 3));
...
...
```

注意，Edge 对象的指针存储在 neighbors 中。这使 neighbors 能够存储 Edge 的任何子类型，以便使用多态性。在下一章中，我们为 neighbors 添加加权边时会看到这样做的好处。

26.4 Graph 类

要点提示：Graph 类定义了图的常用操作。

我们现在设计一个类来对图进行建模。图的常见操作是什么？通常我们需要获取图中的顶点数量、图中的所有顶点、具有指定索引的顶点对象、具有特定名称的顶点的索引、顶点的邻居、邻接矩阵、顶点的度，执行深度优先搜索和广度优先搜索。深度优先搜索和广度优先搜索将在下一节中介绍。图 26.7 在 UML 图中展示了这些函数。

在 Graph 类中定义向量 vertices 用于表示顶点。顶点可以是任何类型：int、string 等。因此，我们使用泛型类型 V 来定义它。neighbors，向量的向量，被定义来表示边。有了这两个数据字段，就足以实现 Graph 类中定义的所有函数。

方便起见，类提供了无参数构造函数。采用无参数构造函数，可以很容易地将该类用作基类和类中数据字段的数据类型。可以使用其他四个构造函数中的某个创建 Graph 对象，以方便的为准。如果有一个边数组，则用 UML 类图中的前两个构造函数。如果有一个 Edge 对象的向量，则用 UML 类图中的后两个构造函数。

泛型类型 V 表示顶点的类型——整数、字符串等。我们可以创建任何类型顶点的图。如果在不指定顶点的情况下创建图，则顶点是整数 0、1、…、n-1，其中 n 是顶点的数量。每个顶点都与一个索引相关联，该索引与顶点数组中顶点的索引相同。

假设该类在 Graph.h 中给出。LiveExample 26.2 给出了一个测试程序，该程序创建了图 26.1 中的图，以及图 26.3a 中的图。

```
                    Graph<V>
#vertices: vector<V>
#neighbors: vector<vector<Edge*>>
+Graph()
+Graph(vertices: const vector<V>&, edges[][2]: const
 int, numberOfEdges: int)
+Graph(numberOfVertices: int, edges[][2]: const int,
 numberOfEdges: int)
+Graph(vertices: const vector<V>&, edges: const
 vector<Edge>&)
+Graph(numberOfVertices: int, edges: const
 vector<Edge>&)
+getSize(): int const
+getDegree(v: int): int const
+getVertex(index: int): V const
+getIndex(v: const V&): int const
+getVertices(): vector<V> const
+getNeighbors(v: int): vector<int> const
+printEdges(): void const
+printAdjacencyMatrix(): void const
+clear(): void
+addVertex(v: const V&): bool
+addEdge(u: int, v: int): bool
#createEdge(e: Edge*): bool
+dfs(v: int): Tree const
+bfs(v: int): Tree const
```

图 26.7　Graph 类定义了图的常用操作

LiveExample 26.2 的互动程序请访问 https://liangcpp.pearsoncmg.com/LiveRunCpp5e/faces/LiveExample.xhtml?header=off&programName=TestGraph&programHeight=1100&resultHeight=1100。

LiveExample26.2　TestGraph.cpp

Source Code Editor:

```cpp
#include <iostream>
#include <string>
#include <vector>
#include "Graph.h" // Defined in LiveExample 26.3
#include "Edge.h" // Defined in LiveExample 26.1
using namespace std;

int main()
{
  // Vertices for graph in Figure 26.1
  string vertices[] = { "Seattle", "San Francisco", "Los Angeles",
    "Denver", "Kansas City", "Chicago", "Boston", "New York",
    "Atlanta", "Miami", "Dallas", "Houston" };

  // Edge array for graph in Figure 26.1
  int edges[][2] = {
    { 0, 1 }, { 0, 3 }, { 0, 5 },
    { 1, 0 }, { 1, 2 }, { 1, 3 },
    { 2, 1 }, { 2, 3 }, { 2, 4 }, { 2, 10 },
    { 3, 0 }, { 3, 1 }, { 3, 2 }, { 3, 4 }, { 3, 5 },
    { 4, 2 }, { 4, 3 }, { 4, 5 }, { 4, 7 }, { 4, 8 }, { 4, 10 },
```

```cpp
22        { 5, 0 }, { 5, 3 }, { 5, 4 }, { 5, 6 }, { 5, 7 },
23        { 6, 5 }, { 6, 7 },
24        { 7, 4 }, { 7, 5 }, { 7, 6 }, { 7, 8 },
25        { 8, 4 }, { 8, 7 }, { 8, 9 }, { 8, 10 }, { 8, 11 },
26        { 9, 8 }, { 9, 11 },
27        { 10, 2 }, { 10, 4 }, { 10, 8 }, { 10, 11 },
28        { 11, 8 }, { 11, 9 }, { 11, 10 }
29    };
30    const int NUMBER_OF_EDGES = 46; // 46 edges in Figure 26.1
31
32    // Create a vector for vertices
33    vector<string> vectorOfVertices(12);
34    copy(vertices, vertices + 12, vectorOfVertices.begin());
35
36    Graph<string> graph1(vectorOfVertices, edges, NUMBER_OF_EDGES); // Create graph1
37    cout << "The number of vertices in graph1: " << graph1.getSize();
38    cout << "\nThe vertex with index 1 is " << graph1.getVertex(1);
39    cout << "\nThe index for Miami is " << graph1.getIndex("Miami");
40
41    cout << "\nedges for graph1: " << endl;
42    graph1.printEdges();
43
44    cout << "\nAdjacency matrix for graph1: " << endl;
45    graph1.printAdjacencyMatrix();
46
47    // vector of Edge objects for graph in Figure 26.3a
48    vector<Edge> edgeVector;
49    edgeVector.push_back(Edge(0, 2));
50    edgeVector.push_back(Edge(1, 2));
51    edgeVector.push_back(Edge(2, 4));
52    edgeVector.push_back(Edge(3, 4));
53    // Create a graph with 5 vertices
54    Graph<int> graph2(5, edgeVector);
55
56    cout << "The number of vertices in graph2: " << graph2.getSize();
57    cout << "\nedges for graph2: " << endl;
58    graph2.printEdges();
59
60    cout << "\nAdjacency matrix for graph2: " << endl;
61    graph2.printAdjacencyMatrix();
62
63    return 0;
64 }
```

Execution Result:

```
command>cl TestGraph.cpp
Microsoft C++ Compiler 2019
Compiled successful (cl is the VC++ compile/link command)

command>TestGraph
The number of vertices in graph1: 12
The vertex with index 1 is San Francisco
The index for Miami is 9
edges for graph1:
Vertex Seattle(0): (Seattle, San Francisco) (Seattle, Denver)
(Seattle, Chicago)
Vertex San Francisco(1): (San Francisco, Seattle)
(San Francisco, Los Angeles) (San Francisco, Denver)
Vertex Los Angeles(2): (Los Angeles, San Francisco) (Los Angeles, Denver)
(Los Angeles, Kansas City) (Los Angeles, Dallas)
Vertex Denver(3): (Denver, Seattle) (Denver, San Francisco)
```

```
(Denver, Los Angeles) (Denver, Kansas City) (Denver, Chicago)
Vertex Kansas City(4): (Kansas City, Los Angeles) (Kansas City, Denver)
(Kansas City, Chicago) (Kansas City, New York) (Kansas City, Atlanta)
(Kansas City, Dallas)
Vertex Chicago(5): (Chicago, Seattle) (Chicago, Denver)
(Chicago, Kansas City) (Chicago, Boston) (Chicago, New York)
Vertex Boston(6): (Boston, Chicago) (Boston, New York)
Vertex New York(7): (New York, Kansas City) (New York, Chicago)
(New York, Boston) (New York, Atlanta)
Vertex Atlanta(8): (Atlanta, Kansas City) (Atlanta, New York)
(Atlanta, Miami) (Atlanta, Dallas) (Atlanta, Houston)
Vertex Miami(9): (Miami, Atlanta) (Miami, Houston)
Vertex Dallas(10): (Dallas, Los Angeles) (Dallas, Kansas City)
(Dallas, Atlanta) (Dallas, Houston)
Vertex Houston(11): (Houston, Atlanta) (Houston, Miami) (Houston, Dallas)

Adjacency matrix for graph1:
0 1 0 1 0 1 0 0 0 0 0 0
1 0 1 1 0 0 0 0 0 0 0 0
0 1 0 1 1 0 0 0 0 0 1 0
1 1 1 0 1 1 0 0 0 0 0 0
0 0 1 1 0 1 0 1 1 0 1 0
1 0 0 1 1 0 1 1 0 0 0 0
0 0 0 0 0 1 0 1 0 0 0 0
0 0 0 0 1 1 1 0 1 0 0 0
0 0 0 0 1 0 0 1 0 1 1 1
0 0 0 0 0 0 0 0 1 0 0 1
0 0 1 0 1 0 0 0 1 0 0 1
0 0 0 0 0 0 0 0 1 1 1 0
The number of vertices in graph2: 5
edges for graph2:
Vertex 0(0): (0, 2)
Vertex 1(1): (1, 2)
Vertex 2(2): (2, 4)
Vertex 3(3): (3, 4)
Vertex 4(4):

Adjacency matrix for graph2:
0 0 1 0 0
0 0 1 0 0
0 0 0 0 1
0 0 0 0 1
0 0 0 0 0

command>
```

程序在第 11 ~ 36 行为图 26.1 中的图创建了 graph1。在第 11 ~ 13 行定义了 graph1 的顶点。在第 16 ~ 29 行定义了 graph1 的边。边用二维数组表示。对于数组中的每行 i, edges[i][0] 和 edges[i][1] 表示存在从顶点 edges[i][0] 到顶点 edges[i][1] 的边。例如，第一行 {0, 1} 表示从顶点 0 (edges[0][0]) 到顶点 1 (edges[0][1]) 的边。行 {0, 5} 表示从顶点 0 (edges[2][0]) 到顶点 5 (edges[2][1]) 的边。该图在第 36 行中创建。第 42 行在 graph1 上调用 printEdges() 函数以显示 graph1 中的所有边。第 45 行在 graph1 上调用 printAdjacencyMatrix() 函数以显示 graph1 的邻接矩阵。

程序在第 48 ~ 54 行为图 26.3a 中的图创建了 graph2。在第 48 ~ 52 行定义了 graph2 的边。在第 54 行用 Edge 对象的向量创建了 graph2。第 58 行在 graph2 上调用 printEdges()

以显示 graph2 中的所有边。第 61 行在 graph2 上调用 printAdjacencyMatrix() 函数以显示 graph2 的邻接矩阵。

注意，graph1 包含字符串的顶点，graph2 包含带有整数 0，1，⋯，n-1 的顶点，其中 n 是顶点的数量。在 graph1 中，顶点与索引 0，1，⋯，n-1 相关联。索引是顶点在 vertices 中的位置。例如，顶点 Miami 的索引为 9。

现在我们将注意力转向 Graph 类的实现，如 LiveExample 26.3 所示。

LiveExample 26.3 的互动程序请访问 https://liangcpp.pearsoncmg.com/LiveRunCpp5e/faces/LiveExample.xhtml?header=off&programName=Graph&fileType=.h&programHeight=6380&resultVisible=false。

LiveExample26.3　Graph.h

Source Code Editor:

```cpp
#ifndef GRAPH_H
#define GRAPH_H

#include "Edge.h" // Defined in LiveExample 26.1
#include "Tree.h" // Defined in LiveExample 26.4
#include <vector>
#include <queue> // For implementing BFS
#include <stdexcept>
#include <sstream> // For converting a number to a string
#include <algorithm> // For the find algorithm
using namespace std;

template<typename V>
class Graph
{
public:
  // Construct an empty graph
  Graph();

  // Construct a graph from vertices in a vector and
  // edges in 2-D array
  Graph(const vector<V>& vertices, const int edges[][2], int numberOfEdges);

  // Construct a graph with vertices 0, 1, ..., n-1 and
  // edges in 2-D array
  Graph(int numberOfVertices, const int edges[][2], int numberOfEdges);

  // Construct a graph from vertices and edges objects
  Graph(const vector<V>& vertices, const vector<Edge>& edges);

  // Construct a graph with vertices 0, 1, ..., n-1 and
  // edges in a vector
  Graph(int numberOfVertices, const vector<Edge>& edges);

  // Return the number of vertices in the graph
  int getSize() const;

  // Return the degree for a specified vertex
  int getDegree(int v) const;

  // Return the vertex for the specified index
  V getVertex(int index) const;

  // Return the index for the specified vertex
  int getIndex(V v) const;
```

```cpp
47    // Return the vertices in the graph
48    vector<V> getVertices() const;
49
50    // Return the neighbors of vertex v
51    vector<int> getNeighbors(int v) const;
52
53    // Print the edges
54    void printEdges() const;
55
56    // Print the adjacency matrix
57    void printAdjacencyMatrix() const;
58
59    // Clear the graph
60    void clear();
61
62    // Adds a vertex to the graph
63    virtual bool addVertex(const V& v);
64
65    // Adds an edge from u to v to the graph
66    bool addEdge(int u, int v);
67
68    // Obtain a depth-first search tree
69    // To be discussed in Section 26.6
70    Tree dfs(int v) const;
71
72    // Starting bfs search from vertex v
73    // To be discussed in Section 26.7
74    Tree bfs(int v) const;
75
76  protected:
77    vector<V> vertices; // Store vertices
78    vector<vector<Edge*>> neighbors; // Adjacency edge lists using vectors
79    bool createEdge(Edge* e); // Add an edge
80
81  private:
82    // Create adjacency lists for each vertex from an edge array
83    void createAdjacencyLists(int numberOfVertices, const int edges[][2],
84      int numberOfEdges);
85
86    // Create adjacency lists for each vertex from an Edge vector
87    void createAdjacencyLists(int numberOfVertices,
88      const vector<Edge>& edges);
89
90    // Recursive function for DFS search
91    void dfs(int v, vector<int>& parent,
92      vector<int>& searchOrders, vector<bool>& isVisited) const;
93  };
94
95  template<typename V>
96  Graph<V>::Graph()
97  {
98  }
99
100 template<typename V>
101 Graph<V>::Graph(const vector<V>& vertices, const int edges[][2],
102   int numberOfEdges)
103 {
104   for (unsigned i = 0; i < vertices.size(); i++)
105     addVertex(vertices[i]);
106
107   createAdjacencyLists(vertices.size(), edges, numberOfEdges);
108 }
109
110 template<typename V>
111 Graph<V>::Graph(int numberOfVertices, const int edges[][2],
```

```cpp
112       int numberOfEdges)
113     {
114       for (int i = 0; i < numberOfVertices; i++)
115         addVertex(i); // vertices is {0, 1, 2, ..., n-1}
116
117       createAdjacencyLists(numberOfVertices, edges, numberOfEdges);
118     }
119
120     template<typename V>
121     Graph<V>::Graph(const vector<V>& vertices, const vector<Edge>& edges)
122     {
123       for (unsigned i = 0; i < vertices.size(); i++)
124         addVertex(vertices[i]);
125
126       createAdjacencyLists(vertices.size(), edges);
127     }
128
129     template<typename V>
130     Graph<V>::Graph(int numberOfVertices, const vector<Edge>& edges)
131     {
132       for (int i = 0; i < numberOfVertices; i++)
133         addVertex(i); // vertices is {0, 1, 2, ..., n-1}
134
135       createAdjacencyLists(numberOfVertices, edges);
136     }
137
138     template<typename V>
139     void Graph<V>::createAdjacencyLists(int numberOfVertices,
140       const int edges[][2], int numberOfEdges)
141     {
142       for (int i = 0; i < numberOfEdges; i++)
143       {
144         int u = edges[i][0];
145         int v = edges[i][1];
146         addEdge(u, v);
147       }
148     }
149
150     template<typename V>
151     void Graph<V>::createAdjacencyLists(int numberOfVertices,
152       const vector<Edge>& edges)
153     {
154       for (unsigned i = 0; i < edges.size(); i++)
155       {
156         int u = edges[i].u;
157         int v = edges[i].v;
158         addEdge(u, v);
159       }
160     }
161
162     template<typename V>
163     int Graph<V>::getSize() const
164     {
165       return vertices.size(); // Return the number of vertices
166     }
167
168     template<typename V>
169     int Graph<V>::getDegree(int v) const
170     {
171       return neighbors[v].size(); // Return the degree of the vertex with index v
172     }
173
174     template<typename V>
175     V Graph<V>::getVertex(int index) const
```

```cpp
176   {
177     return vertices[index]; // Return the vertex with the specified index
178   }
179
180   template<typename V>
181   int Graph<V>::getIndex(V v) const
182   {
183     for (unsigned i = 0; i < vertices.size(); i++)
184     {
185       if (vertices[i] == v)
186         return i;
187     }
188
189     return -1; // Return -1 if vertex is not in the graph
190   }
191
192   template<typename V>
193   vector<V> Graph<V>::getVertices() const
194   {
195     return vertices; // Return vertices in a vector
196   }
197
198   template<typename V>
199   vector<int> Graph<V>::getNeighbors(int u) const
200   {
201     vector<int> result;
202     for (Edge* e: neighbors[u])
203       result.push_back(e->v);
204     return result;
205   }
206
207   template<typename V>
208   void Graph<V>::printEdges() const
209   {
210     for (unsigned u = 0; u < neighbors.size(); u++)
211     {
212       cout << "Vertex " << getVertex(u) << "(" << u << "): ";
213       for (Edge* e: neighbors[u])
214       {
215         cout << "(" << getVertex(e->u) << ", " << getVertex(e->v) << ") ";
216       }
217       cout << endl;
218     }
219   }
220
221   template<typename V>
222   void Graph<V>::printAdjacencyMatrix() const
223   {
224     // Use vector for 2-D array
225     vector<vector<int>> adjacencyMatrix(getSize());
226
227     // Initialize 2-D array for adjacency matrix
228     for (int i = 0; i < getSize(); i++)
229     {
230       adjacencyMatrix[i] = vector<int>(getSize());
231     }
232
233     for (unsigned i = 0; i < neighbors.size(); i++)
234     {
235       for (Edge* e: neighbors[i])
236       {
237         adjacencyMatrix[i][e->v] = 1;
238       }
239     }
240
241     for (unsigned i = 0; i < adjacencyMatrix.size(); i++)
```

```cpp
242      {
243        for (unsigned j = 0; j < adjacencyMatrix[i].size(); j++)
244        {
245          cout << adjacencyMatrix[i][j] << " ";
246        }
247
248        cout << endl;
249      }
250    }
251
252    template<typename V>
253    void Graph<V>::clear()
254    {
255      vertices.clear();
256      for (int i = 0; i < getSize(); i++)
257        for (Edge* e: neighbors[i])
258          delete e; // Delete e
259      neighbors.clear();
260    }
261
262    template<typename V>
263    bool Graph<V>::addVertex(const V& v)
264    {
265      if (find(vertices.begin(), vertices.end(), v) == vertices.end())
266      {
267        vertices.push_back(v);
268        neighbors.push_back(vector<Edge*>(0));
269        return true;
270      }
271      else
272      {
273        return false;
274      }
275    }
276
277    template<typename V>
278    bool Graph<V>::createEdge(Edge* e)
279    {
280      if (e->u < 0 || e->u > getSize() - 1)
281      {
282        stringstream ss;
283        ss << e->u;
284        throw invalid_argument("No such edge: " + ss.str());
285      }
286
287      if (e->v < 0 || e->v > getSize() - 1)
288      {
289        stringstream ss;
290        ss << e->v;
291        throw invalid_argument("No such edge: " + ss.str());
292      }
293
294      vector<int> listOfNeighbors = getNeighbors(e->u);
295      if (find(listOfNeighbors.begin(), listOfNeighbors.end(), e->v)
296          == listOfNeighbors.end())
297      {
298        neighbors[e->u].push_back(e);
299        return true;
300      }
301      else
302      {
303        return false;
304      }
305    }
306
307    template<typename V>
```

```cpp
308    bool Graph<V>::addEdge(int u, int v)
309    {
310      return createEdge(new Edge(u, v));
311    }
312
313    template<typename V>
314    Tree Graph<V>::dfs(int v) const
315    {
316      vector<int> searchOrders;
317      vector<int> parent(vertices.size());
318      for (unsigned i = 0; i < vertices.size(); i++)
319        parent[i] = -1; // Initialize parent[i] to -1
320
321      // Declare isVisited to track all visited vertices
322      vector<bool> isVisited(vertices.size());
323      for (unsigned i = 0; i < vertices.size(); i++)
324      {
325        isVisited[i] = false;
326      }
327
328      // Search by calling the recursive helper function dfs
329      dfs(v, parent, searchOrders, isVisited);
330
331      // Return a search tree
332      return Tree(v, parent, searchOrders);
333    }
334
335    template<typename V>
336    void Graph<V>::dfs(int v, vector<int>& parent,
337      vector<int>& searchOrders, vector<bool>& isVisited) const
338    {
339      // Store the visited vertex
340      searchOrders.push_back(v);
341      isVisited[v] = true; // Vertex v visited
342
343      for (Edge* e: neighbors[v])
344      { // w is a neighbor of v. Note that e->u is v and e->v.
345        int w = e->v;
346        if (!isVisited[w]) // Test if w is not visited
347        {
348          parent[w] = v; // The parent of vertex i is v
349          dfs(w, parent, searchOrders, isVisited); // Recursive search
350        }
351      }
352    }
353
354    template<typename V>
355    Tree Graph<V>::bfs(int v) const
356    {
357      vector<int> searchOrders;
358      vector<int> parent(vertices.size());
359      for (int i = 0; i < getSize(); i++)
360        parent[i] = -1; // Initialize parent[i] to -1
361
362      queue<int> queue; // Stores vertices to be visited
363      vector<bool> isVisited(getSize());
364      queue.push(v); // Enqueue v
365      isVisited[v] = true; // Mark it visited
366
367      while (!queue.empty()) // Test if queue is empty
368      {
369        int u = queue.front(); // Get from the front of the queue
370        queue.pop(); // remove the front element
371        searchOrders.push_back(u); // u searched
372        for (Edge* e: neighbors[u])
```

```
373      { // w is a neighbor of v. Note that e->u is v and e->v.
374        int w = e->v;
375        if (!isVisited[w]) // Test if w is not visited
376        {
377          queue.push(w); // Enqueue w
378          parent[w] = u; // The parent of w is u
379          isVisited[w] = true; // Mark it visited
380        }
381      }
382    }
383
384    return Tree(v, parent, searchOrders); // Return the BFS tree
385  }
386
387  #endif
```

`Answer` `Reset`

要构建一个图，需要创建顶点和边。顶点存储在 vector<V> 中（第 77 行），边存储在 vector<vector<Edge*>> 中（第 78 行），这是 26.3.5 节中描述的邻接边列表。构造函数（第 95 ～ 136 行）创建顶点和边。可以从边数组（26.3.2 节中讨论）或 Edge 对象向量（26.3.3 节中讨论）创建边。私有函数 createAdjacencyList（第 138 ～ 148 行）用于从边数组创建邻接列表，重载的 createAdjacencyList 函数（第 150 ～ 160 行）用于根据 Edge 对象向量创建邻接列表。Edge 类在 LiveExample 26.1 中定义，它只定义了有一条边的两个顶点。

vertices 和 neighbors 被声明为受保护的，以便可以从 Graph 的派生类访问它们以供将来扩展。

函数 getSize 返回图中顶点的数量（第 162 ～ 166 行）。函数 getDegree(int v) 返回索引为 v 的顶点的度（第 168 ～ 172 行）。函数 getVertex(int index) 返回具有指定索引的顶点（第 174 ～ 178 行）。函数 getIndex(T v) 返回指定顶点的索引（第 180 ～ 190 行）。函数 getVertices() 返回顶点的向量（第 192 ～ 196 行）。函数 getNeighbors(u) 返回与 u 相邻的顶点列表（第 198 ～ 205 行）。函数 printEdges() 显示所有顶点以及与每个顶点相邻的边（第 207 ～ 219 行）。函数 printAdjacencyMatrix() 显示邻接矩阵（第 221 ～ 250 行）。

函数 clear 从图中删除所有边和顶点（第 252 ～ 260 行）。函数 addVertex(v) 将一个新顶点添加到图中（第 262 ～ 275 行），如果该顶点不在图中，则返回 true；如果顶点已经在图中，则函数返回 false（第 273 行）。

函数 createEdge 将一条新边添加到图中（第 277 ～ 305 行）。如果边无效，它抛出 invalid_argument 异常（第 280 ～ 292 行）。如果边已经在图中，则返回 false（第 303 行）。注意，此函数是受保护的函数，派生类可以使用它向图中添加不同类型的边。在这个类中，它被 addEdge(u, v) 函数调用以将无权边添加到图中（第 310 行）。

第 313 ～ 385 行中的代码提供了查找深度优先搜索树和广度优先搜索树的函数，这些函数将在后续小节中介绍。

26.5 图遍历

要点提示：深度优先和广度优先是图遍历的两种常见方式。

图遍历是访问图中每个顶点一次的过程。有两种流行的图遍历方式：深度优先遍历（深度优先搜索）和广度优先遍历（**广度优先搜索**）。两个遍历都会产生一个生成树，该树可用一个类对其进行建模，如图 26.8 所示。Tree 类描述树中节点的父子关系，如 LiveExample 26.4 所示。

```
                    Tree
-root: int
-parent: vector<int>
-searchOrders: vector<int>
+Tree()
+Tree(root: int, parent: const vector<int>&,
     searchOrders: const vector<int>&)
+getRoot(): int const
+getSearchOrders(): vector<int>const
+getParent(v: int): int const
+getNumberOfVerticesFound(): int const
+getPath(v: int): vector<int> const
+printTree(): void const
```

图 26.8　Tree 类描述树中节点的父子关系

LiveExample 26.4 的互动程序请访问 https://liangcpp.pearsoncmg.com/LiveRunCpp5e/faces/LiveExample.xhtml?header=off&programName=Tree&programHeight=1430&fileType=.h&resultVisible=false。

LiveExample26.4　Tree.h

Source Code Editor:

```cpp
#ifndef TREE_H
#define TREE_H

#include <vector>
using namespace std;

class Tree
{
public:
  // Construct an empty tree
  Tree()
  {
  }

  // Construct a tree with root, parent, and searchOrder
  Tree(int root, const vector<int>& parent, const vector<int>& searchOrders)
  {
    this->root = root;
    this->parent = parent;
    this->searchOrders = searchOrders;
  }

  // Return the root of the tree
  int getRoot() const
  {
    return root;
  }

  // Return the parent of vertex v
```

```cpp
30      int getParent(int v) const
31      {
32        return parent[v];
33      }
34
35      // Return search order
36      vector<int> getSearchOrders() const
37      {
38        return searchOrders;
39      }
40
41      // Return number of vertices found
42      int getNumberOfVerticesFound() const
43      {
44        return searchOrders.size();
45      }
46
47      // Return the path of vertices from v to the root in a vector
48      vector<int> getPath(int v) const
49      {
50        vector<int> path;
51
52        do
53        {
54          path.push_back(v);
55          v = parent[v];
56        }
57        while (v != -1);
58
59        return path;
60      }
61
62      // Print the whole tree
63      void printTree() const
64      {
65        cout << "Root is: " << root << endl;
66        cout << "Edges: ";
67        for (unsigned i = 0; i < searchOrders.size(); i++)
68        {
69          if (parent[searchOrders[i]] != -1)
70          {
71            // Display an edge
72            cout << "(" << parent[searchOrders[i]] << ", " <<
73              searchOrders[i] << ") ";
74          }
75        }
76        cout << endl;
77      }
78
79    private:
80      int root; // The root of the tree
81      vector<int> parent; // Store the parent of each vertex
82      vector<int> searchOrders; // Store the search order
83    };
84
85    #endif
```

Answer   Reset

Tree 类有两个构造函数。无参数构造函数构造一个空树。另一个构造函数构造一个具有搜索顺序的树（第 16 ~ 21 行）。

Tree 类定义了七个函数。getRoot() 函数返回树的根（第 24 ~ 27 行）。可以调用 getParent(v) 在搜索中查找顶点 v 的父亲（第 30 ~ 33 行）。可以通过调用 getSearchOrders()

函数获得搜索到的顶点的顺序（第 36～39 行）。调用 getNumberOfVerticesFound() 返回搜索到的顶点数（第 42～45 行）。getPath(v) 函数返回从 v 到根的路径（第 48～60 行）。可以用 printTree() 函数（第 63～77 行）显示树中的所有边。

26.6 节和 26.7 节将分别介绍深度优先搜索和广度优先搜索。这两个搜索都将产生 Tree 类的实例。

26.6 深度优先搜索

要点提示：图的深度优先搜索从图中的一个顶点开始，并在回溯之前尽可能地访问图中的所有顶点。

图的深度优先搜索类似于 21.7 节中讨论的树的深度优先遍历。对于树，遍历从根开始。对于图，搜索可以从任何顶点开始。

树的深度优先遍历首先访问根，然后递归地访问根的子树。类似地，图的深度优先搜索首先访问一个顶点，然后递归地访问与该顶点相邻的所有顶点。不同的是，图可能包含环路，这可能导致无限递归。若要避免此问题，需要跟踪已经访问过的顶点，并避免再次访问它们。

这种搜索被称为深度优先，因为它在图中搜索得尽可能"更深"。搜索从某个顶点 v 开始。访问 v 后，访问 v 的未访问邻居。如果 v 没有未访问邻居，则回溯到令搜索到达 v 的顶点。

26.6.1 深度优先搜索算法

深度优先搜索的算法在 Listing 26.1 中描述。

Listing 26.1 深度优先搜索算法

```
输入：G=(V, E) 和起始顶点 v
输出：根为 v 的 DFS 树
1   Tree dfs(vertex v)
2   {
3     访问 v;
4     for v 的每个邻居 w
5       if(w 没有被访问)
6       {
7         parent[w] = v;
8         dfs(w);
9       }
10  }
```

可以用名为 isVisited 的向量来表示是否已访问顶点。开始时每个顶点 i 的 isVisited[i] 都为 false。一旦访问了某个顶点，比如 v，则 isVisited[v] 就设置为 true。

考虑图 26.9a 中的图。假设从顶点 0 开始深度优先搜索。首先访问 0，然后访问它的任意邻居，比如 1。现在访问了 1，如图 26.9b 所示。顶点 1 有三个邻居：0、2 和 4。由于 0 已被访问，所以将访问 2 或 4。我们选择 2。现在 2 被访问了，如图 26.9c 所示。2 有三个邻居：0、1 和 3。由于已经访问了 0 和 1，所以选择 3。现在已访问 3，如图 26.9d 所示。此时，已按以下顺序访问顶点：

```
0, 1, 2, 3
```

由于 3 的所有邻居都已访问过，所以回溯到 2。由于已经访问了 2 的所有邻居，因此回溯到 1。4 与 1 相邻，但 4 尚未被访问。因此，访问 4，如图 26.9e 所示。由于 4 的所有邻居都已访问过，所以回溯到 1。由于已经访问了 1 的所有邻居，因此回溯到 0。由于 0 的所有邻居都已访问过，因此搜索结束。

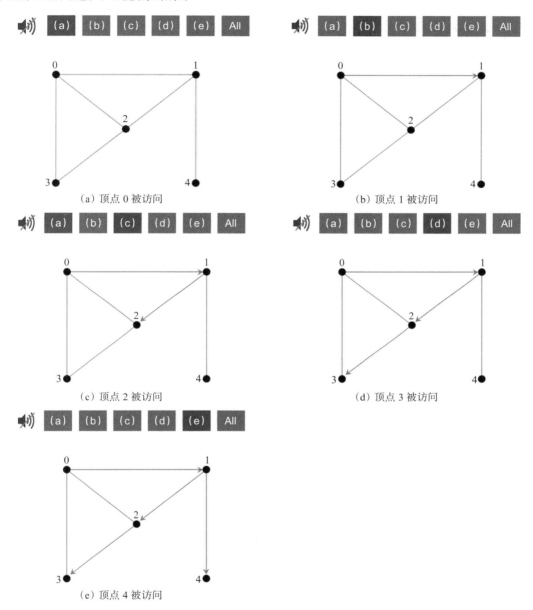

图 26.9 深度优先搜索递归地访问节点及其邻居

由于每条边和每个顶点只访问一次，因此 dfs 函数的时间复杂度为 $O(|E|+|V|)$，其中 $|E|$ 表示边的数量，$|V|$ 表示顶点的数量。

26.6.2 深度优先搜索的实现

Listing 26.1 中用递归描述了该算法，虽然可以用递归实现它。我们也可以用栈实现它（参见编程练习 26.4）。

dfs(int v) 函数在 LiveExample 26.3 中的第 313 ~ 333 行中实现。它返回一个以顶点 u 为根的 Tree 类实例。该函数把搜索到的顶点存储在列表 searchOrders 中（第 316 行），把每个顶点的父亲存储数组 parent 中（第 317 行），用 isVisited 数组指示顶点是否已被访问（第 322 行）。它调用辅助函数 dfs(v, parent, searchOrders, isVisited) 执行深度优先搜索（第 329 行）。

在递归辅助函数中，搜索从顶点 v 开始。v 被添加到 searchOrders（第 340 行）并被标记为已被访问（第 341 行）。对于 v 的每个未访问邻居 w（w 是 e->v），递归调用该函数执行深度优先搜索。当访问顶点 w 时，w 的父节点被存储在 parent[w] 中（第 348 行）。当连通图或连通分量中的所有顶点都被访问完，函数返回。

LiveExample 26.5 给出了一个测试程序，该程序从 Chicago 开始显示图 26.1 中图的 DFS。

LiveExample 26.5 的互动程序请访问 https://liangcpp.pearsoncmg.com/LiveRunCpp5e/faces/LiveExample.xhtml?header=off&programName=TestDFS&fileType=.cpp&programHeight=910&resultHeight=400。

LiveExample 26.5　TestDFS.cpp

Source Code Editor:

```cpp
#include <iostream>
#include <string>
#include <vector>
#include "Graph.h" // Defined in LiveExample 26.3
#include "Edge.h" // Defined in LiveExample 26.1
#include "Tree.h" // Defined in LiveExample 26.4
using namespace std;

int main()
{
  // Vertices for graph in Figure 26.1
  string vertices[] = {"Seattle", "San Francisco", "Los Angeles",
    "Denver", "Kansas City", "Chicago", "Boston", "New York",
    "Atlanta", "Miami", "Dallas", "Houston"};

  // Edge array for graph in Figure 26.1
  int edges[][2] = {
    {0, 1}, {0, 3}, {0, 5},
    {1, 0}, {1, 2}, {1, 3},
    {2, 1}, {2, 3}, {2, 4}, {2, 10},
    {3, 0}, {3, 1}, {3, 2}, {3, 4}, {3, 5},
    {4, 2}, {4, 3}, {4, 5}, {4, 7}, {4, 8}, {4, 10},
    {5, 0}, {5, 3}, {5, 4}, {5, 6}, {5, 7},
    {6, 5}, {6, 7},
    {7, 4}, {7, 5}, {7, 6}, {7, 8},
    {8, 4}, {8, 7}, {8, 9}, {8, 10}, {8, 11},
    {9, 8}, {9, 11},
    {10, 2}, {10, 4}, {10, 8}, {10, 11},
    {11, 8}, {11, 9}, {11, 10}
  };
  const int NUMBER_OF_EDGES = 46; // 46 edges in Figure 26.1

  // Create a vector for vertices
  vector<string> vectorOfVertices(12);
  copy(vertices, vertices + 12, vectorOfVertices.begin());

  Graph<string> graph(vectorOfVertices, edges, NUMBER_OF_EDGES);
  Tree dfs = graph.dfs(graph.getIndex("Chicago")); // DFS starting from Chicago
```

```
40    vector<int> searchOrders = dfs.getSearchOrders();
41    cout << dfs.getNumberOfVerticesFound() <<
42      " vertices are searched in this DFS order:" << endl;
43    for (unsigned i = 0; i < searchOrders.size(); i++)
44      cout << graph.getVertex(searchOrders[i]) << " ";
45    cout << endl << endl;
46
47    for (unsigned i = 0; i < searchOrders.size(); i++)
48      if (dfs.getParent(i) != -1)
49        cout << "parent of " << graph.getVertex(i) <<
50          " is " << graph.getVertex(dfs.getParent(i)) << endl;
51
52    return 0;
53  }
```

Execution Result:
```
command>cl TestDFS.cpp
Microsoft C++ Compiler 2019
Compiled successful (cl is the VC++ compile/link command)

command>TestDFS
12 vertices are searched in this DFS order:
Chicago Seattle San Francisco Los Angeles Denver Kansas City
New York Boston Atlanta Miami Houston Dallas

parent of Seattle is Chicago
parent of San Francisco is Seattle
parent of Los Angeles is San Francisco
parent of Denver is Los Angeles
parent of Kansas City is Denver
parent of Boston is New York
parent of New York is Kansas City
parent of Atlanta is New York
parent of Miami is Atlanta
parent of Dallas is Houston
parent of Houston is Miami

command>
```

该程序在第 37 行为图 26.1 创建了一个图，在第 38 行获得一个从顶点 Chicago 开始的 DFS 树。在第 40 行中获得搜索顺序。从 Chicago 开始的 DFS 的图如图 26.10 所示。

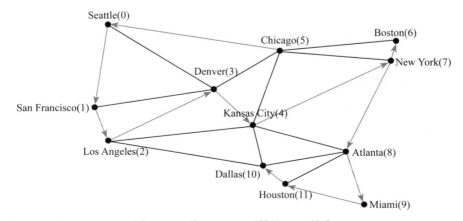

图 26.10　从 Chicago 开始的 DFS 搜索

注意，不要包含 Tree.h 和 Edge.h，因为这两个头文件已经包含在 Graph.h 中。

26.6.3 DFS 的应用

深度优先搜索可以解决许多问题，例如：
- 检测图是否连通。从任意顶点开始搜索图。如果搜索到的顶点数与图的顶点数相同，则表示图是连通的。否则，图不是连通的。
- 检测两个顶点之间是否存在路径。
- 在两个顶点之间查找路径。
- 查找所有连通的分量。连通分量是一个极大连通子图，其中每对顶点都通过一条路径连接。
- 检测图中是否存在环路。
- 在图中查找环路。

使用 LiveExample 26.3 中的 dfs 函数可以很容易地解决前四个问题。要检测或找到图中的环路，必须稍微修改 dfs 函数。

26.7 广度优先搜索

要点提示：图的广度优先搜索逐层访问顶点。第一层由起始顶点组成。每个下一层由与上一层中的顶点相邻的顶点组成。

图的广度优先搜索类似于 21.7 节中讨论的树的广度优先遍历。通过树的广度优先遍历，节点被逐层访问。首先访问根，然后访问根的所有子节点，然后从左到右访问根的孙子节点，以此类推。类似地，图的广度优先搜索首先访问一个顶点，然后访问它的所有相邻顶点，然后再访问与这些顶点相邻的所有顶点，以此类推。要确保每个顶点只访问一次，因此跳过已经访问过的顶点。

26.7.1 广度优先搜索算法

从图中的顶点 v 开始的广度优先搜索的算法如 Listing 26.2 所示。

Listing 26.2 广度优先搜索算法

```
输入：G=(V, E) 和起始顶点 v
输出：根为 v 的 BFS 树

1   Tree bfs(vertex v)
2   {
3     创建一个空队列用于存储要访问的顶点；
4     把 v 加入队列；
5     标记 v 已被访问；
6
7     while 队列不为空
8     {
9       从队列取出一个节点，比如 u
10      访问 u;
11      for u 的每个邻居 w
12        if w 没有被访问过
13        {
14          把 w 加入队列；
15          标记 w 已被访问；
16          parent[w] = u;
```

```
17       }
18     }
19   }
```

考虑图 26.11a 中的图。假设从顶点 0 开始广度优先搜索。首先访问 0，然后访问其所有未被访问的邻居 1、2 和 3，如 26.11b 所示。顶点 1 有三个邻居：0、2 和 4。由于 0 和 2 已经被访问，所以现在只访问 4，如图 26.11c 所示。顶点 2 有三个邻居：0、1 和 3，它们都已被访问过。顶点 3 有两个邻居：0、2，它们都已被访问过。顶点 4 有一个邻居：1，它已被访问过。因此，搜索结束。

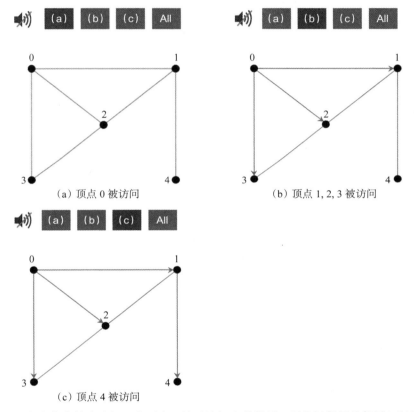

图 26.11 广度优先搜索访问一个顶点，然后访问它的邻居，再访问邻居的邻居，以此类推

由于每条边和顶点只访问一次，bfs 函数的时间复杂度为 $O(|E|+|V|)$，其中 |E| 表示边的数量，|V| 表示顶点的数量。

26.7.2 广度优先搜索的实现

bfs(int v) 函数在 LiveExample 26.3（第 354 ~ 385 行）中定义。它返回以顶点 v 为根的 Tree 类的实例（第 384 行）。该函数把搜索到的顶点存储在列表 searchOrders 中（第 357 行），把每个顶点的父亲存储在数组 parent 中（第 358 行），将要访问的顶点存储在 queue 中（第 362 行），用 isVisited 数组指示顶点是否已被访问（第 363 行）。搜索从顶点 v 开始。v 被添加到队列（第 364 行）并且被标记为已访问（第 365 行）。该函数现在检查队列中的每个顶点 u（第 369 行），并将其添加到 searchOrders 中（第 371 行）。该函数将 u 的每个未访问的邻居 w 添加到队列中（第 377 行），将其父亲设置为 u（第 378 行）并标记其已访问（第 379 行）。

LiveExample 26.6 给出了一个测试程序，该程序从 Chicago 开始显示图 26.1 中图的 BFS。

LiveExample 26.6 的互动程序请访问 https://liangcpp.pearsoncmg.com/LiveRunCpp5e/faces/LiveExample.xhtml?header=off&programName=TestBFS&fileType=.cpp&programHeight=880&resultHeight=400。

LiveExample 26.6　TestBFS.cpp

Source Code Editor:

```cpp
#include <iostream>
#include <string>
#include <vector>
#include "Graph.h" // Defined in LiveExample 26.3
using namespace std;

int main()
{
  // Vertices for graph in Figure 26.1
  string vertices[] = {"Seattle", "San Francisco", "Los Angeles",
    "Denver", "Kansas City", "Chicago", "Boston", "New York",
    "Atlanta", "Miami", "Dallas", "Houston"};

  // Edge array for graph in Figure 26.1
  int edges[][2] = {
    {0, 1}, {0, 3}, {0, 5},
    {1, 0}, {1, 2}, {1, 3},
    {2, 1}, {2, 3}, {2, 4}, {2, 10},
    {3, 0}, {3, 1}, {3, 2}, {3, 4}, {3, 5},
    {4, 2}, {4, 3}, {4, 5}, {4, 7}, {4, 8}, {4, 10},
    {5, 0}, {5, 3}, {5, 4}, {5, 6}, {5, 7},
    {6, 5}, {6, 7},
    {7, 4}, {7, 5}, {7, 6}, {7, 8},
    {8, 4}, {8, 7}, {8, 9}, {8, 10}, {8, 11},
    {9, 8}, {9, 11},
    {10, 2}, {10, 4}, {10, 8}, {10, 11},
    {11, 8}, {11, 9}, {11, 10}
  };
  const int NUMBER_OF_EDGES = 46; // 46 edges in Figure 26.1

  // Create a vector for vertices
  vector<string> vectorOfVertices(12);
  copy(vertices, vertices + 12, vectorOfVertices.begin());

  Graph<string> graph(vectorOfVertices, edges, NUMBER_OF_EDGES);
  Tree bfs = graph.bfs(graph.getIndex("Chicago")); // BFS starting from Chicago

  vector<int> searchOrders = bfs.getSearchOrders();
  cout << bfs.getNumberOfVerticesFound() <<
    " vertices are searched in this BFS order:" << endl;
  for (unsigned i = 0; i < searchOrders.size(); i++)
    cout << graph.getVertex(searchOrders[i]) << " ";
  cout << endl << endl;

  for (unsigned i = 0; i < searchOrders.size(); i++)
    if (bfs.getParent(i) != -1)
      cout << "parent of " << graph.getVertex(i) <<
        " is " << graph.getVertex(bfs.getParent(i)) << endl;
```

```
50      return 0;
51  }
```

Execution Result:

```
command>cl TestBFS.cpp
Microsoft C++ Compiler 2019
Compiled successful (cl is the VC++ compile/link command)
command>TestBFS
12 vertices are searched in this BFS order:
Chicago Seattle Denver Kansas City Boston New York
San Francisco Los Angeles Atlanta Dallas Miami Houston

parent of Seattle is Chicago
parent of San Francisco is Seattle
parent of Los Angeles is Denver
parent of Denver is Chicago
parent of Kansas City is Chicago
parent of Boston is Chicago
parent of New York is Chicago
parent of Atlanta is Kansas City
parent of Miami is Atlanta
parent of Dallas is Kansas City
parent of Houston is Atlanta

command>
```

该程序在第 35 行为图 26.1 创建了一个图，在第 36 行获得一个从顶点 Chicago 开始的 BFS 树。在第 38 行中获得搜索顺序。从 Chicago 开始的 BFS 的图如图 26.12 所示。

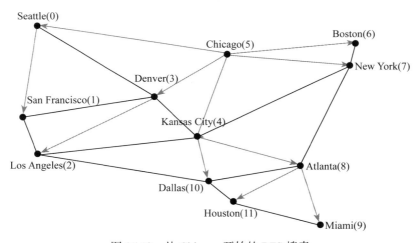

图 26.12　从 Chicago 开始的 BFS 搜索

26.7.3　BFS 的应用

DFS 解决的许多问题也可以用广度优先搜索来解决。具体而言，BFS 可以解决以下问题：
- 检测图是否连通。如果图中任意两个顶点之间存在路径，则图是连通的。
- 检测两个顶点之间是否存在路径。
- 查找两个顶点之间的最短路径。可以证明根和 BFS 树中任意顶点之间的路径是根和顶点之间的最短路径。
- 查找所有连通分量。连通分量是一个极大连通子图，其中每对顶点都通过一条路径连接。
- 检测图中是否存在环路。
- 在图中查找环路。
- 测试图是否是二分图。如果图的顶点可以划分为两个不相交的集合，使得同一集合中的顶点之间不存在边，则图是二分图。

26.8 案例研究：九枚硬币翻转问题

要点提示：九枚硬币翻转问题可以归结为最短路径问题。

DFS 和 BFS 算法有许多应用。本节应用 BFS 来解决九枚硬币翻转问题。

问题陈述如下。九枚硬币被放置在一个 3×3 的矩阵中，有些正面朝上（H），有些正面朝下（T）。合法的移动是取任意面朝上的硬币，并将其与相邻的硬币一起翻转（不包括对角相邻的硬币）。你的任务是找到让所有硬币正面朝下的最小移动次数。例如，从图 26.13a 所示的九枚硬币开始。翻转最后一行的第二枚硬币后，九枚硬币的状态如图 26.13b 所示。翻转第一行的第二枚硬币后，九枚硬币都面朝下，如图 26.13c 所示。

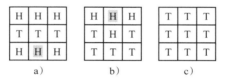

图 26.13 所有硬币都面朝下时，问题解决了

我们编写一个 C++ 程序，提示用户输入九枚硬币的初始状态并显示解决方案，如下示例运行所示。

下面 LiveExample 互动程序请访问 https://liangcpp.pearsoncmg.com/LiveRunCpp5e/faces/LiveExampleResultOnly.xhtml?header=off&programName=NineTail&fileType=.cpp&programHeight=0&resultHeight=370。

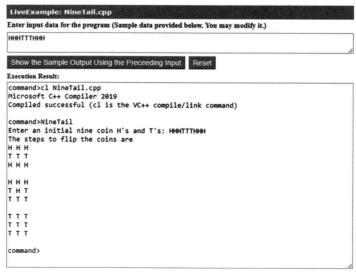

九枚硬币的每一种状态都表示图中的一个节点。例如，图 26.13 中的三种状态对应于图的三个节点。直观地说，可以使用 3×3 矩阵来表示所有节点，并使用 0 表示正面，用 1 表示背面。由于有九个单元格，并且每个单元格是 0 或 1，因此总共有 2^9（512）个节点，分别标记为 0，1，…，511，如图 26.14 所示。

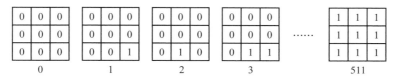

图 26.14 总共有 512 个节点，按此顺序标记为 0, 1, 2, …, 511

如果存在从 v 到 u 的合法移动，则我们分配一条从节点 u 到 v 的边。图 26.15 显示了到节点 56 的有向边。

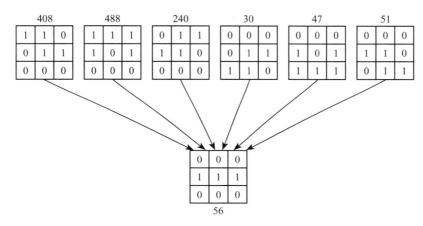

图 26.15 如果翻转单元格后节点 v 变成 u，则分配一条从 u 到 v 的边

图 26.14 的最后一个节点表示九枚正面朝下硬币的状态。方便起见，我们将最后一个节点称为目标节点。因此，目标节点被标记为 511。假设九枚硬币翻转问题的初始状态对应于节点 s。该问题简化为在以目标节点为根的 BFS 树中找到从目标节点到 s 的最短路径。

现在的任务是构建一个由 512 个标记为 0，1，2，…，511 的节点以及节点之间的边组成的图。创建了图，接下来就是获得以节点 511 为根的 BFS 树。从 BFS 树中，可以找到从根到任意节点的最短路径。我们创建一个名为 NineTailModel 的类，其中包含获取从目标节点到任意其他节点的最短路径的函数。类 UML 图如图 26.16 所示。

直观上，节点用字母 H 和 T 表示在 3×3 矩阵中。在 C++ 程序中，可以用 9 个字符的向量来表示节点。getNode(index) 函数返回指定索引的节点。例如，getNode(0) 返回包含 9 个 H 的节点。getNode(511) 返回包含 9 个 T 的节点。getIndex(node) 函数返回节点的索引。printNode(node) 函数在控制台上直观地显示节点。

注意，数据字段 tree 和函数 getEdges()、getFlippedNode(node, position) 和 flipACell(&node, row, column) 被定义为受保护的，以便在下一章中可以从子类访问它们。

getEdges() 函数返回 Edge 对象的向量。

```
                    NineTailModel
#tree: Tree*
+NineTailModel()
+getShortestPath(nodeIndex: int): vector<int>
+getNode(index: int): vector<char>
+getIndex(node: const vector<char>&): int
+printNode(node: const vector<char>&):void
#getEdges(): vector<Edge> const
#getFlippedNode(node: vector<char>&, position: int):
  int const
#flipACell(node: vector<char>&, row: int, column:
  int): void
```

图 26.16 NineTailModel 类用图对九枚硬币翻转问题进行建模

getFlippedNode(node, position) 函数翻转指定位置的节点，并返回新节点的索引。位置 position 的值从 0 到 8，它指向节点中的一个硬币，如下图所示。

例如，对于图 26.15 中的节点 56，翻转其位置 0，将得到节点 408。如果在位置 1 翻转节点 56，将得到节点 488。

flipACell(&node, row, column) 函数用于翻转指定行和列处的节点。例如，如果翻转节点 56 的第 0 行和第 0 列，则新节点为 408。如果翻转节点 56 的第 2 行和第 0 列，则新节点为 30。

LiveExample 26.7 显示了 NineTailModel.h 的源代码。

LiveExample 26.7 的互动程序请访问 https://liangcpp.pearsoncmg.com/LiveRunCpp5e/faces/LiveExample.xhtml?header=off&programName=NineTailModel&fileType=.h&programHeight=2520&resultVisible=false。

LiveExample 26.7　NineTailModel.h

Source Code Editor:

```
1   #ifndef NINETAILMODEL_H
2   #define NINETAILMODEL_H
3
4   #include <iostream>
5   #include "Graph.h" // Defined in LiveExample 26.3
6
7   using namespace std;
8
9   const int NUMBER_OF_NODES = 512;
10
11  class NineTailModel
12  {
13  public:
14    // Construct a model for the Nine Tail problem
15    NineTailModel();
```

```cpp
16
17      // Return the index of the node
18      int getIndex(const vector<char>& node) const;
19
20      // Return the node for the index
21      vector<char> getNode(int index) const;
22
23      // Return the shortest path of vertices from the specified
24      // node to the root
25      vector<int> getShortestPath(int nodeIndex) const;
26
27      // Print a node to the console
28      void printNode(const vector<char>& node) const;
29
30    protected:
31      Tree* tree;
32
33      // Return a vector of Edge objects for the graph
34      // Create edges among nodes
35      vector<Edge> getEdges() const;
36
37      // Return the index of the node that is the result of flipping
38      // the node at the specified position
39      int getFlippedNode(vector<char>& node, int position) const;
40
41      // Flip a cell at the specified row and column
42      void flipACell(vector<char>& node, int row, int column) const;
43    };
44
45    NineTailModel::NineTailModel()
46    {
47      // Create edges
48      vector<Edge> edges = getEdges();
49
50      // Build a graph
51      Graph<int> graph(NUMBER_OF_NODES, edges);
52
53      // Build a BFS tree rooted at the target node
54      tree = new Tree(graph.bfs(511));
55    }
56
57    vector<Edge> NineTailModel::getEdges() const
58    {
59      vector<Edge> edges;
60
61      for (int u = 0; u < NUMBER_OF_NODES; u++)
62      {
63        for (int k = 0; k < 9; k++)
64        {
65          vector<char> node = getNode(u);
66          if (node[k] == 'H')
67          {
68            int v = getFlippedNode(node, k);
69            // Add edge (v, u) for a legal move from node u to node v
70            edges.push_back(Edge(v, u));
71          }
72        }
73      }
74
75      return edges;
76    }
77
78    int NineTailModel::getFlippedNode(vector<char>& node, int position)
79      const
```

```cpp
 80  {
 81    int row = position / 3;
 82    int column = position % 3;
 83
 84    flipACell(node, row, column);
 85    flipACell(node, row - 1, column);
 86    flipACell(node, row + 1, column);
 87    flipACell(node, row, column - 1);
 88    flipACell(node, row, column + 1);
 89
 90    return getIndex(node);
 91  }
 92
 93  void NineTailModel::flipACell(vector<char>& node,
 94    int row, int column) const
 95  {
 96    if (row >= 0 && row <= 2 && column >= 0 && column <= 2)
 97    { // Within boundary
 98      if (node[row * 3 + column] == 'H')
 99        node[row * 3 + column] = 'T'; // Flip from H to T
100      else
101        node[row * 3 + column] = 'H'; // Flip from T to H
102    }
103  }
104
105  int NineTailModel::getIndex(const vector<char>& node) const
106  {
107    int result = 0;
108
109    for (int i = 0; i < 9; i++)
110      if (node[i] == 'T') // Test if it is a tail
111        result = result * 2 + 1;
112      else
113        result = result * 2 + 0;
114
115    return result;
116  }
117
118  vector<char> NineTailModel::getNode(int index) const
119  {
120    vector<char> result(9);
121
122    for (int i = 0; i < 9; i++)
123    {
124      int digit = index % 2;
125      if (digit == 0)
126        result[8 - i] = 'H';
127      else
128        result[8 - i] = 'T';
129      index = index / 2;
130    }
131
132    return result;
133  }
134
135  vector<int> NineTailModel::getShortestPath(int nodeIndex) const
136  {
137    return tree->getPath(nodeIndex);
138  }
139
140  void NineTailModel::printNode(const vector<char>& node) const
141  {
142    for (int i = 0; i < 9; i++)
143      if (i % 3 != 2)
144        cout << node[i];
```

```
145        else
146          cout << node[i] << endl;
147
148      cout << endl;
149    }
150
151    #endif
```

构造函数（第 45 ～ 55 行）创建一个具有 512 个节点的图，每条边对应于从一个节点到另一个节点的移动（第 48 行）。从该图中，获得以目标节点 511 为根的 BFS 树，并将其赋值给指针变量 tree（第 54 行）。我们使用 tree 的指针使 tree 能够引用任何类型的 Tree。在下一章，我们将把 ShortestPathTree 赋值给 tree，ShortestPathTree 是 Tree 的一个子类型。

为了创建边，getEdges 函数（第 57 ～ 76 行）检查每个节点 u，看看它是否可以翻转到另一个节点 v。如果可以，将 (v, u) 添加到 Edge 向量（第 70 行）。getFlippedNode(node, position) 函数通过翻转节点中的 H 单元格及其邻居来查找翻转的节点（第 78 ～ 91 行）。flipACell(node, row, column) 函数实际翻转节点中的 H 单元格及其邻居（第 93 ～ 103 行）。

注意，getFlippedNode(&node, position) 的参数 node 是通过值传递的。如果它被错误地通过引用传递，会发生什么？在第 68 行调用 getFlippedNode(node, k) 之后，节点的内容将被更改，并且节点不再对应于顶点 u。

getIndex(node) 函数的实现方式与将二进制数转换为十进制数相同（第 105 ～ 116 行）。getNode(index) 函数返回一个由字母 H 和 T 组成的节点（第 118 ～ 133 行）。

getShortestpath(nodeIndex) 函数调用 getPath(nodeIndex) 函数，以获取从指定节点到目标节点的最短路径中的顶点（第 135 ～ 138 行）。

LiveExample 26.8 给出一个程序，提示用户输入初始节点并显示到达目标节点的步骤。

LiveExample 26.8 的互动程序请访问 https://liangcpp.pearsoncmg.com/LiveRunCpp5e/faces/LiveExample.xhtml?header=off&programName=NineTail&fileType=.cpp&programHeight=430&resultHeight=340。

LiveExample 26.8　NineTail.cpp

```cpp
1   #include <iostream>
2   #include <vector>
3   #include "NineTailModel.h"
4   using namespace std;
5
6   int main()
7   {
8     // Prompt the user to enter nine coins H and T's
9     cout << "Enter an initial nine coin H's and T's: ";
10    vector<char> initialNode(9);
11
12    for (int i = 0; i < 9; i++)
13      cin >> initialNode[i];
14
15    cout << "The steps to flip the coins are " << endl;
16    NineTailModel model; // Create a model
17    vector<int> path =
```

该程序在第 9 ~ 13 行提示用户输入一个包含 9 个 H 和 T 字母的初始节点，创建一个图模型和 BFS 树（第 16 行），获得从初始节点到目标节点的最短路径（第 17 ~ 18 行），显示路径中的节点（第 20 ~ 21 行）。

关键术语

adjacency list（邻接列表）
adjacency matrix（邻接矩阵）
adjacent vertices（邻接顶点）
breadth-first search（广度优先搜索）
complete graph（完全图）
degree（度）
depth-first search（深度优先搜索）
directed graph（有向图）
graph（图）

incident edges（关联边）
parallel edge（平行边）
Seven Bridges of Königsberg（柯尼斯堡七桥）
simple graph（简单图）
spanning tree（生成树）
undirected graph（无向图）
unweighted graph（无权图）
weighted graph（加权图）

章节总结

1. 图是一种有用的数学结构，表示现实世界中实体之间的关系。
2. 图可以是有向的，也可以是无向的。在有向图中，每条边都有一个方向，表示可以通过边从一个顶点移动到另一个顶点。
3. 边可以加权也可以不加权。加权图具有加权边。
4. 可以用类和接口对图进行建模。
5. 可以用数组和链表来表示顶点和边。
6. 图遍历是访问图中每个顶点一次的过程。遍历图的两种常用方式是深度优先搜索和广度优先搜索。
7. 图的深度优先搜索，首先访问一个顶点，然后递归地访问与该顶点相邻的所有未访问的顶点。
8. 图的广度优先搜索，首先访问一个顶点，然后访问其所有相邻的未访问顶点，再访问与这些顶点相邻的所有未访问顶点，以此类推。
9. DFS 和 BFS 可以用来解决许多问题，例如检测图是否连通、检测图中是否存在环路，以及找到两个顶点之间的最短路径。

编程练习

26.6～26.7 节

*26.1（测试图是否连通）编写一个程序，从文件中读取图并确定图是否连通。文件中的第一行包含一个数字，表示顶点的数量（n）。顶点标记为 0, 1, …, n-1。后续每一行，格式为 u: v1, v2, …，表示边 (u, v1)、(u, v2) 等。图 26.17 给出了两个文件对应图的示例。

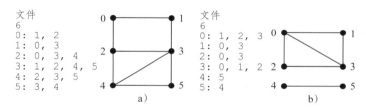

图 26.17 图的顶点和边可以存储在文件中

程序提示用户输入文件名，从文件中读取数据，创建 Graph 的实例 g，调用 g.printEdges() 显示所有边，并调用 dfs(0) 获得一个 Tree 的实例 tree。如果 tree.getNumberOfVerticeFound() 与图中的顶点数相同，则表示该图是连通的。

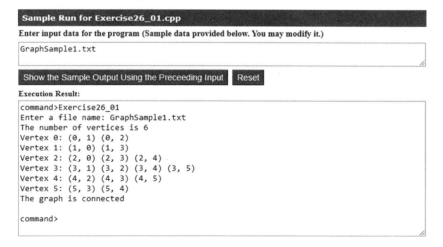

（提示：用 Graph(numberOfVertices, vectorOfEdges) 创建图，其中 vectorOfEdges 包含一个 Edge 对象的向量。用 Edge(u, v) 创建边。读取第一行获得顶点数。读取后续每行，从字符串中提取顶点，并通过顶点创建边。）

*26.2 （为图创建一个文件）修改 LiveExample 26.2，创建一个表示 graph1 的文件。文件格式如编程练习 26.1 所述。根据 LiveExample 26.2 中第 16～29 行定义的图创建文件。图的顶点数为 12，存储在文件的第一行。文件的内容如下所示：

```
12
0: 1, 3, 5
1: 0, 2, 3
2: 1, 3, 4, 10
3: 0, 1, 2, 4, 5
4: 2, 3, 5, 7, 8, 10
5: 0, 3, 4, 6, 7
6: 5, 7
7: 4, 5, 6, 8
8: 4, 7, 9, 10, 11
9: 8, 11
10: 2, 4, 8, 11
11: 8, 9, 10
```

*26.3 （找到最短路径）编写一个程序，从文件中读取连通图。图用编程练习 26.1 中指定的相同格式存储在文件中。程序提示用户输入文件名，然后输入两个顶点，显示两个顶点之间的最短路径。例如，对于图 26.17a 中的图，0 和 5 之间的最短路径显示为 0 1 3 5。

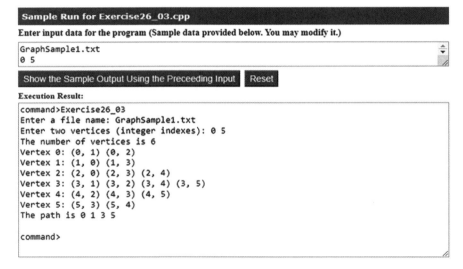

*26.4 （用栈实现 DFS）LiveExample 26.5 中描述的深度优先搜索算法使用了递归。请在不使用递归的情况下实现它。

*26.5 （查找连通分量）在 Graph 类中添加一个有以下函数头的新函数，查找图中的所有连通分量：

    vector<vector<**int**>> getConnectedComponents();

该函数返回一个向量。向量中的每个元素都是另一个包含连通分量所有顶点的向量。例如，如果图有三个连通分量，则函数返回一个包含三个元素的向量，每个元素都包含连通分量的顶点。

*26.6 （查找路径）在 Graph 类中添加一个有以下函数头的新函数，查找两个顶点之间的路径：

    vector<**int**> getPath(**int** u, **int** v);

该函数返回一个向量，向量包含从 u 到 v 的路径中按顺序排列的所有顶点。用 BFS 方法，可以获得从 u 到 v 的最短路径。如果没有从 u 到 v 的路径，则函数返回一个空向量。

*26.7 （检测环路）在 Graph 类中添加一个有以下函数头的新函数，确定图中是否存在环路：

   bool containsCyclic();

*26.8 （查找环路）在 Graph 类中添加一个有以下函数头的新函数，在图中查找环路：

   vector<**int**> getACycle();

该函数返回一个向量，向量包含从 u 到 v 的环路中按顺序排列的所有顶点。如果图没有环路，函数将返回一个空向量。

**26.9 （检测二分图）回想一下，如果图的顶点可以划分为两个不相交的集合，使得同一集合中的顶点之间不存在边，那么该图就是二分图。在 Graph 中添加一个新函数，以检测图是否为二分图：

   bool isBipartite();

**26.10 （获取二分图集合）如果图是二分图，则在 Graph 中添加一个新函数，以返回两个二分图集合：

   vector<vector<**int**>> getBipartiteSets();

该函数返回一个向量。向量中的每个元素都是包含一组顶点的另一个向量。

**26.11 （九枚硬币翻转问题的变体）在九枚硬币翻转问题中，当翻转正面朝上硬币时，水平和垂直相邻单元格也会翻转。重写程序，假设包括对角邻居在内的所有相邻单元格也翻转。

**26.12 （4×4 的 16 枚硬币翻转问题）前文中的九枚硬币翻转问题使用 3×3 矩阵。假设你将 16 枚硬币放在一个 4×4 的矩阵中。创建一个名为 TailModel16 的新模型类。编写一个程序，提示用户输入由 H 和 T 组成的 16 个字母的字符串，并显示将其全部翻转成 T 的步骤。

**26.13 （诱导子图）给定一个无向图 $G = (V, E)$ 和一个整数 k，找到 G 的一个最大的诱导子图 H，使得 H 的所有顶点的度 $\geq k$，或者得出不存在这样的诱导子图的结论。用以下函数头实现该函数：

   Graph<V> maxInducedSubgraph(Graph<V> g, int k)

如果这样的子图不存在，函数将返回一个空图。

（提示：一种直观的方法是删除度数小于 k 的顶点。当顶点与其相邻边一起删除时，其他顶点的度数可能会减少。继续该过程，直到无法删除任何顶点或者删除所有顶点。）

第 27 章

Introduction to C++ Programming and Data Structures, Fifth Edition

加权图及其应用

学习目标

1. 用邻接矩阵和邻接列表表示加权边（27.2 节）。
2. 用扩展 `Graph` 类的 `WeightedGraph` 类对加权图进行建模（27.3 节）。
3. 设计并实现寻找最小生成树的算法（27.4 节）。
4. 定义扩展 `Tree` 类的 `MST` 类（27.4 节）。
5. 设计并实现寻找单源最短路径的算法（27.5 节）。
6. 定义扩展 `Tree` 类的 `ShortestPathTree` 类（27.5 节）。
7. 用最短路径算法解决加权九枚硬币翻转问题（27.6 节）。

27.1 简介

要点提示：如果图的每个边都被赋予了权重，则该图就是加权图。加权图有许多实际应用。

假设图 26.1 中的图表示城市间的航班数量。可以应用 BFS 来查找两个城市之间的最少航班数。假设边代表城市之间的航行距离，如图 27.1 所示。如何找到连接所有城市的最小总距离？如何找到两个城市之间的最短路径？本章将讨论这些问题。前者被称为**最小生成树（MST）**问题，后者被称为最短路径问题。

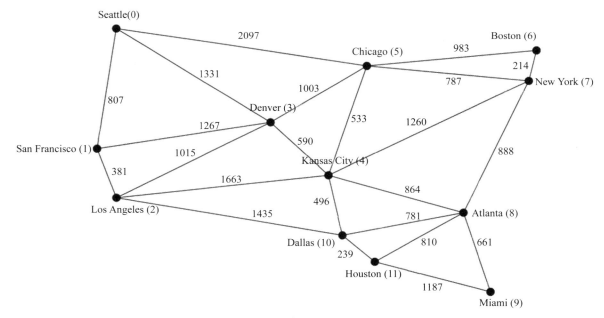

图 27.1 用图模拟城市之间的距离

前一章介绍了图的概念。我们学习了如何用边数组、边向量、邻接矩阵和邻接列表表示边，以及如何用 Graph 类对图建模。该章还介绍了遍历图的两种重要技术：深度优先搜索和广度优先搜索，并用遍历解决实际问题。本章将介绍加权图。我们将在 27.4 节中学习寻找最小生成树的算法，在 27.5 节中学习寻找最短路径的算法。

27.2 加权图的表示

要点提示：加权边可以存储在邻接列表中。

有两种类型的加权图：顶点加权图和边加权图。在**顶点加权图**中，为每个顶点指定权重。在**边加权图**中，为每条边指定权重。在这两种类型中，边加权图的应用更多。本章主要介绍边加权图。

加权图可以用与无权图相同的方式表示，只是必须要表示边上的权重。与无权图一样，加权图中的顶点可以存储在向量中。本节介绍加权图中边的三种表示形式。

27.2.1 加权边的表示：边数组

加权边可以用二维数组来表示。例如，可以用以下数组存储图 27.2 中图的所有边：

```
int edges[][3] =
{
  {0, 1, 2}, {0, 3, 8},
  {1, 0, 2}, {1, 2, 7}, {1, 3, 3},
  {2, 1, 7}, {2, 3, 4}, {2, 4, 5},
  {3, 0, 8}, {3, 1, 3}, {3, 2, 4}, {3, 4, 6},
  {4, 2, 5}, {4, 3, 6}
};
```

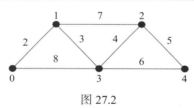

图 27.2

27.2.2 加权相邻列表

表示边的另一种方法是将边定义为对象。定义 Edge 类以表示无权图中的边。对于加权边，我们定义 WeightedEdge 类，如 LiveExample 27.1 所示。

LiveExample 27.1 的互动程序请访问 https://liangcpp.pearsoncmg.com/LiveRunCpp5e/faces/LiveExample.xhtml?header=off&programName=WeightedEdge&fileType=.h&programHeight=830&resultVisible=false。

LiveExample 27.1　WeightedEdge.h

```
Source Code Editor:
 1  #ifndef WEIGHTEDEDGE_H
 2  #define WEIGHTEDEDGE_H
 3
 4  #include "Edge.h"
 5
 6  class WeightedEdge : public Edge
 7  {
 8  public:
```

```
9   double weight; // The weight on edge (u, v)
10
11  // Create a weighted edge on (u, v)
12  WeightedEdge(int u, int v, double weight): Edge(u, v)
13  {
14    this->weight = weight;
15  }
16
17  // Compare edges based on weights
18  bool operator<(const WeightedEdge& edge) const
19  {
20    return (*this).weight < edge.weight;
21  }
22
23  bool operator<=(const WeightedEdge& edge) const
24  {
25    return (*this).weight <= edge.weight;
26  }
27
28  bool operator>(const WeightedEdge& edge) const
29  {
30    return (*this).weight > edge.weight;
31  }
32
33  bool operator>=(const WeightedEdge& edge) const
34  {
35    return (*this).weight >= edge.weight;
36  }
37
38  bool operator==(const WeightedEdge& edge) const
39  {
40    return (*this).weight == edge.weight;
41  }
42
43  bool operator!=(const WeightedEdge& edge) const
44  {
45    return (*this).weight != edge.weight;
46  }
47  };
48  #endif
```

Answer   Reset

在 LiveExample 26.1 中定义的 Edge 类表示从顶点 u 到 v 的边。WeightedEdge 继承自 Edge，并加入了新的属性 weight。

创建 WeightedEdge 对象要用 WeightedEdge(i, j, w)，其中 w 是边 (i, j) 上的权重。边的权重常常会被比较。因此，我们在 WeightedEdge 类中定义了运算符函数 (<, <=, ==, !=, >, >=)。

对于无权图，我们用邻接边列表来表示边。对于加权图，我们仍然用邻接边列表。由于 WeightedEdge 是从 Edge 派生的。无权图的邻接边列表可以用于加权图。例如，图 27.2 中图的顶点的邻接边列表可以表示为：

```
vector<vector<Edge*>> neighbors;
for(unsigned i = 0; i < numberOfVertices; i++)
{
  neighbors.push_back(vector<Edge*>());
}
neighbors[0].push_back(new WeightedEdge(0, 1, 2));
neighbors[0].push_back(new WeightedEdge(0, 3, 8));
...
```

neighbors[0]	WeightedEdge(0, 1, 2)	WeightedEdge(0, 3, 8)		
neighbors[1]	WeightedEdge(1, 0, 2)	WeightedEdge(1, 2, 3)	WeightedEdge(1, 2, 7)	
neighbors[2]	WeightedEdge(2, 3, 4)	WeightedEdge(2, 4, 5)	WeightedEdge(2, 1, 7)	
neighbors[3]	WeightedEdge(3, 1, 3)	WeightedEdge(3, 2, 4)	WeightedEdge(3, 4, 6)	WeightedEdge(3, 0, 8)
neighbors[4]	WeightedEdge(4, 2, 5)	WeightedEdge(4, 3, 6)		

27.3 `WeightedGraph` 类

要点提示：`WeightedGraph` 类继承自 `Graph` 类。

前一章设计了对图进行建模的 `Graph` 类。按照这个模式，我们设计从 `Graph` 类派生的 `WeightedGraph` 类，如图 27.3 所示。

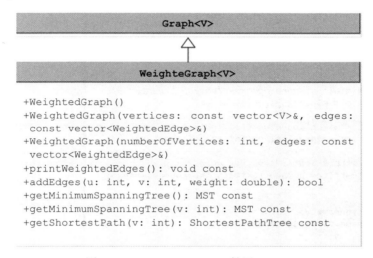

图 27.3 `WeightedGraph` 扩展 `Graph`

`WeightedGraph` 简单地扩展 `Graph`。`WeightedGraph` 继承了 `Graph` 的所有函数，并引入了新函数，来获得最小生成树和查找单源所有最短路径。最小生成树和最短路径将分别在 27.4 节和 27.5 节中介绍。

该类包含三个构造函数。第一个是无参数构造函数，用于创建一个空图。第二个用向量中的指定顶点和边构造 `WeightedGraph`。第三个用顶点 0，1，⋯，n-1 和边向量构造 `WeightedGraph`。`printWeightedEdges` 函数显示每个顶点的所有边。LiveExample 27.2 实现了 `WeightedGraph`。

LiveExample 27.2 的互动程序请访问 https://liangcpp.pearsoncmg.com/LiveRunCpp5e/faces/LiveExample.xhtml?header=off&programName=WeightedGraph&fileType=.h&programHeight=3510&resultVisible=false。

LiveExample 27.2 WeightedGraph.h

Source Code Editor:

```
1  #ifndef WEIGHTEDGRAPH_H
2  #define WEIGHTEDGRAPH_H
3
4  #include "Graph.h"
5  #include "WeightedEdge.h" // Defined in LiveExample 27.1
```

```cpp
 6  #include "MST.h" // Defined in LiveExample 27.4
 7  #include "ShortestPathTree.h" // Defined in LiveExample 27.6
 8
 9  template<typename V>
10  class WeightedGraph : public Graph<V>
11  {
12  public:
13    // Construct an empty graph
14    WeightedGraph();
15
16    // Construct a graph from vertices and edges objects
17    WeightedGraph(const vector<V>& vertices, const vector<WeightedEdge>& edges);
18
19    // Construct a graph with vertices 0, 1, ..., n-1 and
20    // edges in a vector
21    WeightedGraph(int numberOfVertices, const vector<WeightedEdge>& edges);
22
23    // Print all edges in the weighted tree
24    void printWeightedEdges();
25
26    // Add a weighted edge
27    bool addEdge(int u, int v, double w);
28
29    // Get a minimum spanning tree rooted at vertex 0
30    MST getMinimumSpanningTree();
31
32    // Get a minimum spanning tree rooted at a specified vertex
33    MST getMinimumSpanningTree(int startingVertex);
34
35    // Find single-source shortest paths
36    ShortestPathTree getShortestPath(int sourceVertex);
37  };
38
39  template<typename V>
40  WeightedGraph<V>::WeightedGraph()
41  {
42  }
43
44  template<typename V>
45  WeightedGraph<V>::WeightedGraph(const vector<V>& vertices,
46    const vector<WeightedEdge>& edges)
47  {
48    // Add vertices to the graph
49    for (unsigned i = 0; i < vertices.size(); i++)
50    {
51      addVertex(vertices[i]);
52    }
53
54    // Add edges to the graph
55    for (unsigned i = 0; i < edges.size(); i++)
56    {
57      addEdge(edges[i].u, edges[i].v, edges[i].weight);
58    }
59  }
60
61  template<typename V>
62  WeightedGraph<V>::WeightedGraph(int numberOfVertices,
63    const vector<WeightedEdge>& edges)
64  {
65    // Add vertices to the graph
66    for (int i = 0; i < numberOfVertices; i++)
67      addVertex(i); // vertices is {0, 1, 2, ..., n-1}
68
69    // Add edges to the graph
```

```cpp
70      for (unsigned i = 0; i < edges.size(); i++)
71      {
72        addEdge(edges[i].u, edges[i].v, edges[i].weight);
73      }
74    }
75
76    template<typename V>
77    void WeightedGraph<V>::printWeightedEdges()
78    {
79      for (int i = 0; i < getSize(); i++)
80      {
81        // Display all edges adjacent to vertex with index i
82        cout << "Vertex " << getVertex(i) << "(" << i << "): ";
83
84        // Display all weighted edges
85        for (Edge* e: neighbors[i])
86        {
87          cout << "(" << e->u << ", " << e->v << ", "
88            << static_cast<WeightedEdge*>(e)->weight << ") ";
89        }
90        cout << endl;
91      }
92    }
93
94    template<typename V>
95    bool WeightedGraph<V>::addEdge(int u, int v, double w)
96    {
97      return createEdge(new WeightedEdge(u, v, w));
98    }
99
100   template<typename V>
101   MST WeightedGraph<V>::getMinimumSpanningTree()
102   {
103     return getMinimumSpanningTree(0);
104   }
105
106   template<typename V>
107   MST WeightedGraph<V>::getMinimumSpanningTree(int startingVertex)
108   {
109     vector<int> T; // T contains the vertices in the tree
110
111     vector<int> parent(getSize()); // Parent of a verte
112     parent[startingVertex] = -1; // startingVertex is the root
113     double totalWeight = 0; // Total weight of the tree thus far
114
115     // cost[v] stores the cost by adding v to the tree
116     vector<double> cost(getSize());
117     for (unsigned i = 0; i < cost.size(); i++)
118     {
119       cost[i] = INFINITY; // Initial cost set to infinity
120     }
121     cost[startingVertex] = 0; // Cost of source is 0
122
123     // Expand T
124     while (T.size() < getSize())
125     {
126       // Find smallest cost v in V - T
127       int u = -1; // Vertex to be determined
128       double currentMinCost = INFINITY;
129       for (int i = 0; i < getSize(); i++)
130       {
131         if (find(T.begin(), T.end(), i) == T.end()
132           && cost[i] < currentMinCost)
133         {
```

```cpp
                currentMinCost = cost[i];
                u = i;
            }
        }

        T.push_back(u); // Add a new vertex to T
        totalWeight += cost[u]; // Add cost[u] to the tree

        // Adjust cost[v] for v that is adjacent to u and v in V - T
        for (Edge* e: neighbors[u])
        {
            if (find(T.begin(), T.end(), e->v) == T.end()
                && cost[e->v] > static_cast<WeightedEdge*>(e)->weight)
            {
                cost[e->v] = static_cast<WeightedEdge*>(e)->weight;
                parent[e->v] = u;
            }
        }
    } // End of while

    return MST(startingVertex, parent, T, totalWeight);
}

template<typename V>
ShortestPathTree WeightedGraph<V>::getShortestPath(int sourceVertex)
{
    // T stores the vertices whose path found so far
    vector<int> T;

    // parent[v] stores the previous vertex of v in the path
    vector<int> parent(getSize());
    parent[sourceVertex] = -1; // The parent of source is set to -1

    // cost[v] stores the cost of the path from v to the source
    vector<double> cost(getSize());
    for (unsigned i = 0; i < cost.size(); i++)
    {
        cost[i] = INFINITY; // Initial cost set to infinity
    }
    cost[sourceVertex] = 0; // Cost of source is 0

    // Expand T
    while (T.size() < getSize())
    {
        // Find smallest cost v in V - T
        int u = -1; // Vertex to be determined
        double currentMinCost = INFINITY;
        for (int i = 0; i < getSize(); i++)
        {
            if (find(T.begin(), T.end(), i) == T.end()
                && cost[i] < currentMinCost)
            {
                currentMinCost = cost[i];
                u = i;
            }
        }

        if (u == -1) break;

        T.push_back(u); // Add a new vertex to T

        // Adjust cost[v] for v that is adjacent to u and v in V - T
        for (Edge* e:  neighbors[u])
        {
```

```
198            if (find(T.begin(), T.end(), e->v) == T.end() &&
199              cost[e->v] > cost[u] + static_cast<WeightedEdge*>(e)->weight)
200            {
201              cost[e->v] = cost[u] + static_cast<WeightedEdge*>(e)->weight;
202              parent[e->v] = u;
203            }
204          }
205        } // End of while
206
207        // Create a ShortestPathTree
208        return ShortestPathTree(sourceVertex, parent, T, cost);
209      }
210
211      #endif
```

Answer Reset

WeightedGraph派生自Graph（第10行）。属性vertices和neighbors在父类Graph中定义为受保护的，因此可以在子类WeightedGraph中访问它们。

构造WeightedGraph时，通过调用addVertex函数（第51、67行）和addEdge函数（第57、72行）创建其顶点和邻接边列表。addVertex函数在LiveExample 26.3中的Graph类中定义。addEdge函数调用createEdge函数将加权边添加到图中（第94～98行）。createEdge函数在LiveExample 26.3中定义为受保护的。此函数检查边是否有效，如果边无效，则抛出invalid_argument异常。

LiveExample 27.3给出了一个测试程序，该程序为图27.1中的图创建了一个图，为图27.2中的图创建了另一个图。

LiveExample 27.3的互动程序请访问https://liangcpp.pearsoncmg.com/LiveRunCpp5e/faces/LiveExample.xhtml?header=off&programName=TestWeightedGraph&fileType=.cpp&programHeight=1370&resultHeight=680。

LiveExample 27.3 TestWeightedGraph.cpp

Source Code Editor:

```cpp
1   #include <iostream>
2   #include <string>
3   #include <vector>
4   #include "WeightedGraph.h"
5   #include "WeightedEdge.h"
6   using namespace std;
7
8   int main()
9   {
10    // Vertices for graph in Figure 27.1
11    string vertices[] = {"Seattle", "San Francisco", "Los Angeles",
12      "Denver", "Kansas City", "Chicago", "Boston", "New York",
13      "Atlanta", "Miami", "Dallas", "Houston"};
14
15    // Edge array for graph in Figure 27.1
16    int edges[][3] = {
17      {0, 1, 807}, {0, 3, 1331}, {0, 5, 2097},
18      {1, 0, 807}, {1, 2, 381}, {1, 3, 1267},
19      {2, 1, 381}, {2, 3, 1015}, {2, 4, 1663}, {2, 10, 1435},
20      {3, 0, 1331}, {3, 1, 1267}, {3, 2, 1015}, {3, 4, 599},
21        {3, 5, 1003},
22      {4, 2, 1663}, {4, 3, 599}, {4, 5, 533}, {4, 7, 1260},
23        {4, 8, 864}, {4, 10, 496},
```

```cpp
      {5, 0, 2097}, {5, 3, 1003}, {5, 4, 533},
        {5, 6, 983}, {5, 7, 787},
      {6, 5, 983}, {6, 7, 214},
      {7, 4, 1260}, {7, 5, 787}, {7, 6, 214}, {7, 8, 888},
      {8, 4, 864}, {8, 7, 888}, {8, 9, 661},
        {8, 10, 781}, {8, 11, 810},
      {9, 8, 661}, {9, 11, 1187},
      {10, 2, 1435}, {10, 4, 496}, {10, 8, 781}, {10, 11, 239},
      {11, 8, 810}, {11, 9, 1187}, {11, 10, 239}
    };

    // 23 undirected edges in Figure 27.1
    const int NUMBER_OF_EDGES = 46;

    // Create a vector for vertices
    vector<string> vectorOfVertices(12);
    copy(vertices, vertices + 12, vectorOfVertices.begin());

    // Create a vector for edges
    vector<WeightedEdge> edgeVector;
    for (int i = 0; i < NUMBER_OF_EDGES; i++)
      edgeVector.push_back(WeightedEdge(edges[i][0],
      edges[i][1], edges[i][2]));

    WeightedGraph<string> graph1(vectorOfVertices, edgeVector);
    cout << "The number of vertices in graph1: " << graph1.getSize();
    cout << "\nThe vertex with index 1 is " << graph1.getVertex(1);
    cout << "\nThe index for Miami is " << graph1.getIndex("Miami");

    cout << "\nThe edges for graph1: " << endl;
    graph1.printWeightedEdges();

    // Create a graph for Figure 27.2
    int edges2[][3] =
    {
      {0, 1, 2}, {0, 3, 8},
      {1, 0, 2}, {1, 2, 7}, {1, 3, 3},
      {2, 1, 7}, {2, 3, 4}, {2, 4, 5},
      {3, 0, 8}, {3, 1, 3}, {3, 2, 4}, {3, 4, 6},
      {4, 2, 5}, {4, 3, 6}
    }; // 14 edges in Figure 27.2

    // 7 undirected edges in Figure 27.2
    const int NUMBER_OF_EDGES2 = 14;

    vector<WeightedEdge> edgeVector2;
    for (int i = 0; i < NUMBER_OF_EDGES2; i++)
      edgeVector2.push_back(WeightedEdge(edges2[i][0],
      edges2[i][1], edges2[i][2]));

    WeightedGraph<int> graph2(5, edgeVector2); // 5 vertices in graph2

    cout << "The number of vertices in graph2: " << graph2.getSize();
    cout << "\nThe edges for graph2: " << endl;
    graph2.printWeightedEdges();

    return 0;
}
```

Execution Result:
```
command>cl TestWeightedGraph.cpp
```

```
Microsoft C++ Compiler 2019
Compiled successful (cl is the VC++ compile/link command)

command>TestWeightedGraph
The number of vertices in graph1: 12
The vertex with index 1 is San Francisco
The index for Miami is 9
The edges for graph1:
Vertex Seattle(0): (0, 1, 807) (0, 3, 1331) (0, 5, 2097)
Vertex San Francisco(1): (1, 0, 807) (1, 2, 381) (1, 3, 1267)
Vertex Los Angeles(2): (2, 1, 381) (2, 3, 1015) (2, 4, 1663)
(2, 10, 1435)
Vertex Denver(3): (3, 0, 1331) (3, 1, 1267) (3, 2, 1015)
(3, 4, 599) (3, 5, 1003)
Vertex Kansas City(4): (4, 2, 1663) (4, 3, 599) (4, 5, 533)
(4, 7, 1260) (4, 8, 864) (4, 10, 496)
Vertex Chicago(5): (5, 0, 2097) (5, 3, 1003) (5, 4, 533)
(5, 6, 983) (5, 7, 787)
Vertex Boston(6): (6, 5, 983) (6, 7, 214)
Vertex New York(7): (7, 4, 1260) (7, 5, 787) (7, 6, 214)
(7, 8, 888)
Vertex Atlanta(8): (8, 4, 864) (8, 7, 888) (8, 9, 661)
(8, 10, 781) (8, 11, 810)
Vertex Miami(9): (9, 8, 661) (9, 11, 1187)
Vertex Dallas(10): (10, 2, 1435) (10, 4, 496)
(10, 8, 781) (10, 11, 239)
Vertex Houston(11): (11, 8, 810) (11, 9, 1187) (11, 10, 239)
The number of vertices in graph2: 5
The edges for graph2:
Vertex 0(0): (0, 1, 2) (0, 3, 8)
Vertex 1(1): (1, 0, 2) (1, 2, 7) (1, 3, 3)
Vertex 2(2): (2, 1, 7) (2, 3, 4) (2, 4, 5)
Vertex 3(3): (3, 0, 8) (3, 1, 3) (3, 2, 4) (3, 4, 6)
Vertex 4(4): (4, 2, 5) (4, 3, 6)

command>
```

程序在第 11～48 行为图 27.1 中的图创建 graph1。在第 11～13 行中定义 graph1 的顶点。在第 16～33 行中定义 graph1 的边。使用二维数组表示边。对于数组中的第 i 行，edges[i][0] 和 edges[i][1] 表明存在从顶点 edges[i][0] 到顶点 edges[i][1] 的边，且该边的权重是 edges[i][2]。例如，第一行 {0, 1, 807} 表示从顶点 0 (edges[0][0]) 到顶点 1 (edges[0][1]) 的边，权重为 807 (edges[0][2])。行 {0, 5, 2097} 表示从顶点 0 (edges[2][0]) 到顶点 5 (edges[2][1]) 的边，权重为 2097 (edges[2][2])。要创建 WeightedGraph，必须获得 WeightedEdge 的向量（第 43～46 行）。

第 54 行调用 graph1 上的 printWeightedEdges() 函数显示 graph1 中的所有边。

程序在第 57～74 行为图 27.2 中的图创建一个加权图 graph2。第 78 行调用 graph2 上的 printWeightedEdges() 函数显示 graph2 中的所有边。

27.4 最小生成树

要点提示：图的最小生成树是具有最小总权重的生成树。

一个图可能有许多生成树。假设边是加权的。最小生成树是具有最小总权重的生成树。

例如，图27.4b、图27.4c 和图27.4d 中的树是图27.4a 中图的生成树。图27.4c 和图27.4d 中的树是最小生成树，总权重为38。

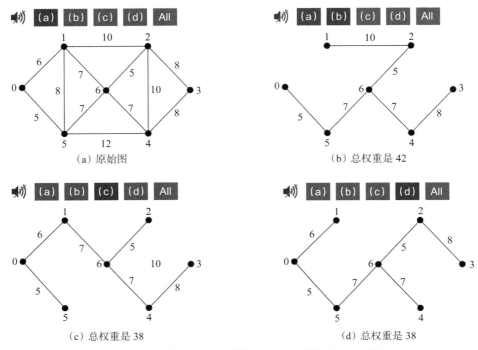

(a) 原始图
(b) 总权重是42
(c) 总权重是38
(d) 总权重是38

图27.4 （c）和（d）中的树是（a）中的图的最小生成树

求最小生成树问题有很多应用。以一家在许多城市都有分支机构的公司为例。该公司想租用电话线把所有分支机构连接起来。电话公司为连接不同的城市收取不同的费用。有很多方法可以将所有分支机构连接在一起。最便宜的方法是找到一个总费用最低的生成树。

27.4.1 最小生成树算法

如何找到最小生成树？有几种著名的算法可以做到。本节介绍 **Prim 算法**。Prim 算法从一个包含任意顶点的生成树 T 开始。该算法通过向树中添加与树中顶点相关连、且边的权重最小的顶点来扩展树。该算法如 Listing 27.1 所示。

Listing 27.1　Prim 最小生成树算法

```
输入：连通无向加权图 G=(V,E)
输出：MST( 最小生成树 )
1   MST minimumSpanningTree()
2   {
3     设 T 是生成树所有顶点集合 ;
4     对所有顶点 v 设置 cost[v]= 无穷大 ;
5     取任意顶点，比如 s，并设置 cost[s]=0 且 parent[s]=-1;
6
7     while(T 的大小 < n)
8     {
9       找到不在 T 中且具有最小 cost[u] 的 u;
10      将 u 加入 T;
11      for 每个不在 T 中的 v 和在 E 中的 (u,v)
12        if(cost[v] > w(u, v))
```

```
13        {
14            cost[v] = w(u, v); parent[v] = u;
15        }
16    }
17 }
```

该算法从选择任意顶点 s 开始,并将 0 赋值给 cost[s] (第 5 行),将无穷大赋值给所有其他顶点的 cost[v] (第 4 行)。然后,在 while 循环的第一次迭代中,它将 s 添加到 T 中 (第 7～16 行)。将 s 添加到 T 中后,调整与 s 相邻的每个 v 的 cost[v]。cost[v] 是 v 与 T 中的某个相邻顶点的所有边中的最小权重。如果 cost[v] 是无穷大,则表明 v 此时不与 T 中任何顶点相邻。在 while 循环的后续每次迭代中,具有最小 cost[u] 的顶点 u 被选取并添加到 T 中 (第 9～10 行),如图 27.5a 所示。然后,对于 V-T 中每个 v,如果 cost[v]>w(u, v),且 v 与 u 相邻,则 cost[v] 更新为 w(u, v),parent[v] 更新为 u (第 13～14 行),如图 27.5b 所示。这里,w(u, v) 表示边 (u, v) 上的权重。

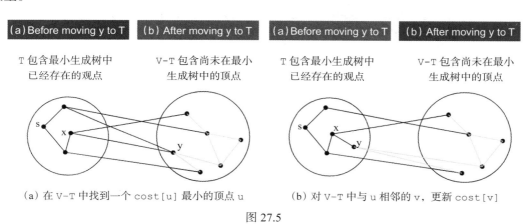

(a) 在 V-T 中找到一个 cost[u] 最小的顶点 u　　(b) 对 V-T 中与 u 相邻的 v,更新 cost[v]

图 27.5

图 27.6 演示了如何用该算法以交互方式找到最小生成树。

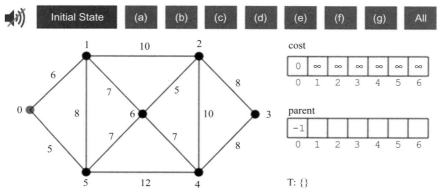

让我们选择 0 作为起始顶点。设定 cost[0] 为 0,对于所有其他顶点,cost[i] 为 ∞。parent[0] 设置为 -1,表示顶点 0 是根。最初,T 是空的。注意,算法会选择代价最小的顶点,并在下一步将其添加到 T 中。在这种情况下,顶点是 0,它将被加到 T 中。

初始状态

图 27.6　具有最小权重的相邻顶点依次添加到 T 中

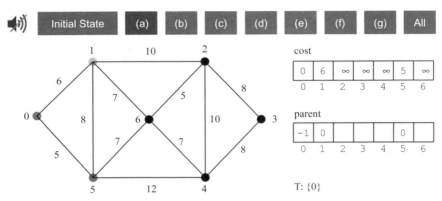

该算法将 0 加到 T,然后调整每个与 0 相邻的顶点的代价。顶点 1 和顶点 5 及它们父顶点的代价会被更新。下一步,算法将在 V-T 中选择具有最小代价的顶点,并将其添加到 T 中。该顶点将为 5。

(a)

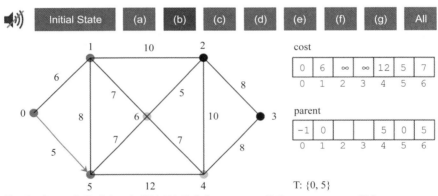

将 5 加到 T,更新每个与 5 相邻的顶点的代价。cost[6] 设为 7,parent[6] 设为 5。cost[4] 设为 12,parent[4] 设为 5。下一步,算法将在 V-T 中选择具有最小代价的顶点,并将其添加到 T 中。该顶点将为 1。

(b)

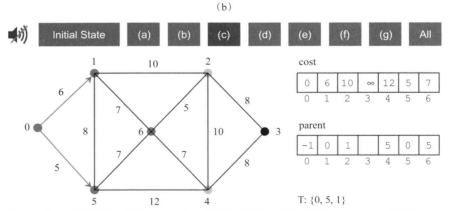

将 1 加到 T。如果合适,更新 V-T 中与 1 相邻的顶点的代价和父顶点。cost[2] 更新为 10,parent[2] 更新为 1。V-T 中的顶点 2、6、4 连接到 T 中的顶点。由于顶点 6 有最小的代价,在下一步将其添加到 T 中。

(c)

图 27.6 具有最小权重的相邻顶点依次添加到 T 中(续)

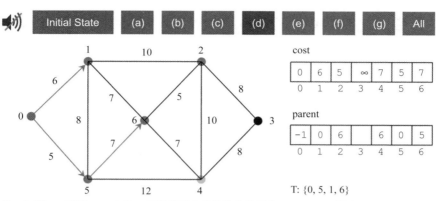

将6加到T。更新V-T中与6相邻的顶点的代价和父顶点。cost[2]设为5，parent[2]设为6。cost[4]设为7，parent[4]设为6。V-T中的顶点2、4连接到T中的顶点。由于cost[2]更小，在下一步将2添加到T中。

(d)

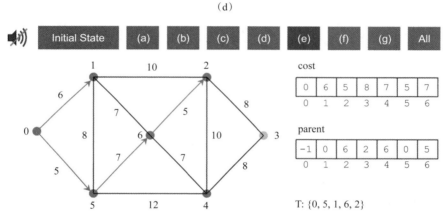

将2加到T。更新V-T中与2相邻的顶点的代价和父顶点。cost[3]设为8，parent[3]设为2。V-T中的顶点3、4连接到T中的顶点。由于cost[4]更小，在下一步将4添加到T中。

(e)

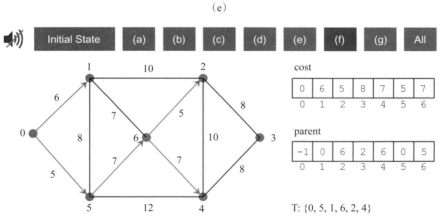

将4加到T。如果合适，更新V-T中与4相邻的顶点的代价和父顶点。在这种情况下，已经无更新。该算法在下一步将3添加到T中。

(f)

图27.6 具有最小权重的相邻顶点依次添加到T中（续）

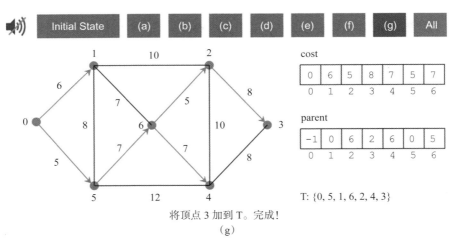

将顶点 3 加到 T。完成！
(g)

图 27.6 具有最小权重的相邻顶点依次添加到 T 中（续）

注意：最小生成树不是唯一的。例如，图 27.4c 和图 27.4d 都是图 27.6a 中图的最小生成树。但如果总权重不同，则图有唯一的最小生成树。

注意：这里假设图是连通的且无向的。如果一个图不是连通的或是有向的，那程序可能无法工作。你可以修改程序，为任意无向图找到一个生成树。

`getMinimumSpanningTree(int v)` 函数是在 `WeightedGraph` 类中定义的。它返回 `MST` 类的一个实例。`MST` 类被定义为 `Tree` 的一个子类，如图 27.7 所示。`Tree` 类在 LiveExample 26.4 中定义。LiveExample 27.4 实现了 `MST` 类。

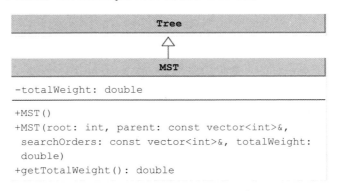

图 27.7 `MST` 类扩展了 `Tree` 类

LiveExample 27.4 的互动程序请访问 https://liangcpp.pearsoncmg.com/LiveRunCpp5e/faces/LiveExample.xhtml?header=off&programName=MST&programHeight=550&fileType=.h&resultVisible=false。

LiveExample 27.4 MST.h

Source Code Editor:

```
1  #ifndef MST_H
2  #define MST_H
3
4  #include "Tree.h" // Defined in LiveExample 26.4
5
6  class MST : public Tree
7  {
```

```cpp
 8    public:
 9      // Create an empty MST
10      MST()
11      {
12      }
13
14      // Construct a tree with root, parent, searchOrders,
15      // and total weight
16      MST(int root, const vector<int>& parent, const vector<int>& searchOrders,
17        double totalWeight) : Tree(root, parent, searchOrders)
18      {
19        this->totalWeight = totalWeight;
20      }
21
22      // Return the total weight of the tree
23      double getTotalWeight()
24      {
25        return totalWeight;
26      }
27
28    private:
29      double totalWeight;
30    };
31    #endif
```

getMinimumSpanningTree 函数在 LiveExample 27.2 的第 100～155 行中实现。get-MinimumSpanningTree() 函数选择顶点 0 作为起始顶点，并调用 getMinimum-SpanningTree(0) 函数返回从 0 开始的 MST（第 103 行）。

getMinimumSpanningTree(startingVertex) 函数将 cost[startingVertex] 设置为 0（第 121 行），并将所有其他顶点的 cost[v] 设置为无穷大（第 117～120 行）。然后，它反复找到 V-T 中代价最小的顶点（第 129～137 行），并将其添加到 T 中（第 139 行）。将顶点 u 添加到 T 后，可以调整 V-T 中每个与 u 相邻的顶点 v 的代价和父顶点（第 143～151 行）。

在新顶点被添加到 T 之后（第 139 行），totalWeight 被更新（第 140 行）。一旦将所有顶点添加到 T，MST 的实例就创建成功（第 154 行）。

MST 类扩展了 Tree 类。要创建 MST 的实例，需要传递 root、parent、searchOrders 和 totalWeight（第 154 行）。数据字段 root、parent 和 searchOrders 在 Tree 类中定义。

由于 T 是向量，通过调用 STL find 算法（第 131 行）测试顶点 k 是否在 T 中需要 $O(|V|)$ 时间，其中 $|V|$ 表示顶点的数量。通过引入一个数组（比如 isInT）来跟踪顶点 k 是否在 T 中，可以将测试时间减少到 $O(1)$。每当将顶点 k 添加到 T 时，将 isInT[k] 设置为 true。有关使用 isInT 实现算法的信息，参见编程练习 27.11。因此，找到要添加到 T 中的顶点的总时间是 $O(|V|^2)$。每次将顶点 u 添加到 T 中后，函数都会更新 V-T 中的每个与 u 相邻的顶点 v 的 cost[v] 和 parent[v]。如果 cost[v]>w(u, v)，则更新顶点 v 的代价和父顶点。每条和 u 相关联的边只检查一次。因此，检查边的总时间是 $O(|E|)$，其中 $|E|$ 表示边的数量。因此，该实现的时间复杂度是 $O(|V|^2+|E|) = O(|V|^2)$。使用优先级队列，我们可以在 $O(|E|\log|V|)$ 时间内实现算法（参见编程练习 27.9），这对稀疏图更高效。

LiveExample 27.5 给出了一个测试程序，该程序分别显示了图 27.1 和图 27.2 中图的最小生成树。

LiveExample 27.5 的互动程序请访问 https://liangcpp.pearsoncmg.com/LiveRunCpp5e/faces/LiveExample.xhtml?header=off&programName=TestMinimumSpanningTree&fileType=.cpp&programHeight=1550&resultHeight=310。

LiveExample 27.5   TestMinimumSpanningTree.cpp

Source Code Editor:

```cpp
#include <iostream>
#include <string>
#include <vector>
#include "WeightedGraph.h" // Defined in LiveExample 27.2
#include "WeightedEdge.h" // Defined in LiveExample 27.1
using namespace std;

// Print tree
template<typename T>
void printTree(const Tree& tree, const vector<T>& vertices)
{
  cout << "\nThe root is " << vertices[tree.getRoot()];
  cout << "\nThe edges are:";
  for (unsigned i = 0; i < vertices.size(); i++)
  {
    if (tree.getParent(i) != -1)
      cout << " (" << vertices[i] << ", "
        << vertices[tree.getParent(i)] << ")";
  }
  cout << endl;
}

int main()
{
  // Vertices for graph in Figure 27.1
  string vertices[] = {"Seattle", "San Francisco", "Los Angeles",
    "Denver", "Kansas City", "Chicago", "Boston", "New York",
    "Atlanta", "Miami", "Dallas", "Houston"};

  // Edge array for graph in Figure 27.1
  int edges[][3] = {
    {0, 1, 807}, {0, 3, 1331}, {0, 5, 2097},
    {1, 0, 807}, {1, 2, 381}, {1, 3, 1267},
    {2, 1, 381}, {2, 3, 1015}, {2, 4, 1663}, {2, 10, 1435},
    {3, 0, 1331}, {3, 1, 1267}, {3, 2, 1015}, {3, 4, 599},
    {3, 5, 1003},
    {4, 2, 1663}, {4, 3, 599}, {4, 5, 533}, {4, 7, 1260},
    {4, 8, 864}, {4, 10, 496},
    {5, 0, 2097}, {5, 3, 1003}, {5, 4, 533},
    {5, 6, 983}, {5, 7, 787},
    {6, 5, 983}, {6, 7, 214},
    {7, 4, 1260}, {7, 5, 787}, {7, 6, 214}, {7, 8, 888},
    {8, 4, 864}, {8, 7, 888}, {8, 9, 661},
    {8, 10, 781}, {8, 11, 810},
    {9, 8, 661}, {9, 11, 1187},
    {10, 2, 1435}, {10, 4, 496}, {10, 8, 781}, {10, 11, 239},
    {11, 8, 810}, {11, 9, 1187}, {11, 10, 239}
  };

  // 23 undirected edges in Figure 27.1
  const int NUMBER_OF_EDGES = 46;

  // Create a vector for vertices
  vector<string> vectorOfVertices(12);
```

```cpp
55      copy(vertices, vertices + 12, vectorOfVertices.begin());
56
57      // Create a vector for edges
58      vector<WeightedEdge> edgeVector;
59      for (int i = 0; i < NUMBER_OF_EDGES; i++)
60        edgeVector.push_back(WeightedEdge(edges[i][0],
61          edges[i][1], edges[i][2]));
62
63      WeightedGraph<string> graph1(vectorOfVertices, edgeVector);
64      MST tree1 = graph1.getMinimumSpanningTree();
65      cout << "The spanning tree weight is " << tree1.getTotalWeight();
66      printTree<string>(tree1, graph1.getVertices());
67
68      // Create a graph for Figure 27.2
69      int edges2[][3] =
70      {
71        {0, 1, 2}, {0, 3, 8},
72        {1, 0, 2}, {1, 2, 7}, {1, 3, 3},
73        {2, 1, 7}, {2, 3, 4}, {2, 4, 5},
74        {3, 0, 8}, {3, 1, 3}, {3, 2, 4}, {3, 4, 6},
75        {4, 2, 5}, {4, 3, 6}
76      }; // 14 edges in Figure 27.2
77
78      // 7 undirected edges in Figure 27.2
79      const int NUMBER_OF_EDGES2 = 14;
80
81      vector<WeightedEdge> edgeVector2;
82      for (int i = 0; i < NUMBER_OF_EDGES2; i++)
83        edgeVector2.push_back(WeightedEdge(edges2[i][0],
84          edges2[i][1], edges2[i][2]));
85
86      WeightedGraph<int> graph2(5, edgeVector2); // 5 vertices in graph2
87      MST tree2 = graph2.getMinimumSpanningTree();
88      cout << "\nThe spanning tree weight is " << tree2.getTotalWeight();
89      printTree<int>(tree2, graph2.getVertices());
90
91      return 0;
92    }
```

Execution Result:

```
command>cl TestMinimumSpanningTree.cpp
Microsoft C++ Compiler 2019
Compiled successful (cl is the VC++ compile/link command)

command>TestMinimumSpanningTree
The spanning tree weight is 6513
The root is Seattle
The edges are: (San Francisco, Seattle) (Los Angeles, San Francisco)
(Denver, Los Angeles) (Kansas City, Denver) (Chicago, Kansas City)
(Boston, New York) (New York, Chicago) (Atlanta, Dallas)
(Miami, Atlanta) (Dallas, Kansas City) (Houston, Dallas)

The spanning tree weight is 14
The root is 0
The edges are: (1, 0) (2, 3) (3, 1) (4, 2)

command>
```

程序在第 63 行为图 27.1 创建一个加权图。然后，它调用 getMinimumSpanningTree() （第 64 行）返回一个 MST，该 MST 表示图的最小生成树。对 MST 对象调用 getTotalWeight()

返回最小生成树的总权重（第 65 行）。MST 对象上的 `printTree()` 函数（第 9～21 行）显示树中的边。注意，MST 是 Tree 的一个子类。

最小生成树的图示如图 27.8 所示。顶点按以下顺序添加到树中：Seattle、San Francisco、Los Angeles、Denver、Kansas City、Dallas、Houston、Chicago、New York、Boston、Atlanta 和 Miami。

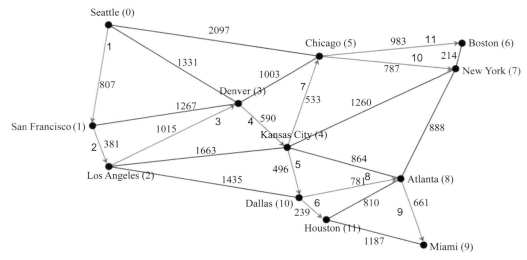

图 27.8 城市的最小生成树中的边被加粗显示

27.5 寻找最短路径

要点提示：两个顶点之间的最短路径是具有最小总权重的路径。

给定一个边上具有非负权重的图，荷兰计算机科学家 Edsger Dijkstra 发现了一种著名的寻找单源最短路径的算法。为了找到从顶点 s 到顶点 v 的最短路径，**Dijkstra 算法**找到从 s 到所有顶点的最短路径。因此 Dijkstra 算法被称为**单源最短路径**算法。该算法使用 `cost[v]` 存储从顶点 v 到源顶点 s 的**最短路径**的代价。`cost[s]` 为 0。初始时，所有其他顶点的 `cost[v]` 都设定为无穷大。让 V 表示图中的所有顶点，T 表示代价已知的顶点集。该算法反复找到 V-T 中具有最小 `cost[u]` 的顶点 u，并将 u 移动到 T。然后，调整 V-T 中每个满足与 u 相邻且 `cost[v]>cost[u]+w(u, v)` 的顶点 v 的代价和父节点。

该算法如 Listing 27.2 所示。

Listing 27.2　Dijkstra 单源最短路径算法

```
输入：具有非负权重和源顶点 s 的图 G=(V,E)
输出：以源顶点 s 为根的最短路径树
1    ShortestPathTree getShortestPath(s)
2    {
3
4        设 T 是包含所有到 s 的路径已知的顶点的集合；
5        设置所有节点 cost[v]= 无穷大；
6        设置 cost[s]=0 且 parent[s]=-1;
7
8        while(T 的大小 <n)
```

```
9    {
10       找到不在 T 中且 cost[u] 最小的 u;
11       将 u 加到 T;
12       for( 每个不在 T 中的 v 和在 E 中的 (u,v))
13           if(cost[v]>cost[u]+w(u,v))
14           {
15                    cost[v]=cost[u]+w(u,v);parent[v]=u;
16           }
17    }
18 }
```

该算法与 Prim 算法在寻找最小生成树方面非常相似。两种算法都将顶点划分为两个集合：T 和 V-T。在 Prim 算法中，集合 T 包含已经添加到树中的顶点。在 Dijkstra 算法中，集合 T 包含已找到到源的最短路径的顶点。这两种算法都反复地从 V-T 中找到一个顶点，并将其添加到 T 中。在 Prim 算法中，该顶点与集合中的某个顶点相邻且其边上具有最小权重。在 Dijkstra 算法中，该顶点与集合中的某个顶点相邻，并具有到源的最小总代价。

该算法首先将 cost[s] 设置为 0，将 parent[s] 设为 -1，将所有其他顶点的 cost[v] 设为无穷大（第 5～6 行）。我们用 parent[i] 表示路径中 i 的父顶点。方便起见，将源节点的父顶点设置为 -1。如图 27.9a 所示，该算法连续将具有最小 cost[u] 的顶点（比如 u）从 V-T 添加到 T 中（第 10～11 行）。将 u 添加到 T 后，对每个不在 T 中的 v，如果 (u, v) 在 T 中，并且 cost[v]>cost[u]+w(u, v)，则该算法更新 cost[v] 和 parent[v]（第 12～16 行）。

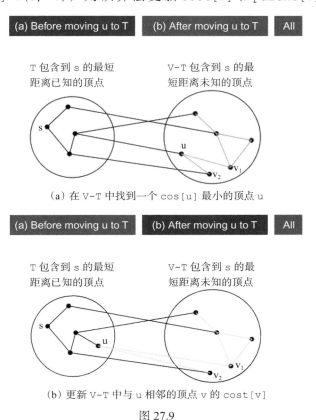

(a) 在 V-T 中找到一个 cos[u] 最小的顶点 u

(b) 更新 V-T 中与 u 相邻的顶点 v 的 cost[v]

图 27.9

图 27.10 演示了如何交互式地使用 Dijkstra 算法找到最短路径。

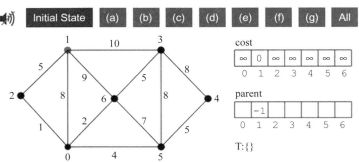

假设 1 是源顶点。设定 cost[1] 为 0，对于所有其他顶点，cost[i] 设置为 ∞。parent[1] 设置为 -1，表示顶点 1 是根。最初，T 是空的。注意，算法会选择代价最小的顶点，并在下一步将其添加到 T 中。在这种情况下，顶点是 1，它将被加到 T。

初始状态

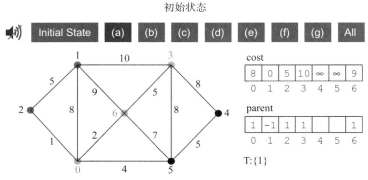

算法将 1 添加到 T。然后调整每个与 1 相邻的顶点的代价。更新顶点 2、0、6、3 的代价及它们的父顶点。在下一步，算法会选择 V-T 中代价最小的顶点，并将其添加到 T 中。该顶点为 2。

(a)

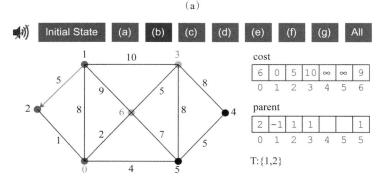

在 V-T 中的所有顶点中，顶点 2 具有最小的代价。将 2 添加到 T。更新与 2 相邻的顶点的代价及其父顶点。注意，代价是与源顶点的距离。cost[0] 更新为 6，它的父顶点设为 2。现在 T 包含 {1,2}。顶点 0 在 V-T 中具有最小的代价，因此在下一步，将其添加到 T 中。

(b)

图 27.10 该算法将找到从源顶点 1 开始的所有最短路径

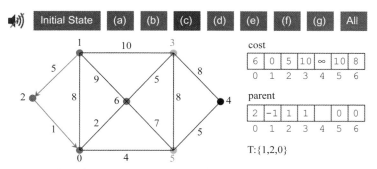

将 0 添加到 T。如果合适，更新 V-T 中与 0 相邻的顶点的代价及其父顶点。cost[5] 更新为 10，它的父顶点设为 0。cost[6] 更新为 8，它的父顶点设为 0。现在 T 包含 {1,2,0}。顶点 6 在 V-T 中具有最小的代价，因此在下一步，将其添加到 T 中。

(c)

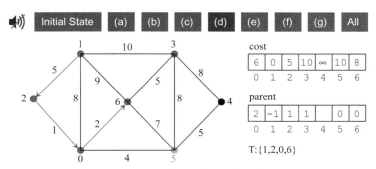

将 6 添加到 T。如果合适，更新 V-T 中与 6 相邻的顶点的代价及其父顶点。在这种情况下，无变化。现在 T 包含 {1,2,0,6}。顶点 3 或 5 在 V-T 中具有最小的代价，在下一步我们选择将 3 添加到 T 中。

(d)

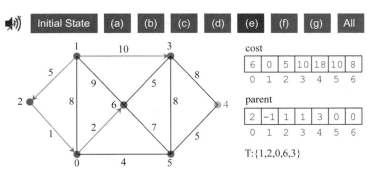

将 3 添加到 T。如果合适，更新 V-T 中与 3 相邻的顶点的代价及其父顶点。cost[4] 更新为 18，它的父顶点设为 3。现在 T 包含 {1,2,0,6,3}。顶点 5 在 V-T 中具有最小的代价，在下一步将其添加到 T 中。

(e)

图 27.10 该算法将找到从源顶点 1 开始的所有最短路径（续）

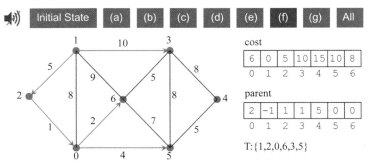

将 5 添加到 T。如果合适，更新 V-T 中与 5 相邻的顶点的代价及其父顶点。cost[4] 更新为 15，它的父顶点设为 5。现在 T 包含 {1,2,0,6,3,5}。顶点 4 在 V-T 中具有最小的代价，在下一步将其添加到 T 中。

(f)

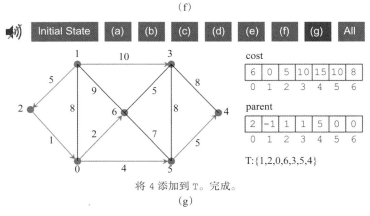

将 4 添加到 T。完成。

(g)

图 27.10 该算法将找到从源顶点 1 开始的所有最短路径（续）

正如所见，该算法本质上是找到从源顶点开始的所有最短路径，从而生成一个以源顶点为根的树。我们将此树称为单源最短路径树（或简称为最短路径树）。为了给此树建模，定义一个名为 ShortestPathTree 的类，该类扩展了 Tree 类，如图 27.11 所示。LiveExample 27.6 实现了 ShortestPathTree 类。

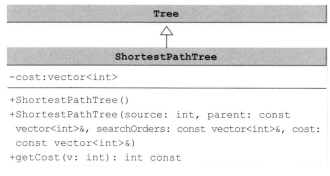

图 27.11 ShortestPathTree 扩展 Tree

LiveExample 27.6 的互动程序请访问 https://liangcpp.pearsoncmg.com/LiveRunCpp5e/faces/LiveExample.xhtml?header=off&programName=ShortestPathTree&fileType=.h&programHeight=560&resultVisible=false。

LiveExample 27.6 ShortestPathTree.h

Source Code Editor:

```cpp
#ifndef SHORTESTPATHTREE_H
#define SHORTESTPATHTREE_H

#include "Tree.h" // Defined in LiveExample 26.4

class ShortestPathTree : public Tree
{
public:
  // Create an empty ShortestPathTree
  ShortestPathTree()
  {
  }

  // Construct a tree with root, parent, searchOrders,
  // and cost
  ShortestPathTree(int root, const vector<int>& parent,
    const vector<int>& searchOrders, const vector<double>& cost)
    : Tree(root, parent, searchOrders)
  {
    this->cost = cost;
  }

  // Return the cost for the path from the source to vertex v.
  double getCost(int v) const
  {
    return cost[v];
  }

private:
  vector<double> cost;
};
#endif
```

Answer   Reset

在 LiveExample 27.2 的第 157 ~ 209 行中实现了 getShortestPath(int sourceVertex) 函数。该函数首先将 cost[sourceVertex] 设置为 0（第 173 行），并将所有其他顶点的 cost[v] 设置为无穷大（第 169 ~ 172 行）。

向量 parent 存储每个顶点的父顶点，以使我们能够识别从源到图中所有其他顶点的路径（第 164 行）。根的父顶点设置为 -1（第 165 行）。

T 是存储已经找到路径的顶点的集合（第 161 行）。函数通过执行以下操作扩充 T：

1. 找到 cost[u] 最小的顶点 u（第 179 ~ 189 行），并将其添加到 T 中（第 193 行）。如果没有找到这样的顶点，则图不连通，并且源顶点不能到达图中的所有顶点（第 191 行）。

2. 在 T 中添加 u 后，对 V-T 中每个与 u 相邻的 v，如果 cost[v]>cost[u]+w(u, v)，则更新 cost[v] 和 parent[v]（第 196 ~ 204 行）。

一旦把所有从 s 可到达的顶点添加到 T，一个 ShortestPathTree 的实例就创建成功了（第 208 行）。

ShortestPathTree 类扩展了 Tree 类。为了创建 ShortestPathTree 的实例，需要传递 sourceVertex、parent、searchOrders 和 cost（第 208 行）。sourceVertex 成为树的根。数据字段 root、parent 和 searchOrders 在 Tree 类中定义。

正如前面在 getMinimumSpanningTree 函数中讨论的那样，我们可以通过引入一

个数组（比如 isInT）来跟踪顶点 k 是否在 T 中，从而将测试顶点 k 是否位于 T 中的时间减少到 $O(1)$。每当将顶点 k 添加到 T 时，将 isInT[k] 设置为 true。用线性搜索找到 cost[u] 最小的 u 需要 $O(|V|)$ 时间（第 181 ~ 189 行），其中 |V| 表示顶点数。因此，找到要添加到 T 中的顶点的总时间是 $O(|V|^2)$。每次将顶点 u 添加到 T 中后，函数都会更新 V-T 中与 u 相邻的每个顶点 v 的 cost[v] 和 parent[v]。如果 cost[v]>w(u,v)，则更新顶点 v 的代价和其父顶点。每个与 u 关联的边只检查一次。因此，检查边的总时间是 $O(|E|)$，其中 |E| 表示边的数量。因此，该实现的时间复杂度是 $O(|V|^2+|E|)=O(|V|^2)$。用优先级队列，我们可以在 $O(|E|\log|V|)$ 时间内实现算法（参见编程练习 27.10），这对稀疏图更高效。

Dijkstra 算法是贪婪算法和动态规划的结合。它是一种贪婪算法，因为它总是添加一个到源距离最短的新顶点。它存储每个已知顶点到源的最短距离，并在以后用它来避免冗余计算，因此 Dijkstra 算法也使用动态规划。

LiveExample 27.7 给出了一个测试程序，该程序分别显示了图 27.1 中从 Chicago 到所有其他城市的最短路径和图 27.2 中从顶点 3 到所有顶点的最短路径。

LiveExample 27.7 的互动程序请访问 https://liangcpp.pearsoncmg.com/LiveRunCpp5e/faces/LiveExample.xhtml?header=off&programName=TestShortestPath&fileType=.cpp&programHeight=1610&resultHeight=490。

LiveExample 27.7　TestShortestPath.cpp

Source Code Editor:

```cpp
#include <iostream>
#include <string>
#include <vector>
#include "WeightedGraph.h" // Defined in LiveExample 27.2
#include "WeightedEdge.h" // Defined in LiveExample 27.1
using namespace std;

// Print paths from all vertices to the source
template<typename T>
void printAllPaths(const ShortestPathTree& tree, vector<T> vertices)
{
  cout << "All shortest paths from " <<
    vertices[tree.getRoot()] << " are:" << endl;
  for (unsigned i = 0; i < vertices.size(); i++)
  {
    cout << "To " << vertices[i] << ": ";
    
    // Print a path from i to the source
    vector<int> path = tree.getPath(i);
    for (int j = path.size() - 1; j >= 0; j--)
    {
      cout << vertices[path[j]] << " ";
    }
    
    cout << "(cost: " << tree.getCost(i) << ")" << endl;
  }
}

int main()
{
  // Vertices for graph in Figure 27.1
  string vertices[] = {"Seattle", "San Francisco", "Los Angeles",
    "Denver", "Kansas City", "Chicago", "Boston", "New York",
```

```cpp
       "Atlanta", "Miami", "Dallas", "Houston"};

  // Edge array for graph in Figure 27.1
  int edges[][3] = {
    {0, 1, 807}, {0, 3, 1331}, {0, 5, 2097},
    {1, 0, 807}, {1, 2, 381}, {1, 3, 1267},
    {2, 1, 381}, {2, 3, 1015}, {2, 4, 1663}, {2, 10, 1435},
    {3, 0, 1331}, {3, 1, 1267}, {3, 2, 1015}, {3, 4, 599},
      {3, 5, 1003},
    {4, 2, 1663}, {4, 3, 599}, {4, 5, 533}, {4, 7, 1260},
      {4, 8, 864}, {4, 10, 496},
    {5, 0, 2097}, {5, 3, 1003}, {5, 4, 533},
      {5, 6, 983}, {5, 7, 787},
    {6, 5, 983}, {6, 7, 214},
    {7, 4, 1260}, {7, 5, 787}, {7, 6, 214}, {7, 8, 888},
    {8, 4, 864}, {8, 7, 888}, {8, 9, 661},
      {8, 10, 781}, {8, 11, 810},
    {9, 8, 661}, {9, 11, 1187},
    {10, 2, 1435}, {10, 4, 496}, {10, 8, 781}, {10, 11, 239},
    {11, 8, 810}, {11, 9, 1187}, {11, 10, 239}
  };

  // 23 undirected edges in Figure 27.1
  const int NUMBER_OF_EDGES = 46;

  // Create a vector for vertices
  vector<string> vectorOfVertices(12);
  copy(vertices, vertices + 12, vectorOfVertices.begin());

  // Create a vector for edges
  vector<WeightedEdge> edgeVector;
  for (int i = 0; i < NUMBER_OF_EDGES; i++)
    edgeVector.push_back(WeightedEdge(edges[i][0],
    edges[i][1], edges[i][2]));

  WeightedGraph<string> graph1(vectorOfVertices, edgeVector);
  ShortestPathTree tree = graph1.getShortestPath(5);
  printAllPaths<string>(tree, graph1.getVertices());

  // Create a graph for Figure 27.1
  int edges2[][3] =
  {
    {0, 1, 2}, {0, 3, 8},
    {1, 0, 2}, {1, 2, 7}, {1, 3, 3},
    {2, 1, 7}, {2, 3, 4}, {2, 4, 5},
    {3, 0, 8}, {3, 1, 3}, {3, 2, 4}, {3, 4, 6},
    {4, 2, 5}, {4, 3, 6}
  }; // 7 undirected edges in Figure 27.2

  // 7 undirected edges in Figure 27.2
  const int NUMBER_OF_EDGES2 = 14;

  vector<WeightedEdge> edgeVector2;
  for (int i = 0; i < NUMBER_OF_EDGES2; i++)
    edgeVector2.push_back(WeightedEdge(edges2[i][0],
    edges2[i][1], edges2[i][2]));

  WeightedGraph<int> graph2(5, edgeVector2); // 5 vertices in graph2
  ShortestPathTree tree2 = graph2.getShortestPath(3);
  printAllPaths<int>(tree2, graph2.getVertices());

  return 0;
}
```

```
command>cl TestShortestPath.cpp
Microsoft C++ Compiler 2019
Compiled successful (cl is the VC++ compile/link command)

command>TestShortestPath
All shortest paths from Chicago are:
To Seattle: Chicago Seattle (cost: 2097)
To San Francisco: Chicago Denver San Francisco (cost: 2270)
To Los Angeles: Chicago Denver Los Angeles (cost: 2018)
To Denver: Chicago Denver (cost: 1003)
To Kansas City: Chicago Kansas City (cost: 533)
To Chicago: Chicago (cost: 0)
To Boston: Chicago Boston (cost: 983)
To New York: Chicago New York (cost: 787)
To Atlanta: Chicago Kansas City Atlanta (cost: 1397)
To Miami: Chicago Kansas City Atlanta Miami (cost: 2058)
To Dallas: Chicago Kansas City Dallas (cost: 1029)
To Houston: Chicago Kansas City Dallas Houston (cost: 1268)
All shortest paths from 3 are:
To 0: 3 1 0 (cost: 5)
To 1: 3 1 (cost: 3)
To 2: 3 2 (cost: 4)
To 3: 3 (cost: 0)
To 4: 3 4 (cost: 6)

command>
```

程序在第 69 行为图 27.1 创建一个加权图。然后调用 getShortestPath(5) 函数返回一个 ShortestPathTree 对象，该对象包含从顶点 5（即 Chicago）开始的所有最短路径（第 70 行）。printAllPaths 函数显示所有路径（第 71 行）。

从 Chicago 开始的所有最短路径的图示如图 27.12 所示。从 Chicago 到各个城市的最短路径按以下顺序发现：Kansas City、New York、Boston、Denver、Dallas、Houston、Atlanta、Los Angeles、Miami、Seattle 和 San Francisco。

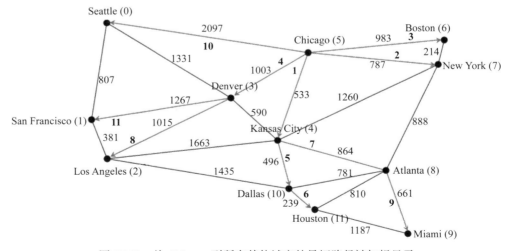

图 27.12　从 Chicago 到所有其他城市的最短路径被加粗显示

27.6 案例研究：加权九枚硬币翻转问题

要点提示：加权九枚硬币翻转问题可以归结为加权最短路径问题。

26.8 节介绍了九枚硬币翻转问题，并使用 BFS 算法进行了求解。本节介绍问题的一个变体，并使用最短路径算法进行求解。

九枚硬币翻转问题是找出使所有硬币正面朝下的最小翻转次数。每次移动都会翻转一枚硬币及其邻居。加权九枚硬币翻转问题将翻转次数指定为每次移动的权重。例如，通过翻转三枚硬币，可以从图 27.13a 中的硬币状态转移到图 27.13b。所以这次移动的权重是 3。通过翻转五枚硬币，可以从图 27.13c 中的硬币状态转移到图 27.13d。所以这次移动的权重是 5。

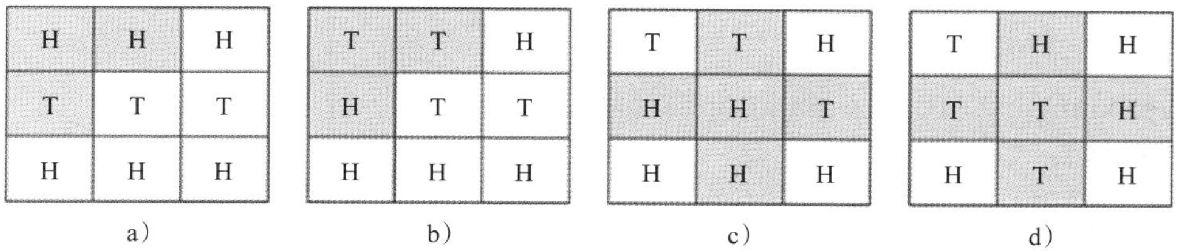

图 27.13 每次移动的权重是该移动的翻转次数

加权九枚硬币翻转问题是找到使所有硬币正面朝下的最小翻转次数。该问题可以简化为在边加权图中找到从起始节点到目标节点的最短路径。该图有 512 个节点。如果从节点 u 移动到节点 v，则创建从节点 v 到节点 u 的边。将翻转次数指定为边的权重。

回想一下，我们在 26.8 节中定义了一个类 NineTailModel，用于对九枚硬币翻转问题进行建模。我们现在定义一个名为 WeightedNineTailModel 的新类，它扩展了 NineTailModel，如图 27.14 所示。

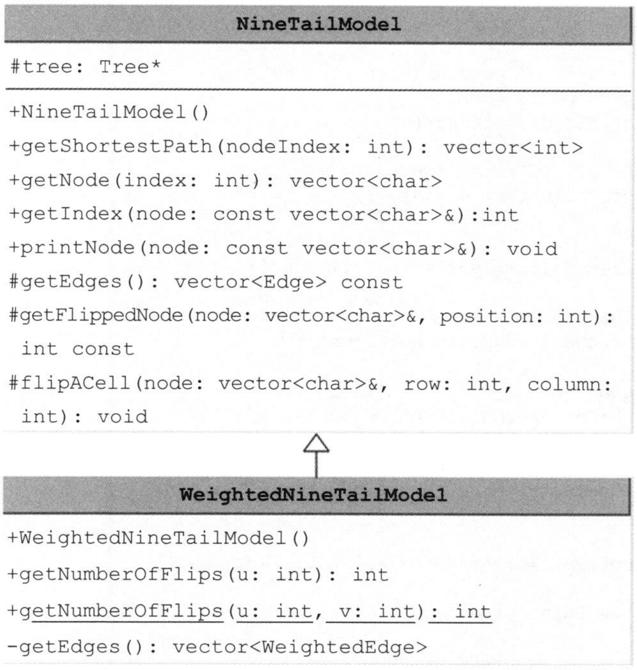

图 27.14 WeightedNineTailModel 扩展 NineTailModel

NineTailModel 类创建 Graph 并获得以目标节点 511 为根的 Tree。WeightedNineTailModel 与 NineTailModel 相同，不同之处在于它创建了 WeightedGraph 并获得了以目标节点 511 为根的 ShortestPathTree。NineTailModel 中定义的函数 getShortestPath(int)、getNode(int) 和 getIndex(node)、getFlippedNode(node,position) 和 flipACell(&node, row, column) 继承自 NineTailModel。getNumberOfFlips(int u) 函数返回从节点 u 到目标节点的翻转次数。

LiveExample 27.8 实现了 WeightedNineTailModel。

LiveExample 27.8 的互动程序请访问 https://liangcpp.pearsoncmg.com/LiveRunCpp5e/faces/LiveExample.xhtml?header=off&programName=WeightedNineTailModel&fileType=.h&programHeight=1320&resultVisible=false。

LiveExample 27.8　WeightedNineTailModel.h

Source Code Editor:

```
 1  #ifndef WEIGHTEDNINETAILMODEL_H
 2  #define WEIGHTEDNINETAILMODEL_H
 3
 4  #include "NineTailModel.h" // Defined in LiveExample 26.7
 5  #include "WeightedGraph.h" // Defined in LiveExample 27.2
 6
 7  using namespace std;
 8
 9  class WeightedNineTailModel : public NineTailModel
10  {
11  public:
12    // Construct a model for the Nine Tail problem
13    WeightedNineTailModel();
14
15    // Get the number of flips from the target to u
16    int getNumberOfFlips(int u);
17
18  private:
19    // Return a vector of Edge objects for the graph
20    // Create edges among nodes
21    vector<WeightedEdge> getEdges();
22
23    // Get the number of flips from u to v
24    int getNumberOfFlips(int u, int v);
25  };
26
27  WeightedNineTailModel::WeightedNineTailModel()
28  {
29    // Create edges
30    vector<WeightedEdge> edges = getEdges();
31
32    // Build a graph
33    WeightedGraph<int> graph(NUMBER_OF_NODES, edges);
34
35    // Build a shortest path tree rooted at the target node
36    tree = new ShortestPathTree(graph.getShortestPath(511));
37  }
38
39  vector<WeightedEdge> WeightedNineTailModel::getEdges()
40  {
41    vector<WeightedEdge> edges;
42
43    for (int u = 0; u < NUMBER_OF_NODES; u++)
44    {
```

```
45        for (int k = 0; k < 9; k++)
46        {
47          vector<char> node = getNode(u);
48          if (node[k] == 'H')
49          {
50            int v = getFlippedNode(node, k);
51            int numberOfFlips = getNumberOfFlips(u, v);
52            // Add edge (v, u) for a legal move from node u to node v
53            // with weight numberOfFlips
54            edges.push_back(WeightedEdge(v, u, numberOfFlips));
55          }
56        }
57      }
58
59      return edges;
60    }
61
62    int WeightedNineTailModel::getNumberOfFlips(int u, int v)
63    {
64      vector<char> node1 = getNode(u);
65      vector<char> node2 = getNode(v);
66
67      int count = 0; // Count the number of different cells
68      for (unsigned i = 0; i < node1.size(); i++)
69        if (node1[i] != node2[i]) count++;
70
71      return count;
72    }
73
74    int WeightedNineTailModel::getNumberOfFlips(int u)
75    {
76      return static_cast<ShortestPathTree*>(tree)->getCost(u);
77    }
78    #endif
```

Answer Reset

WeightedNineTailModel 扩展了 NineTailModel，以构建 WeightedGraph 对加权九枚硬币翻转问题进行建模（第 9 行）。对于每个节点 u，getEdges() 函数会找到一个已翻转的节点 v，并将翻转次数指定为边 (v,u) 的权重（第 50 行）。getNumberOfFlips(int u, int v) 函数返回从节点 u 到节点 v 的翻转次数（第 62～72 行）。翻转次数是两个节点的不同单元格的数量（第 69 行）。

WeightedNineTailModel 构造 WeightedGraph（第 33 行），并获得以目标节点 511 为根的 ShortestPathTree（第 36 行）。然后将其赋值给 tree*（第 36 行）。Tree* 类型的 tree 是在 NineTailModel 中定义的受保护的数据字段。NineTailModel 中定义的函数使用 tree 属性。注意，ShortestPathTree 是 Tree 的一个子类。

getNumberOfFlips(int u) 函数（第 74～77 行）返回从节点 u 到目标节点的翻转次数，这是从节点 u 到目标节点路径的代价。该代价可以通过调用在 ShortestPathTree 类中定义的 getCost(u) 函数获得（第 76 行）。

LiveExample 27.9 给出了一个程序，提示用户输入初始节点，并显示到达目标节点的最小翻转次数。

LiveExample 27.9 的互动程序请访问 https://liangcpp.pearsoncmg.com/LiveRunCpp5e/faces/LiveExample.xhtml?header=off&programName=WeightedNineTail&fileType=.cpp&programHeight=480&resultHeight=410。

LiveExample 27.9   WeightedNineTail.cpp

Source Code Editor:

```cpp
#include <iostream>
#include <vector>
#include "WeightedNineTailModel.h"
using namespace std;

int main()
{
  // Prompt the user to enter nine coins H and T's
  cout << "Enter an initial nine coin H and T's: ";
  vector<char> initialNode(9);

  for (int i = 0; i < 9; i++)
    cin >> initialNode[i];

  cout << "The steps to flip the coins are " << endl;
    WeightedNineTailModel model; // Create a weighted Nine Tail model
  vector<int> path =
    model.getShortestPath(model.getIndex(initialNode));

  for (unsigned i = 0; i < path.size(); i++)
    model.printNode(model.getNode(path[i]));

  cout << "The number of flips is " <<
    model.getNumberOfFlips(model.getIndex(initialNode)) << endl;

  return 0;
}
```

Enter input data for the program (Sample data provided below. You may modify it.)

```
HHHTTTHHH
```

[Automatic Check] [Compile/Run] [Reset] [Answer]   Choose a Compiler: VC++

Execution Result:

```
command>cl WeightedNineTail.cpp
Microsoft C++ Compiler 2019
Compiled successful (cl is the VC++ compile/link command)

command>WeightedNineTail
Enter an initial nine coin H and HHHTTTHHH
The steps to flip the coins are
HHH
TTT
HHH

HHH
THT
TTT

TTT
TTT
TTT

The number of flips is 8

command>
```

该程序在第 9 ～ 13 行提示用户输入一个包含 9 个 H 和 T 字母的初始节点，创建一个模型（第 16 行），获得从初始节点到目标节点的最短路径（第 17 ～ 18 行），显示路径中的节点（第 20 ～ 21 行），并调用 getNumberOfFlips 获取到达目标节点所需的翻转次数（第 23 ～ 24 行）。

关键术语

Dijkstra's algorithm（Dijkstra 算法）
edge-weighted graph（边加权图）
minimum spanning tree（最小生成树）
Prim's algorithm（Prim 算法）

shortest path（最短路径）
single-source shortest path（单源最短路径）
vertex-weighted graph（顶点加权图）

章节总结

1. 可以用邻接矩阵或邻接列表存储图中的加权边。
2. 图的生成树是一个子图，是连接图中所有顶点的一棵树。
3. 寻找最小生成树的 Prim 算法如下：该算法从一个包含任意顶点的生成树 T 开始，通过将最小权重边的顶点添加到树中来扩展树，其中最小权重边与树中已有顶点相关联。
4. Dijkstra 算法从源顶点开始搜索，并不断寻找到源的路径最短的顶点，直到找到所有顶点。

编程练习

*27.1 （Kruskal 算法）本章介绍了求最小生成树的 Prim 算法。Kruskal 算法是另一种众所周知的寻找最小生成树的算法。该算法反复找到最小权重边，如果不形成环路，则将其添加到树中。当所有顶点都在树中时，该过程结束。用 Kruskal 算法设计并实现一种查找 MST 的算法。

*27.2 （用邻接矩阵实现 Prim 算法）本书使用相邻边的向量实现 Prim 算法。请用加权图的邻接矩阵实现算法。

*27.3 （用邻接矩阵实现 Dijkstra 算法）本书使用相邻边的向量实现 Dijkstra 算法。请用加权图的邻接矩阵实现算法。

*27.4 （修改九枚硬币翻转问题中的权重）在前文中，我们将翻转次数指定为每次移动的权重。修改程序，假定权重是翻转次数的三倍。

*27.5 （证明或证伪）猜想 NineTailModel 和 WeightedNineTailModel 可能会产生相同的最短路径。写一个程序来证明或证伪它。
（提示：在 Tree 类中添加一个名为 depth(int v) 的新函数，以返回树中 v 的深度。设 tree1 和 tree2 分别表示从 NineTailModel 和 WeightedNineTailModel 获得的树。如果节点 u 的深度在 tree1 和 tree2 中相同，则从 u 到目标的路径长度相同。）

*27.6 （旅行推销员问题）旅行推销员问题（TSP）是找到一条最短的往返路线，该路线只访问每个城市一次，然后返回起始城市。这个问题等价于找到一个最短的哈密顿环路。在 WeightedGraph 类中添加以下函数：

```
// Return a shortest cycle
vector<int> getShortestHamiltonianCycle()
```

*27.7 （寻找最小生成树）编写一个程序，从文件中读取连通图并显示其最小生成树。文件中的第一行包含一个数字，表示顶点的数量（n）。顶点标记为 0, 1, …, n-1。后续每一行指定边格式为 u1, v1, w1|u2, v2, w2|…。每个三元组描述一条边及其权重。图 27.15 显示相应图的文件示例。注意，我们假设图是无向的。如果图有边 (u, v)，它也有边 (v, u)。文件中只表示一

条边。构造图时，需要添加两条边。

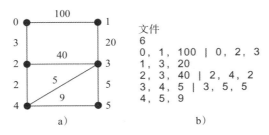

图 27.15 加权图的顶点和边可以存储在文件中

程序提示用户输入文件的名称，从文件中读取数据，创建 `WeightedGraph` 的实例 g，调用 `g.printWeightedEdges()` 显示所有边，调用 `getMinimumSpanningTree()` 获得 MST 的实例 tree，调用 `tree.getTotalWeight()` 显示最小生成树的权重，调用 `tree.printTree()` 显示树。

```
Sample Run for Exercise27_07.cpp
Enter input data for the program (Sample data provided below. You may modify it.)
WeightedGraphSample.txt

Show the Sample Output Using the Preceeding Input    Reset
Execution Result:
command>Exercise27_07
Enter a file name: WeightedGraphSample.txt
The number of vertices is 6
Vertex 0: (0, 2, 3) (0, 1, 100)
Vertex 1: (1, 3, 20) (1, 0, 100)
Vertex 2: (2, 4, 2) (2, 3, 40) (2, 0, 3)
Vertex 3: (3, 4, 5) (3, 5, 5) (3, 1, 20) (3, 2, 40)
Vertex 4: (4, 2, 2) (4, 3, 5) (4, 5, 9)
Vertex 5: (5, 3, 5) (5, 4, 9)
Total weight is 35
Root is: 0
Edges: (3, 1) (0, 2) (4, 3) (2, 4) (3, 5)

command>
```

*27.8 （为图创建一个文件）修改 LiveExample 27.3，创建一个表示 graph1 的文件。编程练习 27.7 中介绍了文件格式。根据 LiveExample 27.3 中第 16～33 行定义的图去创建文件。图的顶点数为 12，将它存储在文件的第一行。如果 u<v，则存储边 (u, v)。文件的内容如下所示：

```
12
0, 1, 807  | 0, 3, 1331 | 0, 5, 2097
1, 2, 381  | 1, 3, 1267
2, 3, 1015 | 2, 4, 1663 | 2, 10, 1435
3, 4, 599  | 3, 5, 1003
4, 5, 533  | 4, 7, 1260 | 4, 8, 864 | 4, 10, 496
5, 6, 983  | 5, 7, 787
6, 7, 214
7, 8, 888
8, 9, 661  | 8, 10, 781 | 8, 11, 810
9, 11, 1187
10, 11, 239
```

***27.9 （Prim 算法的可替代版本）Prim 算法的可替代版本描述如下：

输入：连通无向加权图 G=(V,E)

输出：MST（最小生成树）
```
1  MST minimumSpanningTree()
2  {
3      设T是生成树中所有顶点的集合;
4      初始，将起始顶点加入T;
5
6      while(T 的大小 <n)
7      {
8
9          找到具有最小权重的边(u,v), u在T中且v在V-T中, 如图 27.16 所示;
10         将v加入T并设置parent[v]=u;
11     }
12 }
```

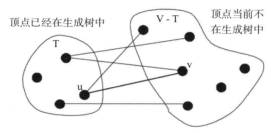

图 27.16　在 T 中找到一个连接 V-T 中顶点 v 的顶点 u, 其边具有最小权重

该算法首先将起始顶点添加到 T 中。然后将 V-T 中的顶点（比如 v）连续添加到 T。v 是与 T 中某个顶点相邻且边权重最小的顶点。例如，五条边连接 T 和 V-T 中的顶点, 如图 27.16 所示, (u, v) 是权重最小的一条。该算法可以用优先级队列来实现, 以获得 $O(|E|\log|V|)$ 的时间复杂度。用 LiveExample 27.5 测试新算法。

***27.10 (Dijkstra 算法的可替代版本) Dijkstra 算法的可替代版本描述如下:

输入：具有非负权重的加权图 G=(V,E)
输出：来自源顶点 s 的最短路径树
```
1  ShortestPathTree getShortestPath(s)
2  {
3
4      设T是包含所有到s的路径已知的顶点的集合;
5      初始, 令T包含源顶点s, cost[s]=0;
6      对每个V-T中的u, 设置cost[u]=无穷大;
7      while(T 的大小 <n)
8      {
9
10         找到V-T中的v, 对于T中的所有u, 具有最小的cost[u]+w(u,v) 值;
11         将v加到T中, 并设置cost[v]=cost[u]+w(u,v);
12           parent[v]=u;
13     }
14 }
```

该算法用 cost[v] 存储从顶点 v 到源顶点 s 的最短路径的代价。cost[s] 为 0。初始时, 给 cost[v] 赋值无穷大, 表示从 v 到 s 没有找到路径。设 V 表示图中的所有顶点, T 表示其代价已知的顶点集。初始时, 源顶点 s 在 T 中。该算法反复地在 T 中找到一个顶点 u, 在 V-T 中找到顶点 v, 使得 cost[u]+w(u, v) 最小, 并将 v 移动到 T。本书给出的最短路

径算法不断更新 V-T 中顶点的代价和父顶点。该算法将每个顶点的代价初始化为无穷大，然后仅在将顶点添加到 T 中时更改一次顶点的代价。该算法可以用优先级队列来实现，以获得 $O(|E|\log|V|)$ 的时间复杂度。使用 LiveExample 27.7 测试新算法。

***27.11 （检测 T 中的顶点是否有效）由于 T 是用 LiveExample 27.2 中 getMinimumSpanningTree 和 getShortestPath 函数中的列表实现的，因此通过调用 STL find 算法检测顶点 u 是否在 T 中需要 $O(n)$ 时间。通过引入一个名为 isInT 的数组来修改这两个函数。当将顶点 u 添加到 T 时，将 isInT[u] 设置为 true。现在可以在 $O(1)$ 时间内检测顶点 u 是否在 T 中。使用以下代码编写测试程序，其中 graph1、printTree 和 printAllPath 与 LiveExample 27.5 和 LiveExample 27.7 中的相同：

```
WeightedGraph<string> graph1(vectorOfVertices, edgeVector);
MST tree1 = graph1.getMinimumSpanningTree();
cout << "The spanning tree weight is " << tree1.getTotalWeight();
printTree<string>(tree1, graph1.getVertices());
ShortestPathTree tree = graph1.getShortestPath(5);
printAllPaths<string>(tree, graph1.getVertices());
```

*27.12 （查找最短路径）编写一个程序，从文件中读取连通图。使用编程练习 27.7 中指定的相同格式将图存储在文件中。程序提示用户输入文件名，然后输入两个顶点，程序显示两个顶点之间的最短路径。例如，对于图 27.15 中的图，0 和 1 之间的最短路径显示为 0 2 4 3 1。

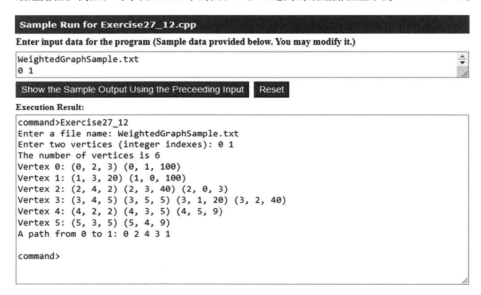

附录 A

C++ 关键字

以下关键字是为供 C++ 语言使用而保留的。它们不应用于 C++ 中预定义用途之外的任何其他用途。

asm	double	int	return	typeid
auto	dynamic_cast	log	short	typename
bool	else	long	signed	union
break	enum	mutable	sizeof	unsigned
case	explicit	namespace	static	using
catch	extern	new	static_cast	virtual
char	false	nullptr	struct	void
class	final	operator	switch	volatile
const	float	override	template	wchar_t
const_cast	for	private	this	while
continue	friend	protected	throw	
default	goto	public	true	
delete	if	register	try	
do	inline	reinterpret_cast	typedef	

注意，以下 11 个 C++ 关键字不是必需的。并不是所有的 C++ 编译器都支持它们。但它们为一些 C++ 运算符提供了更可读的替代方案。

关键字	等价运算符
and	&&
and_eq	&=
bitand	&
bitor	|
compl	~
not	!
not_eq	!=
or	||
or_eq	||=
xor	^
xor_eq	^=

附录 B
Introduction to C++ Programming and Data Structures, Fifth Edition

ASCII 字符集

表 B.1 和表 B.2 显示 ASCII 字符及其各自的十进制和十六进制编码。字符的十进制或十六进制编码是其行索引和列索引的组合。例如，在表 B.1 中，字母 A 位于第 6 行和第 5 列，因此其十进制等效值为 65；在表 B.2 中，字母 A 位于第 4 行和第 1 列，因此其十六进制等效值为 41。

表 B.1　以十进制索引表示的 ASCII 字符集

	0	1	2	3	4	5	6	7	8	9
0	nul	soh	stx	etx	eot	enq	ack	bel	bs	ht
1	nl	vt	ff	cr	so	si	dle	dc1	dc2	dc3
2	dc4	nak	syn	etb	can	em	sub	esc	fs	gs
3	rs	us	sp	!	"	#	$	%	&	'
4	(	)	*	+	,	-	.	/	0	1
5	2	3	4	5	6	7	8	9	:	;
6	<	=	>	?	@	A	B	C	D	E
7	F	G	H	I	J	K	L	M	N	O
8	P	Q	R	S	T	U	V	W	X	Y
9	Z	[	\	]	^	_	`	a	b	c
10	d	e	f	g	h	i	j	k	l	m
11	n	o	p	q	r	s	t	u	v	w
12	x	y	z	{	|	}	~	del		

表 B.2　以十六进制索引表示的 ASCII 字符集

	0	1	2	3	4	5	6	7	8	9	A	B	C	D	E	F
0	nul	soh	stx	etx	eot	enq	ack	bel	bs	ht	nl	vt	ff	cr	so	si
1	dle	dc1	dc2	dc3	dc4	nak	syn	etb	can	em	sub	esc	fs	gs	rs	us
2	sp	!	"	#	$	%	&	'	(	)	*	+	,	-	.	/
3	0	1	2	3	4	5	6	7	8	9	:	;	<	=	>	?
4	@	A	B	C	D	E	F	G	H	I	J	K	L	M	N	O
5	P	Q	R	S	T	U	V	W	X	Y	Z	[	\	]	^	_
6	`	a	b	c	d	e	f	g	h	i	j	k	l	m	n	o
7	P	q	r	s	t	u	v	w	x	y	z	{	|	}	~	del

附录 C

运算符优先级表

运算符按优先级从上到下递减的顺序显示。同一组中的运算符具有相同的优先级，其结合性如下表所示。

运算符	类型	结合性
:: ::	二元作用域解析 一元作用域解析	从左到右
. -> () [] ++ -- typeid dynamic_cast static_cast reinterprete_cast	通过对象访问对象成员 通过指针访问对象成员 函数调用 数组下标 后置递增 后置递减 运行时类型信息 动态转换（运行时） 静态转换（编译时） 非标准转换的强制转换	从左到右
++ -- + - ! ~ sizeof & * new new[] delete delete[]	前置递增 前置递减 一元加 一元减 一元逻辑否定 按位否定 类型的大小 变量的地址 变量的指针 动态内存分配 动态数组分配 动态内存解除分配 动态数组解除分配	从右到左
(type)	C 风格转换	从右到左
* / %	乘 除 取余	从左到右
+ -	加 减	从左到右
<< >>	输出或按位左移 输入或按位右移	从右到左
< <= > >=	小于 小于或等于 大于 大于或等于	从左到右

（续）

运算符	类型	结合性
== !=	等于 不等	从左到右
&	按位与	从左到右
^	按位异或	从左到右
|	按位兼或	从左到右
&&	布尔与	从左到右
||	布尔或	从左到右
?:	三元运算符	从右到左
= += -= *= /= %= &= ^= |= <<= >>=	赋值 加赋值 减赋值 乘赋值 除赋值 取余赋值 按位与赋值 按位异或赋值 按位兼或赋值 按位左移赋值 按位右移赋值	从右到左

附录 D

Introduction to C++ Programming and Data Structures, Fifth Edition

数字系统

D.1 简介

计算机内部使用二进制数,因为计算机天生就可以存储和处理 0 和 1。二进制数字系统有两个数字,0 和 1。数字或字符存储为 0 和 1 的序列。每个 0 或 1 被称为一个位(二进制数字)。

在日常生活中,我们使用十进制数。当我们在程序中写入一个数字(如 20)时,假定它是一个十进制数。计算机软件在内部将十进制数转换为二进制数,反之亦然。

我们用十进制数编写计算机程序。但为了与操作系统打交道,我们需要用二进制数来达到"机器级别"。二进制数往往很冗长。所以通常用十六进制数来缩写它们,每个十六进制数字代表四个二进制数字。十六进制数字系统有 16 个数字:0 ~ 9 和 A ~ F。字母 A、B、C、D、E 和 F 分别对应于十进制数字 10、11、12、13、14 和 15。

十进制的数字是 0、1、2、3、4、5、6、7、8 和 9。十进制数由一个或多个这些数字的序列表示。每个数字表示的值取决于其位置,位置代表 10 的整数幂。例如,十进制数 7423 中的数字 7、4、2 和 3 分别表示 7000、400、20 和 3,如下所示:

$$7423 = 7 \times 10^3 + 4 \times 10^2 + 2 \times 10^1 + 3 \times 10^0 = 7423$$

十进制有 10 个数字,位置值是 10 的整数幂。我们说 10 是十进制的基数。类似地,由于二进制数字系统有两个数字,所以其基数为 2,并且由于十六进制数字系统有 16 个数字,因此其基数为 16。

如果 1101 是二进制字,则数字 1、1、0 和 1 分别表示 1×2^3、1×2^2、0×2^1 和 1×2^0:

$$1101 = 1 \times 2^3 + 1 \times 2^2 + 0 \times 2^1 + 1 \times 2^0 = 13$$

如果 7423 是十六进制数,则数字 7、4、2 和 3 分别表示 7×16^3、4×16^2、2×16^1 和 3×16^0:

$$7423 = 7 \times 16^3 + 4 \times 16^2 + 2 \times 16^1 + 3 \times 16^0 = 29731$$

D.2 二进制数和十进制数之间的转换

给定一个二进制数 $b_n b_{n-1} b_{n-2} \cdots b_2 b_1 b_0$,等效的十进制值为

$$b_n \times 2^n + b_{n-1} \times 2^{n-1} + b_{n-2} \times 2^{n-2} + \cdots + b_2 \times 2^2 + b_1 \times 2^1 + b_0 \times 2^0$$

以下是将二进制数转换为十进制数的一些示例:

输入一个二进制数：1110   显示其十进制值

（二进制）$1110 = 1 \times 2^3 + 1 \times 2^2 + 1 \times 2^1 + 0 \times 2^0 = 14$（十进制）

将十进制数 d 转换为二进制数就是找到位 b_n，b_{n-1}，b_{n-2}，…，b_2，b_1 和 b_0，满足

$$d = b_n \times 2^n + b_{n-1} \times 2^{n-1} + b_{n-2} \times 2^{n-2} + \cdots + b_2 \times 2^2 + b_1 \times 2^1 + b_0 \times 2^0$$

这些位可以通过连续地将 d 除以 2 直到商为 0 来找到。余数就是 b_0，b_1，b_2，…，b_{n-2}，b_{n-1} 和 b_n。

例如，十进制数 123 是二进制数 1111011。转换过程如下：

输入一个十进制数：123   显示其二进制值

二进制值是 1111011（即 $1 \times 2^6 + 1 \times 2^5 + 1 \times 2^4 + 1 \times 2^3 + 0 \times 2^2 + 1 \times 2^1 + 1 \times 2^0$）

```
      0       1       3       7       15       30       61   ← 商
   2⟌1    2⟌3    2⟌7    2⟌15   2⟌30   2⟌61   2⟌123
      0       2       6       14       30       60       122
      ↓       ↓       ↓       ↓        ↓        ↓        ↓
      1       1       1       1        0        1        1   ← 余数
     b₆      b₅      b₄      b₃       b₂       b₁       b₀
```

提示：如图 D.1 所示，Windows 计算器是执行数字转换的有用工具。要运行它，从"开始"按钮搜索"计算器"并启动"计算器"，然后在"视图"下选择"科学"。

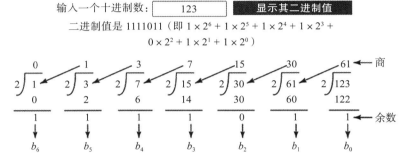

图 D.1 可以用 Windows 计算器进行进制转换

D.3 十六进制数和十进制数之间的转换

给定一个十六进制数 $h_n h_{n-1} h_{n-2} \cdots h_2 h_1 h_0$，等效的十进制值为

$$h_n \times 16^n + h_{n-1} \times 16^{n-1} + h_{n-2} \times 16^{n-2} + \cdots + h_2 \times 16^2 + h_1 \times 16^1 + h_0 \times 16^0$$

以下是将十六进制数转换为十进制数的一些示例：

输入一个十六进制数：　F23　　显示其十进制值

（十六进制）F23 = F × 16^2 + 2 × 16^1 + 3 × 16^0 = 3875（十进制）

将十进制数 d 转换为十六进制数就是找到十六进制数字 h_n, h_{n-1}, h_{n-2}, ⋯, h_2, h_1 和 h_0, 满足

$$d = h_n \times 16^n + h_{n-1} \times 16^{n-1} + h_{n-2} \times 16^{n-2} + \cdots + h_2 \times 16^2 + h_1 \times 16^1 + h_0 \times 16^0$$

这些数字可以通过连续地将 d 除以 16 直到商为 0 来求出。余数是 h_0, h_1, h_2, ⋯, h_{n-2}, h_{n-1} 和 h_n。

例如，十进制数 123 是十六进制数 7B。转换过程如下：

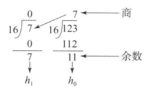

输入一个十进制数：　123　　显示其十六进制值

十六进制值是 7B（即 7 × 16^1 + B × 16^0）

D.4　二进制数和十六进制数之间的转换

要将十六进制数转换为二进制数，只需使用表 D.1 将十六进制数中的每个数字转换为四位数的二进制数。

例如，十六进制数 7B 是 1111011，其中 7 是二进制的 111，而 B 是二进制的 1011。

输入一个十六进制数：　7B　　显示其二进制值

（十六进制）7B = 01111011（二进制）

要将二进制数转换为十六进制数，将二进制数中从右到左的每四个数转换为一个十六进制数字。

例如，二进制数 1110001101 是 38D，因为 1101 是 D，1000 是 8，11 是 3，如下所示。

输入一个二进制数：　1110001101　　显示其十六进制值

（二进制）1110001101 = 38D（十六进制）

表 D.1　将十六进制数转换为二进制数

十六进制	二进制	十进制
0	0000	0
1	0001	1
2	0010	2
3	0011	3
4	0100	4
5	0101	5

(续)

十六进制	二进制	十进制
6	0110	6
7	0111	7
8	1000	8
9	1001	9
A	1010	10
B	1011	11
C	1100	12
D	1101	13
E	1110	14
F	1111	15

注意：八进制数也很有用。八进制有八位数，从 0 到 7。十进制数 8 在八进制中表示为 10。

附录 E

按位运算

要在机器级别编写程序，通常需要直接处理二进制数，并在位级执行运算。C++ 提供了表 E.1 中定义的按位运算符和移位运算符。

表 E.1

运算符	名字	示例（在示例中使用字节）	描述
&	按位与	10101110 & 10010010 的结果是 10000010	如果两个对应的位都是 1，则对其进行与运算得到 1
|	按位兼或	10101110 | 10010010 的结果是 10111110	如果任意一位为 1，则两个对应位的或运算结果为 1
^	按位异或	10101110 ^ 10010010 的结果是 00111100	两个对应位的异或只在两个位不同时结果为 1
~	某数的补	~10101110 的结果是 01010001	该运算符将每个位从 0 到 1 和从 1 到 0 进行切换
<<	左移	10101110 <<2 的结果是 10111000	该运算符将第一个操作数的位向左移动第二个操作数中指定的位数，并在右侧填充 0
>>	带符号扩展的右移	10101110 >> 2 的结果是 11101011 00101110 >> 2 的结果是 00001011	该运算符将第一个操作数的位向右移动第二个操作数中指定的位数，用左边最高的（符号）位填充

位运算符仅适用于整数类型。位运算中涉及的字符被转换为整数。所有二进制按位运算符都可以组成按位赋值运算符，如 &=、|=、^=、<<= 和 >>=。

使用按位运算符的程序比使用算术运算符更高效。例如，要将 int 值 x 乘以 2，可以写入 x<<1，而不是 x*2。

附录 F
Introduction to C++ Programming and Data Structures, Fifth Edition

使用命令行参数

我们可以从命令行向 C++ 程序传递参数。为此，要创建一个具有以下函数头的 main 函数：

```
int main(int argc, char* argv[])
```

其中 argv 指定参数，argc 指定参数的数量。例如，以下命令行使用三个字符串 arg1、arg2 和 arg3 启动程序 TestMain：

```
TestMain arg1 arg2 arg3
```

arg1、arg2 和 arg3 是字符串，并且传递给 argv。argv 是一个 C 字符串数组。在本例中，argc 为 4，因为传递了三个字符串参数，程序名 TestMain 也算作一个参数，该参数被传递给 argv[0]。

参数必须是字符串，但它们不必出现在命令行的双引号中。字符串用空格分隔。包含空格的字符串必须用双引号括起来。考虑以下命令行：

```
TestMain "First num" alpha 53
```

它用三个字符串启动程序："First num"、alpha，以及一个数字字符串 53。注意，53 实际被视为一个字符串。你可以在命令行中使用 "53" 替代 53。

LiveExample F.1 给出了一个对整数执行二元运算的程序。该程序接收三个参数：一个整数、后跟着的运算符和另一个整数。例如，要将两个整数相加，用以下命令：

```
Calculator 1 + 2
```

程序显示以下输出：

```
1 + 2 = 3
```

图 F.1 显示了该程序的示例运行。

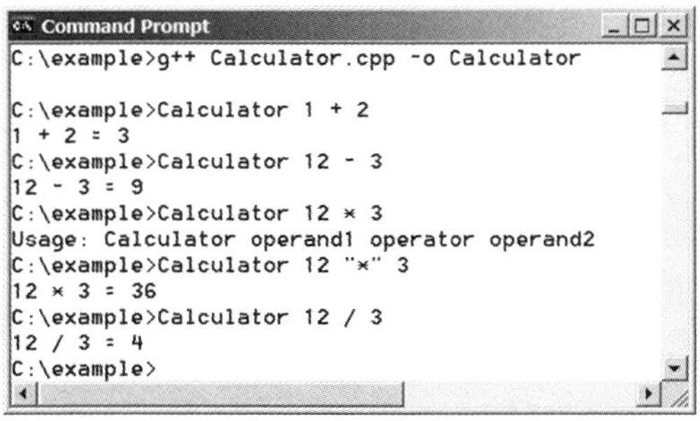

图 F.1 该程序从命令行获取三个参数（operand1、operator、operand2），并显示表达式和算术运算的结果

以下是程序中的步骤：

1. 检查 argc 以确定命令行中是否提供了三个参数。如果没有，用 exit(0) 终止程序。
2. 用 args[2] 中指定的运算符对操作数 args[1] 和 args[3] 执行算术运算。

LiveExample F.1 的互动程序请访问 https://liangcpp.pearsoncmg.com/LiveRunCpp5e/faces/LiveExample.xhtml?header=off&programName=Calculator&fileType=.cpp&programHeight=620&resultHeight=160。

LiveExample F.1 Calculator.cpp

Source Code Editor:

```cpp
#include <iostream>
using namespace std;

int main(int argc, char* argv[])
{
  // Check number of strings passed
  if (argc != 4)
  {
    cout << "Usage: Calculator operand1 operator operand2";
    exit(0);
  }

  // The result of the operation
  int result = 0;

  // Determine the operator
  switch (argv[2][0])
  {
    case '+':
      result = atoi(argv[1]) + atoi(argv[3]);
      break;
    case '-':
      result = atoi(argv[1]) - atoi(argv[3]);
      break;
    case '*':
      result = atoi(argv[1]) * atoi(argv[3]);
      break;
    case '/':
      result = atoi(argv[1]) / atoi(argv[3]);
  }

  // Display result
  cout << argv[1] << ' ' << argv[2] << ' ' << argv[3]
    << " = " << result;
}
```

Enter command arguements (Sample arguments provided below. You may modify it.)

```
99 + 728
```

Automatic Check | Compile/Run | Reset | Answer   Choose a Compiler: VC++

Execution Result:

```
command>cl Calculator.cpp
Microsoft C++ Compiler 2019
Compiled successful (cl is the VC++ compile/link command)

command>Calculator 99 + 728
```

```
99 + 728 = 827
command>
```

`atoi(argv[1])`（第 20 行）将数字字符串转换为整数。字符串必须由数字组成。否则，程序将异常终止。

在图 F.1 中的示例运行中，命令

```
Calculator 12 "*" 3
```

必须使用 "*" 而不是 *。在 C++ 中，当在命令行上使用 * 符号时，它指的是当前目录中的所有文件。因此，为了指定乘法运算符，* 必须在命令行中用引号括起来。下面 LiveExample F.2 中的程序在发出命令 `DisplayAllFiles*` 时显示当前目录中的所有文件。

LiveExample F.2 的互动程序请访问 https://liangcpp.pearsoncmg.com/LiveRunCpp5e/faces/LiveExample.xhtml?header=off&programName=DisplayAllFiles&fileType=.cpp&programHeight=190&resultVisible=false。

LiveExample F.2 DisplayAllFiles.cpp

Source Code Editor:
```cpp
#include <iostream>
using namespace std;

int main(int argc, char* argv[])
{
  for (int i = 0; i < argc; i++)
    cout << argv[i] << endl;

  return 0;
}
```

Answer   Reset

附录 G

枚举类型

G.1 简单枚举类型

C++ 提供了基本数据类型，如 `int`、`long long`、`float`、`double`、`char` 和 `bool`，以及对象类型，如 `string` 和 `vector`。也可以使用类定义自定义类型。此外，还可以定义枚举类型。本附录介绍枚举类型。

一个枚举类型定义一组枚举值。例如

```
enum Day {MONDAY, TUESDAY, WEDNESDAY, THURSDAY, FRIDAY};
```

声明了一个名为 Day 的枚举类型，其值依次为 MONDAY、TUESDAY、WEDNESDAY、THURSDAY 和 FRIDAY。每个值（称为枚举数）都是一个标识符，而不是字符串。枚举数一旦在类型中声明，整个程序就会知道它们。

惯例上，枚举类型以每个单词的大写首字母命名，枚举数命名规则跟常量一样，用所有大写字母命名。

定义了类型，就可以声明该类型的变量：

```
Day day;
```

变量 day 可以包含枚举类型中定义的某个值。例如，以下语句将枚举数 MONDAY 赋值给变量 day：

```
day = MONDAY;
```

与其他类型一样，我们可以在一条语句中声明和初始化变量：

```
Day day = MONDAY;
```

此外，C++ 支持在一条语句中声明枚举类型和变量。例如

```
enum Day {MONDAY, TUESDAY, WEDNESDAY, THURSDAY, FRIDAY} day = MONDAY;
```

警告：枚举数不能被重复声明。例如，以下代码会导致语法错误。

```
enum Day {MONDAY, TUESDAY, WEDNESDAY, THURSDAY, FRIDAY};
const int MONDAY = 0; // Error: MONDAY already declared.
```

G.2 整数和枚举数之间的对应关系

枚举数以整数形式存储在内存中。默认情况下，这些值按照它们在列表中出现的顺序对

应于 0,1,2,…。因此，MONDAY、TUESDAY、WEDNESDAY、THURSDAY 和 FRIDAY 对应于整数值 0,1,2,3 和 4。可以直接用任意整数指定枚举数。例如

```
enum Color {RED = 20, GREEN = 30, BLUE = 40};
```

RED 为整数值 20，GREEN 为 30，BLUE 为 40。

如果在枚举类型声明中为某些值指定整数值，那么其他枚举数将被默认赋值。例如

```
enum City {PARIS, LONDON, DALLAS = 30, HOUSTON};
```

PARIS 被指定为 0，LONDON 为 1，DALLAS 为 30，HOUSTON 为 31。

可以将枚举值赋值给整数变量。例如

```
int i = PARIS;
```

会将 0 赋值给 i。

```
int j = DALLAS;
```

会将 30 赋值给 j。

也可以将枚举数赋值给整数变量。那么可以给如下枚举类型的变量赋值一个整数吗？

```
City city = 1; // Not allowed
```

这是不允许的。但是，以下语法是奏效的：

```
City city = static_cast<City>(1);
```

这相当于

```
City city = LONDON;
```

现在，如果用以下语句显示 city：

```
cout << "city is " << city;
```

输出将是

```
city is 1
```

G.3 使用带有枚举变量的 **if** 或 **switch** 语句

可以用六个比较运算符对枚举数赋值的整数值进行比较。例如，(PARIS<LONDON) 的结果为 true。

枚举变量包含一个值。通常，程序需要根据值执行特定的操作。例如，如果值为 MONDAY，则踢足球；如果值是 TUESDAY，则上钢琴课，以此类推。可以用 if 语句或 switch 语句检测变量中的值，如下面图 G.1a 和 G.1b 所示。

```
        (a)      (b)                    (a)      (b)
    if (day == MONDAY)             switch (day)
    {                              {
      // process Monday              case MONDAY:
    }                                  // process Monday
    else if (day == TUESDAY)           break;
    {                                case TUESDAY:
      // process Tuesday               // process Tuesday
    }                                  break;
    else                             ...
      ...                          }
    (a) 在 if 语句中使用枚举变量      (b) 在 switch 语句中使用枚举变量
```

图 G.1

LiveExample G.1 给出了一个使用枚举类型的示例。

LiveExample G.1 的互动程序请访问 https://liangcpp.pearsoncmg.com/LiveRunCpp5e/faces/LiveExample.xhtml?programName=TestEnumeratedType&programHeight=470&header=off&resultHeight=180。

LiveExample G.1　TestEnumeratedType.cpp

Source Code Editor:

```cpp
#include <iostream>
using namespace std;

int main()
{
  enum Day {MONDAY = 1, TUESDAY, WEDNESDAY, THURSDAY, FRIDAY} day;

  cout << "Enter a day (1 for Monday, 2 for Tuesday, etc): ";
  int dayNumber;
  cin >> dayNumber;

  switch (dayNumber) {
    case MONDAY:
      cout << "Play soccer" << endl;
      break;
    case TUESDAY:
      cout << "Piano lesson" << endl;
      break;
    case WEDNESDAY:
      cout << "Math team" << endl;
      break;
    default:
      cout << "Go home" << endl;
  }

  return 0;
}
```

Enter input data for the program (Sample data provided below. You may modify it.)

2

Automatic Check　Compile/Run　Reset　Answer　　　Choose a Compiler: VC++

Execution Result:

```
command>cl TestEnumeratedType.cpp
Microsoft C++ Compiler 2019
Compiled successful (cl is the VC++ compile/link command)

command>TestEnumeratedType
Enter a day (1 for Monday, 2 for Tuesday, etc): 2
Piano lesson

command>
```

第 6 行声明了一个枚举类型 Day，并在该条语句中声明了一个名为 day 的变量。第 10 行从键盘上读取一个 int 值。第 12～24 行中的 switch 语句检查当天是 MONDAY、TUESDAY、WEDNESDAY 还是其他时间，以显示相应的信息。

G.4 在 C++11 中使用 `enum class`

假设我们需要如下所示定义两个枚举类型 CommonColor 和 BasicColor：

```
enum CommonColor {RED, BLACK, WHITE, YELLOW};
enum BasicColor {RED, GREEN, BLUE};
```

该代码不能编译，因为 CommonColor 和 BasicColor 中都定义了 RED。枚举数就像一个常量。在同一作用域内只能定义一次。

要解决该问题，可以在 C++11 中使用 `enum class`。前面两种类型可以定义如下：

```
enum class CommonColor {RED, BLACK, WHITE, YELLOW};
enum class BasicColor {RED, GREEN, BLUE};
```

用 `enum class` 定义的枚举类型称为强类型枚举。要访问强类型枚举中的枚举数，该枚举数必须以枚举类型为前缀，后跟 :: 运算符，如以下示例所示：

```
CommonColor color1 = CommonColor::BLACK;
BasicColor color2 = BasicColor::RED;
```

强类型枚举数与常规枚举数一样存储为整数。但是，需要显式强制转换才能获得枚举数的整数值。这里有一个例子，

```
int i = static_cast<int>(BasicColor::RED);
```

此语句将枚举数 BasicColor::RED 的整数值赋值给整数变量 i。下面是另一个示例：

```
cout << static_cast<int>(CommonColor::BLACK) <<
    static_cast<int>(BasicColor::RED) << endl;
```

此语句显示枚举数 CommonColor::BLACK 和 BasicColor::RED 的整数值。

附录 H

Introduction to C++ Programming and Data Structures, Fifth Edition

正则表达式

H.1 匹配字符串

我们常常需要编写代码来验证用户输入，例如检查输入是数字、全小写字母的字符串还是社会保障号码。你是如何编写此类代码的？完成此任务的一种简单有效的方法是使用正则表达式。

正则表达式（缩写为 regex）是一个字符串，描述用于匹配一组字符串的模式。正则表达式是字符串操作的强大工具。可以在 C++11 中用正则表达式来匹配、查找、替换和拆分字符串。本附录中涵盖的所有函数都在 `<regex>` 头文件中定义。

要了解正则表达式如何工作，让我们从 `<regex>` 头文件中的 `regex_match` 函数开始。下面是一个示例：

```
bool isMatched = regex_match("John", regex("J.*"));
```

`regex_match("John", regex("J.*"))` 函数接受两个参数：字符串 s 和正则表达式 r。`regex(r)` 从正则表达式 r 创建一个 regex 对象。如果 s 与 r 中指定的模式匹配，则函数返回 `true`。这里的字符串是 `"John"`。`"J.*"` 是一个正则表达式。它描述了一种字符串模式，该模式以字母 J 开头，后跟零个或任意多个字符。子字符串 `.*` 与零个或任意多个字符匹配。前面的函数返回 `true`，因为 `"John"` 与模式 `"J.*"` 匹配。

以下是更多示例：

```
regex_match("Johnson", regex("J.*"));
regex_match("johnson", regex("J.*"));
```

第一个函数调用返回 `true`，因为 `"Johnson"` 与模式 `"J.*"` 匹配，但第二个函数调用返回 `false`，因为 `"johnson"` 与模式 `"J.*"` 不匹配。

H.2 正则表达式语法

正则表达式由字面量字符和特殊符号组成。表 H.1 列出了一些常用的正则表达式语法。

表 H.1 常用正则表达式

正则表达式	匹配	示例
x	一个指定的字符 x	Good 匹配 Good
.	任何单个字符	Good 匹配 G..d
(ab|cd)	ab 或 cd	ten 匹配 t(en|im)
[abc]	a,b 或 c	Good 匹配 Go[opqr]d
[^abc]	除 a,b,c 以外的任何字符	Good 匹配 Go[^ pqr]d
[a-z]	a ~ z	Good 匹配 [A-M]oo[a-d]

(续)

正则表达式	匹配	示例
[^a-z]	除了 a～z 以外的任何字符	Good 匹配 Goo[^e-t]
\d	一个数字，0～9	number2 匹配 "number[\\d]"
\D	一个非数字	$ Good 匹配 "$ [\\D][\\D]od"
\w	一个单词字符	Good1 匹配 "[\\w]ood[\\d]"
\W	一个非单词字符	$ Good 匹配 "[\\W][\\w]ood"
\s	一个空白字符	"Good 2" 匹配 "Good\\s2"
\S	一个非空白字符	Good 匹配 "[\\S]ood"
p*	模式 p 的 0 次或多次出现	aaaa 匹配 "a*" abab 匹配 "(ab)*"
p+	模式 p 的 1 次或多次出现	a 匹配 "a+b" able 匹配 "(ab)+.*"
p?	模式 p 的 0 次或 1 次出现	Give 匹配 "G?Give" ive 匹配 "G?ive"
p{n}	模式 p 的 n 次出现	Good 匹配 "Go{2}d" Good 不匹配 "Go{1}d"
p{n,}	模式 p 的至少 n 次出现	aaaa 匹配 "a{1,}" a 不匹配 "a{2,}"
p{n,m}	出现次数在 n 到 m 之间（包括 n 和 m）	aaaa 匹配 "a{1,9}" abb 不匹配 "a{2,9}bb"

注意：反斜杠是个特殊的字符，用于启动字符串中的转义序列。因此，要用 \\ 来表示字面量 \。

注意：空白字符指的是 ' '、'\t'、'\n'、'\r' 或 '\f'。因此，\s 等同于 [\t\n\r\f]，\S 等同于 [^\t\n\r\f]。

注意：单词字符是任意字母、数字或下划线字符。所以 \w 与 [a-zA-Z0-9_] 相同，\W 与 [^a-zA-Z0-9_] 相同。

注意：表 H.1 中的最后六项 *、+、?、{n}、{n, } 和 {n, m} 被称为量词，它们指定了量词前的模式可以重复的次数。例如，A* 匹配零个或多个 A，A+ 匹配一个或多个 A，A? 匹配零个或一个 A，A{3} 完全匹配 AAA，A{3, } 至少匹配三个 A，A{3, 6} 匹配 3 到 6 个 A。* 与 {0, } 相同，+ 与 {1, } 相同，? 与 {0, 1} 相同。

警告：不要在重复量词中使用空格。例如，A{3, 6} 不能写成逗号后有空格的 A{3, 6}。

注意：可以用括号对模式进行分组。例如，(ab){3} 匹配 ababab，但 ab{3} 与 abbb 匹配。

我们用几个例子来演示如何构造正则表达式。

示例 1：社会保障号码的模式是 xxx-xx-xxxx，其中 x 是一个数字。社会保障号码的正则表达式可以描述为

    [\\d]{3}-[\\d]{2}-[\\d]{4}

例如，

```
regex_match("111-22-3333", regex("[\\d]{3}-[\\d]{2}-[\\d]{4}"))
returns true.
regex_match("11-22-3333", regex("[\\d]{3}-[\\d]{2}-[\\d]{4}"))
returns false.
```

示例 2：偶数以数字 0、2、4、6 或 8 结尾。偶数的模式可以描述为

```
[\\d]*[02468]
```

例如，

```
regex_match("123", regex("[\\d]*[02468]")) returns false.
regex_match("122", regex("[\\d]*[02468]")) returns true.
```

示例 3：电话号码的模式是 (xxx)xxx-xxxx，其中 x 是一个数字，第一个数字不能为零。电话号码的正则表达式可以描述为

```
\\([1-9][\\d]{2}\\) [\\d]{3}-[\\d]{4}
```

注意，括号符号 (和) 是正则表达式中对模式进行分组的特殊字符。要在正则表达式中表示字面量 (或)，必须使用 \\(和 \\)。

例如，

```
regex_match("(912) 921-2728", regex("\\([1-9][\\d]{2}\\)
[\\d]{3}-[\\d]{4}")) returns true.
regex_match("921-2728", regex("\\([1-9][\\d]{2}\\) [\\d]{3}-[\\d]
{4}")) returns false.
```

示例 4：假设姓氏最多由 25 个字母组成，且第一个字母大写。姓氏的模式可以描述为

```
[A-Z][a-zA-Z]{1,24}
```

注意，正则表达式中不能有任何空格。例如，[A-Z][a-zA-Z]{1, 24} 是错误的。

例如

```
regex_match("Smith", regex("[A-Z][a-zA-Z]{1,24}")) returns true.
regex_match("Jones123", regex("[A-Z][a-zA-Z]{1,24}")) returns
false.
```

示例 5：在 2.4 节中定义了 C++ 标识符。标识符必须以字母或美元符号（$）开头。不能以数字开头。

标识符是由字母、数字和下划线（_）组成的一系列字符。标识符的模式可以描述为

```
[a-zA-Z_][\\w]*
```

示例 6：正则表达式 "Welcome to(US|Canada)" 匹配哪些字符串？答案是 Welcome to US 或 Welcome to Canada。

示例 7：正则表达式 ".*" 匹配哪些字符串？答案是任意字符串。

H.3 查找匹配的子字符串

可以用 regex_search 函数查找与正则表达式匹配的子字符串。以下是一个示例：

```
1  string text("Kim Smith 212 Ed Snow 345 Jo James 313");
2  smatch match;
3  regex reg("\\d{3}");
4  regex_search(text, match, reg);
5  cout << match.str() << endl; // Display 212
```

regex_search 函数（第 4 行）接受三个参数：字符串 text、smatch 对象 match 和正

则表达式 reg。调用 match.str()（第 5 行）返回 text 中第一个匹配的字符串，在本例中为 212。

可以用以下循环获取与正则表达式模式匹配的所有子字符串：

```
1  while (regex_search(text, match, reg))
2  {
3    cout << match.str() << endl;
4    text = match.suffix();
5  }
```

该循环显示

```
212
345
313
```

如果 text 中有匹配正则表达式的子字符串，则 regex_search 函数（第 1 行）返回 true。match.str()（第 3 行）返回第一个匹配的子字符串。match.suffix() 函数（第 4 行）返回 text 中第一个匹配子字符串中最后一个字符之后的剩余子字符串。

H.4 替换子字符串

如果字符串与正则表达式匹配，则 regex_match 函数返回 true。如果字符串中的子字符串与正则表达式匹配，则 regex_search 函数返回 true。C++11 还支持 regex_replace 函数来替换子字符串。以下是一个示例：

```
1  string text("abcaabaaaerddd");
2  string newText = regex_replace(text, regex("a+"), "T");
3  cout << newText << endl; // newText is TbcTbTerddd
```

regex_replace(s1, regex(r), s2) 函数返回一个新字符串，该字符串将 s1 中与正则表达式 r 匹配的子字符串替换为 s2。原始字符串 s1 没有更改。

注意：默认情况下，所有的量词都是贪婪的。这意味着它们将匹配尽可能多的出现字符。例如，下面的语句显示 T，因为整个字符串 aaa 与正则表达式 a+ 匹配。

```
cout << regex_replace("aaa", regex("a+"), "T");
```

可以在量词后面加一个问号（?）来改变它的默认行为。量词就会变得懒惰，这意味着它将尽可能少地匹配。例如，下面的语句显示 TTT，因为每个 a 都匹配一个 a+?。

```
cout << regex_replace("aaa", regex("a+?"), "T");
```

H.5 从字符串中抽取元组（token）

可以指定分隔符并用分隔符从字符串中抽取元组。假设字符串是

```
Programming is fun
```

分隔符是空格。则该字符串有三个元组：Programming、is 和 fun。

可以用正则表达式定义分隔符。通过为带有分隔符的字符串创建 sregex_token_iterator 类的实例来提取元组。

LiveExample H.1 给出了一个示例，演示如何从字符串中提取元组。

LiveExample H.1 的互动程序请访问 https://liangcpp.pearsoncmg.com/LiveRunCpp5e/faces/LiveExample.xhtml?header=off&programName=ExtractTokenDemo&programHeight=470&fileType=.cpp&resultHeight=270。

LiveExample H.1　ExtractTokenDemo.cpp

Source Code Editor:

```cpp
#include <iostream>
#include <string>
#include <regex>
using namespace std;

int main()
{
  string text("Good morning. How are you? Hi, Hi");

  // Create a regex for separating words
  regex delimiter("[ ,.?]");

  sregex_token_iterator tokenIterator(text.begin(), text.end(),
    delimiter, -1);
  sregex_token_iterator end;

  while (tokenIterator != end)
  {
    string token = *tokenIterator; // Get a token
    if (token.size() > 0)
      cout << token << endl; // Display token
    tokenIterator++; // Iterate to next token
  }

  return 0;
}
```

Execution Result:

```
command>cl ExtractTokenDemo.cpp
Microsoft C++ Compiler 2019
Compiled successful (cl is the VC++ compile/link command)

command>ExtractTokenDemo
Good
morning
How
are
you
Hi
Hi

command>
```

该程序创建一个名为 delimiter 的正则表达式（第 11 行），该表达式被用作分隔符，把字符串拆分为多个部分。在这种情况下，分隔符是空格、逗号、句点和问号。使用 sregex_token_iterator 类创建一个元组迭代器，用 delimiter 从字符串中提取元组（第 13 ~ 14 行）。sregex_token_iterator 构造函数中的前两个参数是源字符串 text

的起始迭代器和结束迭代器，用于指定字符串的范围。第三个参数是分隔符。最后一个参数 -1 告诉迭代器在提取元组时跳过分隔符。

该程序创建一个名为 end 的空元组迭代器（第 15 行）。如果一个元组迭代器为空，则在该元组迭代器中没有剩下任何元组。该程序使用循环重复显示元组迭代器中的所有元组（第 17～23 行）。*tokenIterator 获得当前元组（第 19 行），tokenIterator++ 移动到下一个元组（第 22 行）。

可以重写 LiveExample H.1 来定义一个名为 split 的可重用函数，将元组存储在一个向量中，如 LiveExample H.2 所示。

LiveExample H.2 的互动程序请访问 https://liangcpp.pearsoncmg.com/LiveRunCpp5e/faces/LiveExample.xhtml?header=off&programName=SplitString&programHeight=630&resultHeight=270。

LiveExample H.2　SplitString.cpp

```cpp
#include <iostream>
#include <string>
#include <regex>
using namespace std;

// Split a string into substrings using the specified delimiter
vector<string> split(const string& text, const string& delim)
{
  regex delimeter(delim); // Create a regex for separating words

  sregex_token_iterator tokenIterator(text.begin(), text.end(),
    delimeter, -1);
  sregex_token_iterator end;

  vector<string> result;
  while (tokenIterator != end)
  {
    string token = *tokenIterator; // Get a token
    if (token.size() > 0)
      result.push_back(token); // Save token in the vector
    tokenIterator++; // Iterate to next token
  }

  return result;
}

int main()
{
  string text("Good morning. How are you? Hi, Hi");
  vector<string> tokens = split(text, "[ ,.?]");

  for (string& s : tokens)
    cout << s << endl;

  return 0;
}
```

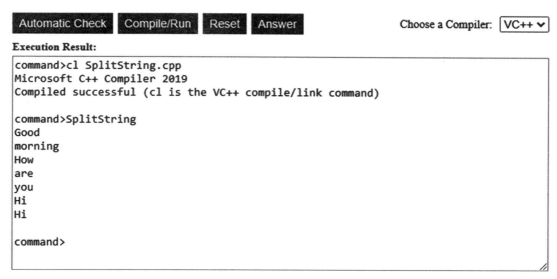

split 函数（第 7 ～ 25 行）用指定的分隔符将字符串拆分为子字符串，并将子字符串保存在向量中后返回向量。

附录 I
Introduction to C++ Programming and Data Structures, Fifth Edition

大 O、大 Omega 和大 Theta 表示法

第 18 章用外行的术语介绍了大 O 表示法。在本附录中，我们给出大 O 表示法的精确数学定义。我们还将给出大 Omega 和大 Theta 表示法。

I.1 大 O 表示法

大 O 表示法是一种渐近表示法，描述函数的自变量接近特定值或无穷大时的行为。设 $f(n)$ 和 $g(n)$ 为两个函数，如果存在一个常数 c（$c>0$）和值 m，对于 $n \geq m$，使得 $f(n) \leq c \times g(n)$，我们说 $f(n)$ 是 $O(g(n))$，读作 "$g(n)$ 的大 O"。

例如，$f(n) = 5n^3 + 8n^2$ 是 $O(n^3)$，因为可以找到 $c = 13$，$m = 1$，对于 $n \geq m$，使 $f(n) \leq cn^3$。$f(n) = 6n\log n + n^2$ 是 $O(n^2)$，因为可以找到 $c = 7$，$m = 2$，对于 $n \geq m$，使 $f(n) \leq cn^2$。$f(n) = 6n\log n + 400n$ 是 $O(n\log n)$，因为可以找到 $c = 406$，$m = 2$，对于 $n \geq m$，使 $f(n) \leq cn\log n$。$f(n^2)$ 是 $O(n^3)$，因为可以找到 $c = 1$，$m = 1$，对于 $n \geq m$，使 $f(n) \leq cn^3$ 的。注意，c 和 m 有无限多的选择，对于 $n \geq m$，使 $f(n) \leq c \times g(n)$。

大 O 表示法表示一个函数 $f(n)$ 渐近地小于或等于另一个函数 $g(n)$。这使你可以通过忽略乘法常数和放弃函数中的非支配项来简化函数。

I.2 大 Omega 表示法

大 Omega 表示法与大 O 表示法相反。它也是一种渐近表示法，表示一个函数 $f(n)$ 大于或等于另一个函数 $g(n)$。设 $f(n)$ 和 $g(n)$ 为两个函数，如果有一个常数 c（$c>0$）和值 m，对于 $n \geq m$，有 $f(n) \geq c \times g(n)$，我们说 $f(n)$ 是 $\Omega(g(n))$，读作 "$g(n)$ 的大欧米伽"。

例如，$f(n) = 5n^3 + 8n^2$ 是 $\Omega(n^3)$，因为可以找到 $c = 5$，$m = 1$，对于 $n \geq m$，使 $f(n) \geq cn^3$。$f(n) = 6n\log n + n^2$ 是 $\Omega(n^2)$，因为可以找到 $c = 1$，$m = 1$，对于 $n \geq m$，使 $f(n) \geq cn^2$。$f(n) = 6n\log n + 400n$ 是 $\Omega(n\log n)$，因为可以找到 $c = 6$，$m = 1$，对于 $n \geq m$，使 $f(n) \geq cn\log n$。$f(n) = n^2$ 是 $\Omega(n)$，因为可以找到 $c = 1$，$m = 1$，对于 $n \geq m$，使 $f(n) \geq cn$。注意，有无限多的 c 和 m，对于 $n \geq m$，使得 $f(n) \geq c \times g(n)$。

I.3 大 Theta 表示法

大 Theta 表示法表示两个函数渐近相同。设 $f(n)$ 和 $g(n)$ 为两个函数，如果 $f(n)$ 是 $O(g(n))$ 且 $f(n)$ 是 $\Omega(g(n))$，我们说 $f(n)$ 是 $\Theta(g(n))$，读作 "$g(n)$ 的大西塔"。

例如，$f(n) = 5n^3 + 8n^2$ 是 $\Theta(n^3)$，因为 $f(n)$ 是 $O(n^3)$ 且 $f(n)$ 是 $\Omega(n^3)$。$f(n) = 6n\log n + 400n$ 是 $\Theta(n\log n)$，因为 $f(n)$ 是 $O(n\log n)$ 且 $f(n)$ 是 $\Omega(n\log n)$。

注意：大 O 表示法给出了一个函数的上界。大 Omega 表示法给出了一个函数的下界。大 Theta 表示法给出了一个函数的紧密界限。简单起见，通常使用大 O 表示法，尽管大 Theta 表示法可能更符合实际。